Carl Claus, Adam Sedgwick

Elementary text-book of zoology

Carl Claus, Adam Sedgwick

Elementary text-book of zoology

ISBN/EAN: 9783337278298

Printed in Europe, USA, Canada, Australia, Japan

Cover: Foto ©berggeist007 / pixelio.de

More available books at **www.hansebooks.com**

ELEMENTARY TEXT-BOOK

OF

ZOOLOGY.

SPECIAL PART: MOLLUSCA TO MAN.

BY

DR. C. CLAUS.

Professor of Zoology and Comparative Anatomy in the University of Vienna;
Director of the Zoological Station at Trieste.

TRANSLATED AND EDITED BY

ADAM SEDGWICK, M.A.,

Fellow and Lecturer of Trinity College, Cambridge,

WITH THE ASSISTANCE OF

F. G. HEATHCOTE, B.A.,

Trinity College, Cambridge.

WITH 706 WOODCUTS.

NEW YORK:

MACMILLAN & CO.

TABLE OF CONTENTS.

SPECIAL PART:

MOLLUSCA TO MAN.

CHAPTER I.

MOLLUSCA.*

Bilaterally symmetrical unsegmented animals, without a locomotory skeleton; with a ventral foot and usually a calcareous univalve or bivalve shell; with brain (supraœsophageal ganglia), circumœsophageal ring, and subœsophageal group of ganglia.

SINCE Cuvier several different groups of animals, which were placed amongst the worms by Linnæus, have been included in the Mollusca. Of late years, however, the anatomy and development of these forms have been more closely examined, and it seems fairly certain that some of them are allied to the Worms. In any case, the group Mollusca must be looked upon as of more limited extent than has for some time been the case. The bivalved *Brachiopoda*, which in structure and development stand in closer relationship to the Bryozoa, may be removed from the Mollusca and united with the latter under the head *Molluscoidea*. The *Tunicata* also must be constituted an independent group between the Mollusca and the Vertebrata.

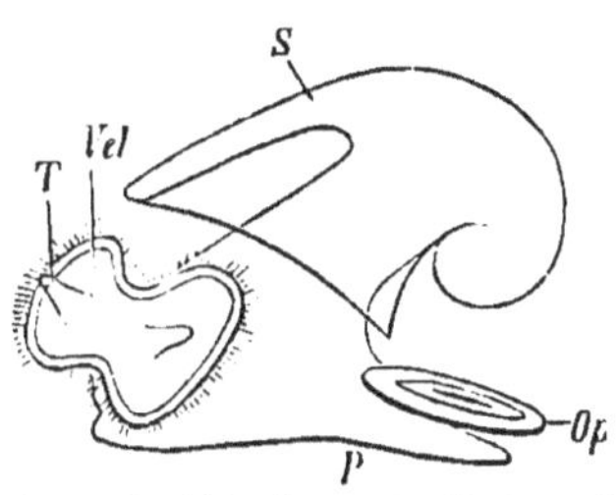

FIG. 192.—Older larva of a *Gasteropod* (after Gegenbaur). *S*, Shell; *P*, foot; *Vel*, velum; *T*, tentacles; *Op*, operculum for the closure of the shell opening.

* G. Cuvier, "Mémoires pour servir à l'histoire et à l'anatomie des Mollusques." Paris, 1817.

R. Leuckart, "Ueber die Morphologie und die Verwandschaftsverhältnisse der wirbellosen Thiere." Braunschweig, 1848.

Huxley, "On the Morphology of the Cephalous Mollusca, as illustrated by the Anatomy of certain Heteropoda and Pteropoda, etc." *Phil. Trans.*, 1853.

The Mollusca are unjointed, unsegmented animals, without jointed appendages. The body is covered by a soft slimy skin. They lack both an internal and external locomotory skeleton, and appear therefore especially suited for life in water. But few of them are terrestrial, and when this is the case the locomotion is always limited and slow; while the aquatic forms, in correspondence with the far more favourable conditions for locomotion presented by water, may be endowed with the power of rapid swimming.

The *dermal muscular system* plays an important part in the locomotion of these animals, especially that part of it placed on the lower, *i.e.*, ventral, surface of the body. In this region it is greatly developed, and gives rise to a more or less projecting locomotory organ of very various shape, the **foot** (figs. 492 and 493). The foot always consists of an unpaired median structure, which is sometimes divided into several parts and may possess in addition lateral paired portions, the **epipodia.** Above the foot there very generally exists on the body a shield-shaped thickening of the integument, the so-called *mantle*, the edges of which, in more advanced development, grow over the body as a fold of the skin and partially or completely cover it. The surface of this fold of skin secretes calcareous and pigmentary substances, and gives rise to the variously shaped and coloured shells which contain and protect the soft body. In addition to the *foot* and *mantle*, the body generally possesses in the anterior region, on either side of the mouth, a pair of lobe-like appendages, the *buccal lobes*, which are the remnants of a largely-developed larval structure, known as the **velum.**

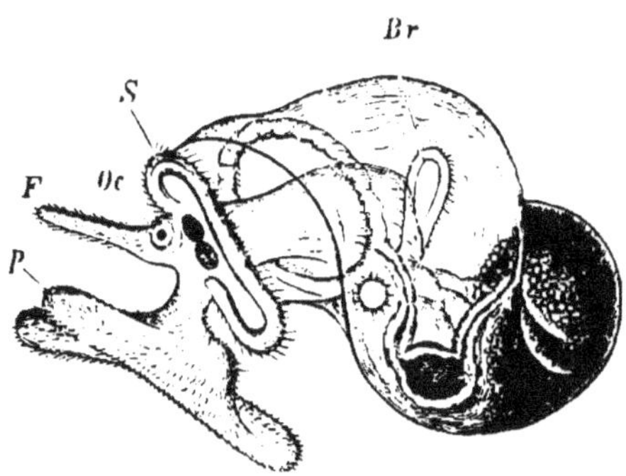

FIG. 493.—Larva of *Vermetus* (after Lacaze-Duthiers). *S*, velum; *Br*, gill; *F*, tentacle; *P*, foot; *Oc*, eye.

In the higher Mollusca (*Cephalophora*) the anterior part of the body bearing the buccal lobes, and containing the central parts of the nervous system and the sense organs, is more or less sharply marked off as a *head*. The part of the body behind the head constitutes the main mass of the animal. Its dorsal portion (*the visceral sac*) contains the viscera and is frequently spirally twisted, as a result of which the bilateral symmetry undergoes externally a remarkable disturbance. The visceral sac may, however, have a flattened or cylindrical form and retain its symmetry. In this group (the *Cephalophora*) the shell may be simply plate-shaped or spirally wound, or remain as a mere flat rudiment hidden under the dorsal integument. In one group of the Cephalophora, viz., the *Cephalopoda*, a circle of arms is attached to the head around the mouth opening. They serve both for swimming and creeping, and for the capture of nourishment. By Lovén and R. Leuckart they were looked upon as modifications of the buccal lobes; by others, perhaps with greater justice, as tentacles, and by others again as

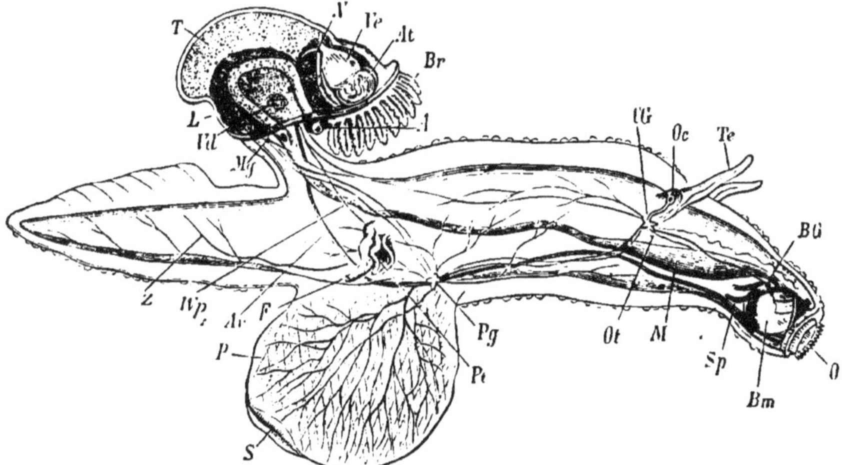

FIG. 494.—Male of *Carinaria mediterranea* (after Gegenbaur). *P*, Foot; *S*, sucker; *O*, mouth; *Bm*, buccal mass; *M*, stomach; *Sp*, salivary gland; *L*, liver; *A*, anus; *CG*, cerebral ganglion; *Te*, tentacle; *Oc*, eye; *Ot*, auditory vesicle; *BG*, buccal ganglion; *Pg*, pedal ganglion; *Mg*, mantle ganglion; *N*, kidney; *Br*, gill; *At*, auricle; *Ve*, ventricle; *Ar*, aorta; *Z*, hinder branch of the same; *T*, testis; *Vd*, vas deferens; *Wp*, ciliated furrow; *Pe*, penis; *F*, flagellum with gland.

modifications of the foot. A perforated funnel-shaped cone, through which the excretory products and water which has passed over the gills is expelled from the large mantle cavity, and which thus serves at the same time as a swimming organ, probably corresponds to the fused folds of the epipodia. Amongst the *Gastropoda* the head is provided with tentacles and buccal lobes, and the ventrally placed foot possesses a large flat plantar surface; more rarely it has the form of a vertically placed fin (*Heteropoda*, fig. 494). In another group, the *Lamellibranchiata* (*Acephala*), there is no independent head, and the laterally compressed body bears two large lateral mantle lobes, each of which secretes a single shell; the two valves so formed are united on the dorsal surface by a ligament.

The internal organization of the Mollusca presents as many differences as does the external form. Like the external form, the internal structure also frequently presents surprising deviations from the bilateral arrangement.

The **nervous system** * (figs. 495, 496, 497) consists of a dorsal pair of ganglia lying on the œsophagus (only exceptionally—fig. 495—dissolved into a general ganglionic investment of the commissure), the *cerebral ganglia* (figs. 496, 497, *Cg*), from which pass off the sense nerves and an œsophageal ring, composed of several fibrous cords. The latter primitively gives off two pairs of nerve-trunks. The nerves of the upper and lateral pair are the *pallial nerves* (fig. 495, *PaSt*); they supply the lateral parts of the body and the mantle. The nerves of the ventral pair are placed nearer the middle line, and are known as the *pedal nerves* (fig. 495, *PeSt*); they are connected together by transverse commissures (fig. 495) and innervate the muscles of the foot. This arrangement, found in the simplest form in *Chiton*, agrees essentially with that of the Gephyrean-like genus, *Neomenia*. At a more advanced stage, two large swellings are found at the origin of the pedal nerves; these are the *pedal ganglia* (figs. 496, 497, *Pg*). In addition, a third group of ganglia, known as the *visceral ganglia*, is also found. The arrangement of the latter ganglia is very various; they are sometimes fused with the cerebral, sometimes with the pedal ganglia, and are sometimes broken up into several groups of ganglia. They are

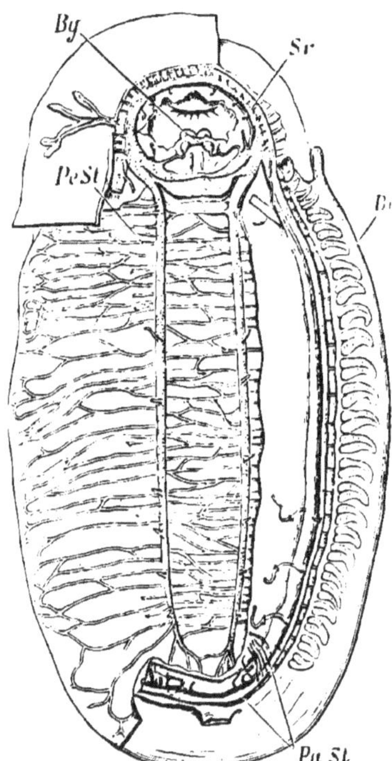

FIG. 495.—Nervous system of *Chiton* (after B. Haller). *Sr*, œsophageal ring; *Bg*, buccal ganglion; *PeSt*, pedal nerve; *PaSt*, pallial nerve; *Br*, gills.

* H. v. Jhering, "Vergleichende Anatomie des Nervensystems und Phylogenie der Mollusken." Leipzig, 1877.

connected with the cerebral ganglia by a longer or shorter commissure, and give off nerve plexuses to the heart, gills, and generative organs. This third pair of ganglia is, therefore, regarded as the equivalent of the sympathetic, but unjustly, as it also gives off nerves to the skin and muscles. Small ganglia (*buccal ganglia*), lying above and below the buccal mass and sending off nerves to the œsophagus and intestine, may more justly be regarded as sympathetic.

Tactile organs are present in the more highly-developed Mollusca, as two or four lobes placed near the mouth, the above-mentioned buccal lobes; in addition to which, tentacles round the edge of the mantle are often found in the *Acephala*, and in the *Cephalophora* two or four retractile tentacles on the head. The **eyes** have almost always a complicated structure, and are provided with lens, iris, choroid, and retina. There are usually two of them on the head; in rare cases—*e.g.*, in some Lamellibranchs—they are more numerous, and are placed on the edge of the mantle. **Auditory organs** are very generally present. They have the form of closed otocysts, provided with hairs on their internal walls. They are usually paired, and lie either on the cerebral or pedal ganglia (fig. 497, *Ot*). They are, however, always innervated from the former.

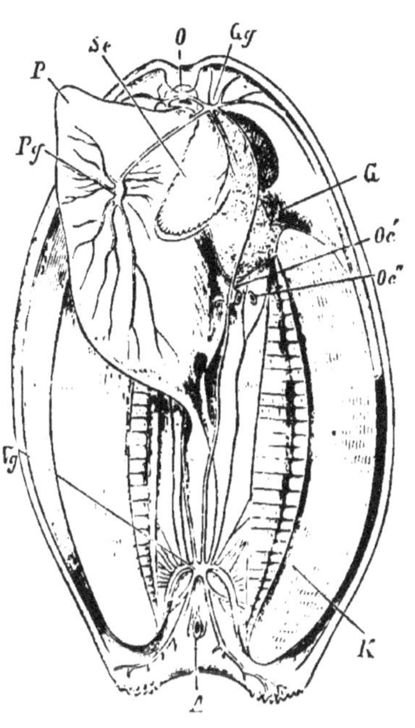

Fig. 496.—Nervous system of the pearl mussel (*Anodonta*)(after Keber). *O*, mouth; *A*, anus; *K*, gills; *P*, foot; *Se*, labial palps; *Gg*, cerebral ganglion; *Pg*, pedal ganglion; *Vg*, splanchnic ganglion; *G*, generative gland; *Oc'*, external opening of kidney; *Oc''*, opening of generative gland.

In the **alimentary canal**, three divisions, at least, can be clearly distinguished—the œsophagus, the stomach and intestine, and the hindgut or rectum. Of these the middle or digesting division (stomach and intestine) is usually characterized by the possession of a very extensive *liver*. **Kidneys** are always present, and are frequently paired and symmetrical in each half of the body. Often, however—principally when the body is asymmetrical—the kidney of one side is smaller (*Patella, Haliotis*) or is entirely absent (*Gastro-*

poda). They usually have the form of sacs with a wide lumen, and open on the one hand into the body cavity (*pericardial sinus*), and on the other to the exterior by a lateral opening. In all probability the molluscan kidney is homologous with an annelidan segmental organ. The internal, funnel-shaped opening is frequently beset with cilia. The anus is very often removed from the middle line, and placed on one side of the body.

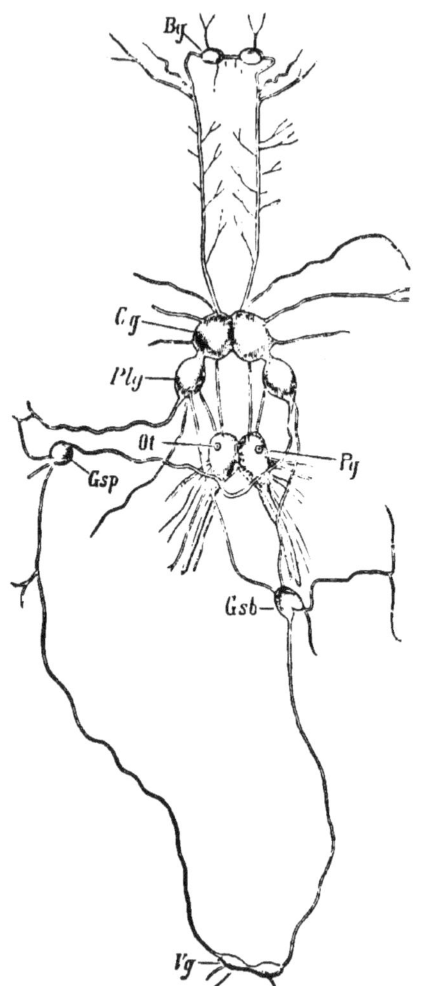

FIG. 197. — Nervous system of *Cassidaria* (after Haller). *Cg*, cerebral ganglion; *Pg*, pedal ganglion; *Plg*, pleural ganglion; *Bg*, buccal ganglion; *Gsp*, supraintestinal ganglion; *Gsb*, subintestinal ganglion; *Vg*, visceral ganglion; *Ot*, otocyst.

A compact **heart** is always present, driving the blood through the vessels into the organs. The vascular system is never completely closed, for, even when the arteries and veins are connected by capillaries, blood sinuses, derived from the body cavity, are inserted into the course of the vessels. The heart is always arterial—*i.e.*, systemic—and receives arterial blood from the respiratory organs.

Respiration is in all cases carried on through the general outer surface of the body; but in addition special respiratory organs, in the form of *branchiæ*, more rarely of *lungs*, are present. The branchiæ are ciliated projections of the body surface, and are usually placed between the mantle and the foot: they may have the form of branched appendages, or of broad lamellæ (*Lamellibranchiata*). The lung, on the other hand, is derived from the mantle cavity, which is filled with air, and the inner surface of which is thrown into a number of complicated folds, so as to expose a large surface for the respiratory blood-vessels: it communicates by an opening with the external medium. The pulmonary and branchial cavities are, therefore, morphologically equivalent.

Reproduction is always sexual. The *hermaphrodite* condition, on the whole, preponderates; nevertheless, not only many marine Gastropods, but also most Lamellibranchs and all Cephalopods are diœcious.

Development usually begins with a total segmentation, which is followed by the formation of a blastoderm surrounding the hinder part of the yolk or the whole yolk. The just hatched young often pass through a complicated metamorphosis, and possess an anterior cutaneous expansion bordered with cilia—*the velum*—which functions as a locomotory organ. In form, disposition of cilia and organisation, many molluscan larvæ permit of a closer comparison with Lovén's worm larva.

By far the majority of the Mollusca are aquatic animals, especially marine: only a few live on land, and these always seek damp localities. When we consider the extraordinarily wide distribution of the Mollusca in past times, the importance of their fossil remains for the determination of the age of the sedimentary formations becomes intelligible.

Class 1.—LAMELLIBRANCHIATA.*

Laterally compressed Mollusca without separated head, with bilobed mantle and bivalve shell, composed of a right and left half and connected by a dorsally-placed ligament: with large gill plates: sexes usually separate.

The Lamellibranchs were formerly united with the Brachiopoda as *Conchifera*. Like the latter, they lack a differentiated cephalic region, and possess a large and usually bilobed mantle and a bivalve shell. Nevertheless, the structural differences between these

* G. Cuvier, "L' histoire et l'anatomie des Mollusques." Paris, 1817.

Bojanus, "Ueber die Athem- und Kreislaufswerkzeuge der zweischaligen Muscheln." *Isis*, 1817, 1820, 1827.

S. Lovén, *K. Vet. Akad. Handlgr.* Stockholm, 1848. Translated in the *Arch. für Naturgesch.*, 1849.

Lacaze-Duthiers, *Ann. des Sc. Nat.*, 1854—1861.

H. and A. Adams, "The Genera of the Recent Mollusca." London, 1853-1858.

L. Reeve, "Conchologia iconica." London, 1846-1858.

two groups are so essential, that a close connection between them cannot be maintained.

The usually strictly symmetrical body is laterally compressed and of considerable extent, and is surrounded by two lateral mantle lobes, which are continuous across the dorsal middle line and secrete a right and left shell valve. To the sides of the mouth opening are found two pairs of leaf- or tentacle-like buccal lobes—the **labial palps**. On the ventral surface a large, usually hatchet-shaped foot (fig. 498, *F*) projects; and two pairs, rarely one pair, of large lamellar gills are always placed in the mantle furrow between the mantle and foot (fig. 498, *K*).

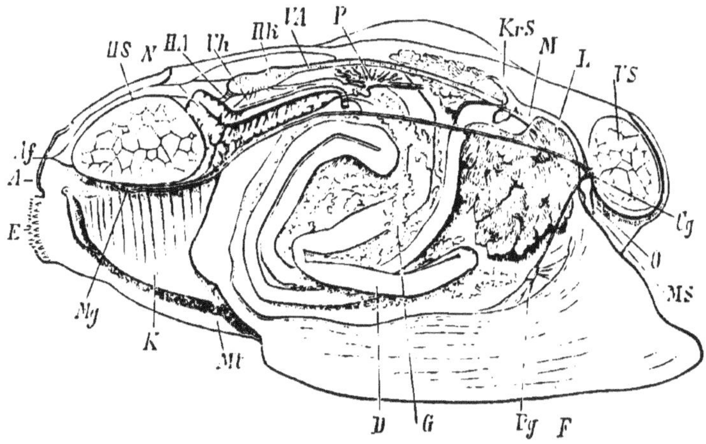

Fig. 498.—Anatomy of *Unio pictorum* (after C. Grobben). *VS*, anterior adductor muscle; *HS*, posterior adductor muscle; *MS*, labial palp; *F*, foot; *Mt*, Mantle; *K*, branchiæ; *Cg*, cerebral ganglion; *Pg*, pedal ganglion; *Mg*, splanchnic ganglion; *O*, mouth; *M*, stomach; *L*, liver; *KrS*, crystalline style; *D*, intestine; *Af*, anus; *G*, generative organs; *A*, region of mantle lobes bounding the exhalent or cloacal orifice; *E*, region of ditto bounding inhalent or branchial orifice; *N*, kidney; *Vk*, auricle; *Hk*, ventricle; *VA*, anterior aorta; *HA*, posterior aorta; *P*, pericardial gland (schematic).

The hind end of the edges of each mantle lobe almost always presents two slight, contiguous excavations (fig. 498, *A* and *E*), the ventral of which is bordered by numerous papillæ. When the two halves of the mantle are above together, these excavations form, with the corresponding structures of the opposite side, two slit-like openings, placed one behind the other. The upper or dorsal of these two openings functions as the cloacal, or exhalent opening; the lower or ventral as the inhalent opening. Through the latter, with a slightly gaping shell, the water is driven by the peculiar arrangement and action of the cilia on the inner surface of the mantle and the gills into the mantle and respiratory chamber. Food materials pass with the water to the labial palps, and so to the mouth. The edges of the mantle lobes do not always remain free through their whole extent, but frequently fuse together, first at the hind end, and then gradually forwards. As a result of this fusion, a posterior opening, including in itself the inhalent and exhalent orifices, becomes separated from the anterior opening into the mantle cavity; and, further, the exhalent and inhalent openings become separated from each other by a transverse bridge of tissue. The long anterior opening or *foot-cleft*, in consequence of the progressive fusion of the mantle edges, often becomes gradually so shortened, that the foot, which is correspondingly reduced, can scarcely be protruded. In this case, the mantle comes to have the form of a saccular investment with two openings. The further forward the fusion of the two mantle lobes proceeds, the more marked becomes a peculiar elongation of the posterior mantle region round the inhalent and exhalent openings—an elongation of such a nature that two contractile tubes, or **siphons**, become formed (fig. 499, *a*). The latter may reach such a size that they can no longer be drawn between the posterior edges of the gaping valves of the shell. The two siphons often fuse with one another; but the two canals, with their openings surrounded by tentacles, remain separate. In the most extreme cases the siphons are enormously enlarged, and the posterior region of the body is peculiarly elongated and uncovered by the rudimentary shell; so that the whole animal acquires a vermiform appearance, the shell-bearing anterior region of the body constituting the head (*Teredo*, fig. 505).

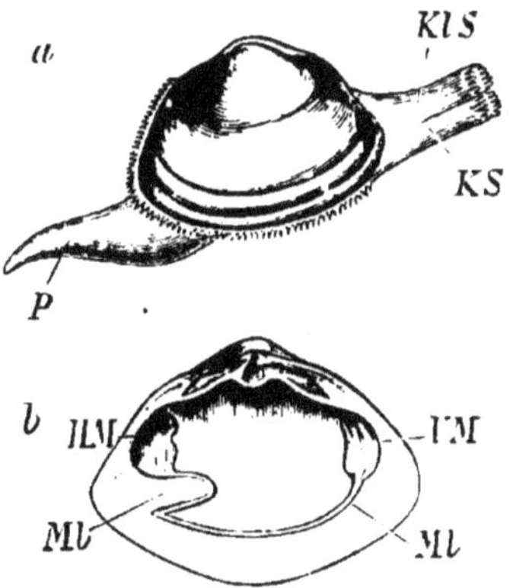

FIG. 499.—*a*, *Mactra elliptica*, animal with shell; *KlS*, cloacal or exhalent siphon; *KS*, branchial or inhalent siphon; *P*, foot. *b*, left valve of *M. solida*; *VM*, anterior adductor muscle; *HM*, posterior adductor muscle; *Ml*, pallial line; *Mb*, pallial indentation.

The mantle and skin consist of a cellular, slimy epidermis, beneath which lies a connective tissue, richly traversed by muscular fibres. The epidermis on the outer surface of the mantle consists of columnar cells; while on the inner surface of the mantle the cells composing it are ciliated. Pigments are present principally upon the edges of the mantle, which are frequently folded or beset by papillæ and tentacles.

The outer surface of the mantle secretes a strong calcareous shell, which is constituted of two valves corresponding to the two mantle lobes. The two valves are united dorsally. They are rarely exactly alike. Nevertheless, the term unequivalve is only applied to those shells in which the asymmetry is very marked, and the valves can be distinguished as upper and lower. The lower valve is the larger and more arched, while the upper is smaller and flatter, closing up the cavity of the lower after the manner of an operculum. The edges of the two valves are generally closely applied to one another, still they may gape more or less widely at various points for the exit of the foot, byssus and siphons. The latter is especially the case for those Molluscs which bore in sand, wood, or hard rock. In extreme cases the shell may, by a wide anterior emargination and an extended docking of its posterior part, be reduced to an annular rudiment (*Teredo*), while to its hinder end is applied a calcareous tube, which may intimately fuse with the shell rudiments and receive the latter entirely into itself (*Aspergillum*).

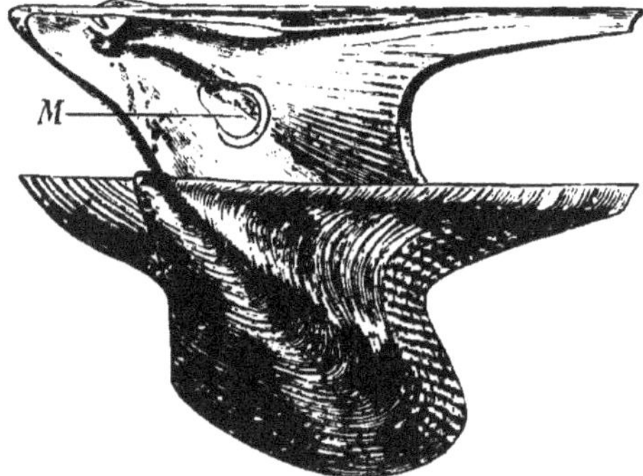

FIG. 500.—*Avicula semisagitta*, the valves are shifted over one another; *M*, muscle impression.

The two valves of the shell are always connected dorsally by an external or internal ligament, which tends to keep the valves open. The two shell valves are also firmly connected together dorsally by interlocking teeth, which constitute the so-called hinge (*cardo*). The hinge edge with the ligament is therefore to be distinguished from the free edge of the shell, which is divided into an anterior, inferior (ventral), and posterior or siphonal edge. The anterior and posterior edge may generally be easily determined by the position of the hinge-ligament with regard to the two *umbones* (*nates*), which have the form of two prominencies projecting over the dorsal edge, and indicate the point (*apex*) where the development of the valves began. The *area* is behind the apex, and includes the dorsal posterior side of the shell. The part of the dorsal edge in front of the apex is usually shorter, and contains, at least in the equivalve species, an excavation, the *lunula*, by means of which the anterior edge can be at once recognised.

While the outer surface of the shell presents various sculpture markings, the inner surface is smooth and shines with the lustre of mother-of-pearl. On a closer examination, impressions and pits become visible on the inner surface. A narrow line, the so-called *mantle* or *pallial line* (the line of attachment of the mantle edge to the shell), is placed near and fairly parallel to the ventral edge of the shell (fig. 499, *Ml*). In the siphoned forms this presents posteriorly a bend directed forwards and upwards (*Mb*), the *pallial bay*, which is due to the siphons. Impressions are usually caused by the insertion of an anterior and posterior adductor muscle which pass through the body of the animal transversely from one side to the other, and are attached to the inner surface of the shell (fig. 499, *HM*, *VM*). While in the equivalve mussels (*Orthoconcha*) the two impressions are usually of equal size, in the unequivalve forms (*Pleuroconcha*) the anterior adductor is reduced, and may completely vanish; the posterior adductor, on the other hand, now a muscle of much larger size, shifts forward to the middle of the shell (fig. 500). Hence the names *Dimyaria* and *Monomyaria*. According to its chemical composition, the shell consists of carbonate of lime and an organic matrix (conchyolin), which usually presents a laminated texture. In addition to this laminated layer there is also a thick external calcareous layer, composed of large, pallisade-like prisms, which are placed side by side and may be compared to the enamel of teeth (fig. 501, *S*). Finally, on the outer surface of the shell there is a horny cuticle, the so-called epidermis (*Cu*). The internal laminated layer is secreted by the whole internal surface of the mantle,

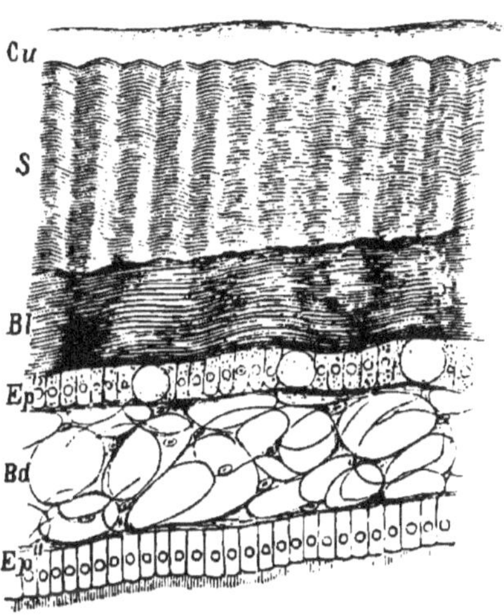

FIG. 501.—Vertical section through the shell and mantle of *Anodonta* (after Leydig). *Cu*, cuticle; *S*, prismatic layer; *Bl*, laminated (mother-of-pearl) layer; *Ep*, external epithelium of mantle; *Bd*, connective tissue substance: *Ep''*, internal epithelium of mantle.

while the two outer layers are formed only by the free edge of the mantle. The growth of the shell is effected in two ways; (1) by additions to the internal laminated layer, whereby the shell increases in thickness; (2) by additions to the prismatic and horny layers, whereby it increases in superficial extent. Accordingly the outer coloured part of the shell, which is composed of vertical prisms and a horny cuticle, when once formed cannot increase in thickness; while new concentric layers are constantly being added to the internal colourless mother-of-pearl layer during the whole life of the animal. The mantle-secretion gives rise in the so-called pearl-mussel (*Meleagrina*, *Unio margaritifer*), to the formation of pearls.

The foot is completely absent in comparatively few of the *Lamellibranchiata*, and only in those which have lost the power of locomotion (*Ostrea*, *Anomia*). In many forms, principally in the larva (*Unio*), less frequently in the adult (*Mytilus*), the foot possesses a **byssus** gland, which secretes silk-like fibres, by which a temporary or permanent attachment of the animal is effected. The form and size of the foot vary very considerably, according to the special kind of locomotion. The foot is most frequently used for creeping in sand, and then is hatchet-shaped; in other cases it is spread out laterally and its creeping surface has the form of a disc. More rarely it is of a large size and bent, in which case it serves for springing movements in the water (*Cardium*). Some Lamellibranchs possess a linear club-shaped or cylindrical foot (*Solen*, *Solenomya*), and move by rapidly retracting the foot and ejecting water through the siphons. Many use the foot for burying themselves in mud; others bore into wood (*Teredo*) or hard rock (*Pholas*, *Lithodomus*, *Saxicava*, etc.), for which purpose they push themselves against the rock with their short blunt foot (*Pholas*, *Teredo*), and use the hard and often finely serrated edge of their shell as a grater, giving it a rotatory movement. According to Hancock, the foot and edge of the mantle at the anterior end of the gaping shell are beset with siliceous crystals, and effect the excavation of the rock after the manner of a file.

The **nervous system** presents three pairs of ganglia, the cerebral, pedal, and visceral ganglia. The visceral ganglia are connected with the cerebral by a longer or shorter commissure on each side (figs. 496 and 498). Since there is never a distinct head, and sense organs do not appear on the anterior region of the body, the brain (cerebral ganglia) is proportionately little developed. Its nerves supply mainly the region round the mouth and the mantle, to which two large nerves are often distributed. The two halves of the brain are frequently (*Unio*) far removed from one another laterally, and are approximated to the anteriorly placed pedal ganglia (*Pecten*), whose nerves are distributed on the ventral side of the body in the foot. The large *visceral ganglia* are placed on the ventral side of the posterior adductor muscle, and supply nerves partly to the gills and partly to the viscera and to the mantle; the latter are two large trunks which run in the edge of the mantle and anastomose with the mantle nerves from the brain, often forming plexuses. Large nerves also pass off from the visceral ganglia to the siphons, at the base of which they form an accessory pair of ganglia.

Sense organs.—Auditory organs, eyes, and tactile organs are present. The former have the form of paired auditory vesicles, and lie beneath the œsophagus attached to the pedal ganglia (their nerve, however, arises from the brain); they are characterised by the large hair cells which line the wall of the vesicle. Eyes may either be simple pigment spots at the end of the respiratory tube (*Solen*, *Venus*), or be much more highly developed and placed on the edge of the mantle of *Arca*, *Pectunculus*, *Tellina*, and especially of *Pecten* and *Spondylus*. In the latter genera they are placed on stalks between the marginal tentacles, and have an emerald green or brown red colour; they consist of an eye-bulb with a corneal lens, choroid, iris, and a well-developed layer of rods into which the optic nerve passes. The sense of touch is provided for by the labial palps, the edges of the respiratory apertures (siphons) with their papillæ and cirri, and also the often numerous tentacles at the edge of the mantle (*Lima*, *Pecten*). In all probability the hair cells found in the mantle are the seat of a special olfactory sense (tracking sense).

The **digestive organs** begin with the mouth, which is placed between the labial palps (fig. 498). The mouth leads into a short œsophagus, into which the cilia of the labial palps drive small nutrient particles received into the mantle cavity with the water. Jaws and tongue are always absent. The œsophagus widens into a spherical stomach, at the pyloric end of which a blind sac, which can be closed up, is attached. A rod-like transparent structure (*crystalline style*) is often found either in the above-mentioned blind diverticulum of the stomach, or in the alimentary canal itself. It is to be regarded as an excretion-product of the alimentary epithelium, and is periodically renewed. The intestine always attains a considerable length, is much coiled and is surrounded by the liver and generative glands; it projects into the foot and then ascends again behind the stomach to the dorsal surface; it then traverses the ventricle of the heart, passes over (dorsal to) the posterior adductor muscle to open at the hind end of the body into the mantle cavity at the end of a projecting papilla.

The **circulation** is effected by an arterial heart, which is enclosed in a pericardium and lies in the dorsal middle line slightly in front of the posterior adductor muscle. The heart consists of a median ventricle, which is perforated by the alimentary canal, and of two lateral auricles, through which the blood enters the ventricle. The ventricle of the heart of *Arca* is peculiar in being double; the efferent aortæ, however, unite to form an unpaired vessel. The ramifications of the anterior and posterior aorta lead the blood into a complicated system of lacunæ in the mantle and in the interspaces between the viscera. These, which coincide with the body cavity, represent the capillaries and finer venous vessels; while, by some observers, they have been regarded as a true capillary and venous system. The chief venous sinuses are two lateral sinuses placed at the base of the gills, and a median sinus into which the lacunæ of the foot lead. From these part of the blood passes direct into the gills; the main part, however, first passes through a network of canals in the walls of the kidney or organ of Bojanus, as through a kind of portal circulation, and thence into the gills, whence it is returned as arterial blood to the auricles of the heart. Water is said to enter the circulation through openings in the foot and to become mixed with the blood. Nevertheless the erectile networks of the foot are blood-lacunæ.

Organs of respiration.—There are usually two pairs of branchial leaflets (gills), which begin behind the labial palps and pass backwards along the sides of the body. The outer surfaces and the interlamellar water-spaces of these branchial leaflets are covered with cilia, which keep up a continuous flow of water over the gills. The outer gill, viz., that lying next the mantle, is usually considerably the smaller of the two. It is often completely absent, so that the number of the gills is reduced to a single pair. Sometimes the gills of the two sides fuse with one another across the middle line in the posterior region, and may in extreme cases represent a sack, like the branchial sack of the Ascidians (*Clavagella*).

The most important of the **excretory organs**—the organ of Bojanus, so-called after its discoverer—is a paired, glandular sac with folded walls, and of an elongated oval form, whose cavity communicates with the pericardium (fig. 498). The substance of this gland, which functions as kidney, is composed of a yellow or brown spongy tissue, which is covered with a closely ciliated layer of cells, from which concrements containing calcareous matter and uric acid (also *guanin*) are excreted. The simple duct often receives the duct of the generative apparatus, or the two organs open together on a common papilla on either side. In the *Siphoniata*, on the other hand, the renal and generative openings are almost always separate.

Generative organs. — The Lamellibranchs are, with a few exceptions (the genera, *Cyclas, Pecten, Ostrea, Clavagella, Pandora*), diœcious. Both kinds of sexual organs lie amongst the viscera, and have the form of lobed or racemose glands, which are placed near the liver, surround the windings of the intestine, and extend into the base of the foot. The testis and ovary can usually be distinguished from one another with the unaided eye by their colour; the ovary being red in consequence of the colour of the ova: the sperm, on the contrary, is milk-white or yellow. The openings of the ducts are placed right and left near the base of the foot. The form, position and opening are exactly the same in the hermaphrodite glands, in which the male and female follicles may be separate and open separately (*Pandora*) or together (*Pecten, Clavagella, Cyclas*); or the same follicles may function sometimes as ovary and sometimes as testis (*Ostrea, Cardium norwegicum*). In the diœcious forms, the male and female animals may differ in the shape of the shell, as is the case in the fresh water *Unionidæ*. Here the outer gills of the female are used for the reception of the eggs (brood pouch), and the shell is more arched. Hermaphrodite individuals are met with among the freshwater mussels, both in *Unio* and in *Anodonta*. The fertilization of the eggs is probably usually effected in the mantle or branchial cavity of the female.

But few Lamellibranchs are viviparous. The fertilized eggs, however, almost always remain for some time between the valves of the shell, or pass into the branchial leaflets, where they undergo the early processes of embryonic development under the protection of the mother. This care of the brood is especially conspicuous in the freshwater forms; in the *Unionidæ* the eggs pass into the great longitudinal canal of the external gill, whence they are distributed into the gill spaces, which become enormously widened and modified into peculiar brood-pouches. In the emptying of these brood-pouches the contents are expelled through the great longitudinal canal as a mass of eggs, united together by mucus and containing ciliated embryos, or as a continuous string of eggs.

The development* of the embryo is introduced by an unequal segmentation. The segments arrange themselves in the form of a blastosphere, on which the archenteron often arises by invagination, while the mesoderm is developed from two cells which are early

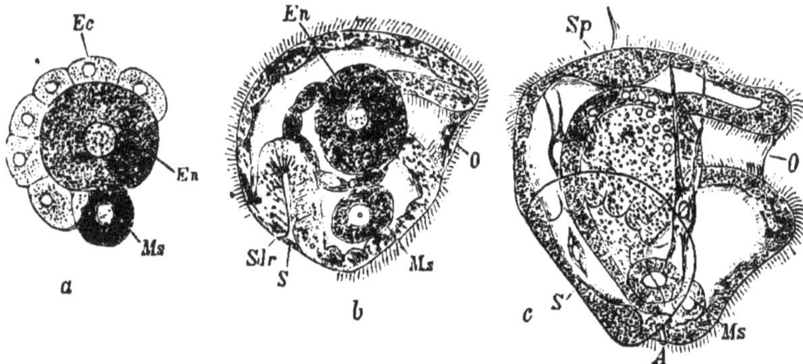

FIG. 502.—Stages in the development of the larva of *Teredo* (after B. Hatschek). *a*, optical median section of an embryo with two mesoderm cells (*Ms*) and two entoderm cells (*En*); *Ec*, ectoderm cells. *b*, Ciliated embryo with mouth (*O*), stomach, intestine, and shell gland (*Sdr*); *S*, shell.—*c*, Later stage; *Sp*, apical plate; *A*, anal invagination.

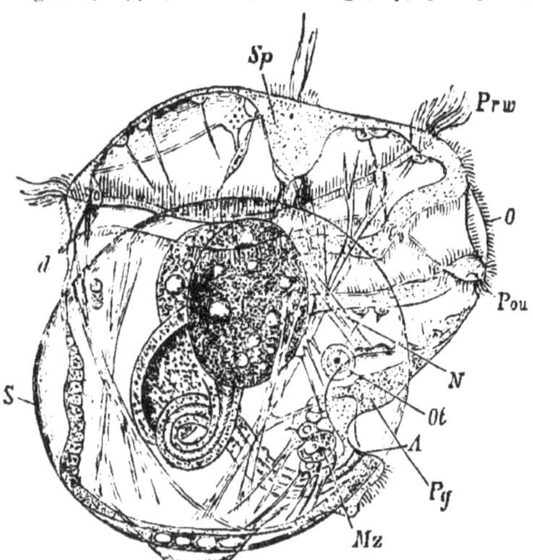

d, Larva of *Teredo*. *O*, mouth; *A*, anus; *Prw*, praeoral ciliated ring; *Pow*, postoral ciliated ring; *N*, pronephros; *Ot*, otocyst; *Pg*, pedal ganglion; *Mz*, mesoderm cells.

separated. The first trace of the endoderm also may have the form of two cells (fig. 502). The embryo, which is partially ciliated and often rotates within the egg membranes, soon acquires a ciliated velum and shell gland. The nervous system, otocysts, and foot are not differentiated till

* Vide especially Lovén. "Bidrag till Kännedomen om Utvecklingen af Mollusca Acephala Lamellibranchiata." Stockholm, 1848.

Flemming, "Studien über die Entwickelungsgeschichte der Najaden." *Sitzungsber. der K. Akad. der Wissensch.* Vienna, 1875.

Carl Rabl, "Ueber die Entwickelungsgeschichte der Malermuschel." Jena. 1876.

B. Hatschek, "Ueber die Entwick-gesch. von Teredo." *Arbeiten aus dem zool. Institute, etc.*, Tom. III. Vienna, 1881.

afterwards; while the heart, kidney, and gills are still later in making their appearance. Among the provisional arrangements the velum, which proceeds from the sides of the præoral ciliated ring, is very generally present, and in the free-swimming larvæ has the form of a large ciliated ring or collar.

The development of the freshwater forms (*Cyclas*, *Unio*, *Anodonta*), in which the eggs and embryos are contained in well-protected brood pouches, may generally be called direct. The marine Lamellibranchs, on the other hand, are born at an early stage, and swim about for a long time as larvæ with large umbrella-like velum, from which the labial palps are developed (fig. 503).

The Lamellibranchs are for the most part marine and live at different depths, sometimes creeping, sometimes swimming and jumping. Many are without the power of changing their position, inasmuch as they fix themselves at an early age by means of the byssus threads to rocks and stones (oysters). Others, as the boring forms, bore passages in the wood of ships and piles and in rocks.

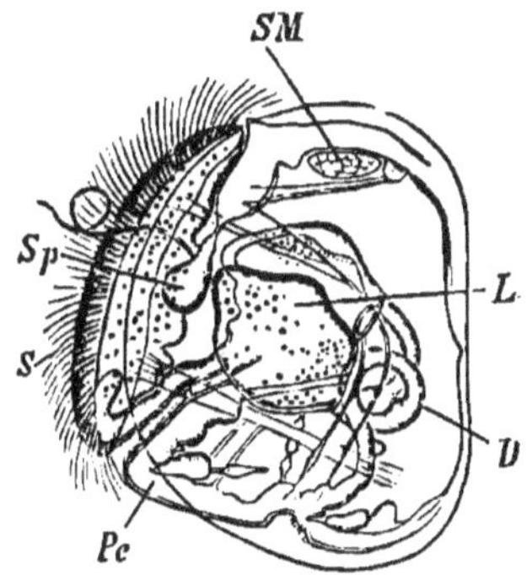

Fig. 503.—Larva of *Montacuta bidentata* (after Lovén). *S*, velum; *Sp*, apical plate with flagellum; *D*, intestine; *L*, liver; *SM*, anterior adductor muscle; *Pc*, foot.

The Lamellibranchs had a wide distribution in the earlier periods of the earth's history, and their fossil shells are most excellently preserved; they are therefore of the greatest importance as characteristic fossils for the determination of the age of formations.

1.—ASIPHONIA.

Mantle without siphons. Pallial impression simple.

Fam. **Ostreidæ.** Oysters. Shell valves unequal, laminated, with weak hinge usually without teeth, and simple central adductor muscle. In the true oysters the more arched left valve is firmly attached, while the right and upper valve, which is fastened by an internal ligament, lies as an operculum on the lower valve. Mantle completely split and fringed at the edge; gill lamellæ, on the contrary, partially fused on their outer edge. Foot absent or rudimentary. They usually live together, like colonies, in the warmer seas, where they may form banks of considerable extent (oyster banks). They were also represented in earlier times, especially in the Jura and in the Chalk. *Ostrea edulis* L., oyster, on the coasts of Europe on rocky ground; probably includes a series of different species according to the locality. According to Davaine, the oysters are said to produce only male sexual products towards the end of the first year.

and it is only later, from the third year onwards, that they become females and produce ova. Moebius, on the contrary, asserts that the sperm is the later formed, and not until after the pregnant beast has got rid of her eggs. The reproduction takes place especially in the months of June and July, at which time, in spite of their extraordinary fertility, the oysters must not be gathered. *O. crista galli* Chemn., in the Indian Ocean. *Anomia ephippium* L., *Placuna placenta* L.

Fam. **Pectinidæ.** Scollops. Shell equivalved or unequivalved, but tolerably equisided, with straight hinge line; often with fan-shaped ribs and bands, with single adductor muscle. The free and completely split mantle edges bear numerous tentacles, and often emerald green eyes in great number. The small foot often secretes byssus fibres for attachment. Some are attached by their arched shell valve (*Spondylus*), others swim about by rapidly opening and closing the shell (*Pecten*). Many are edible and are even more esteemed than

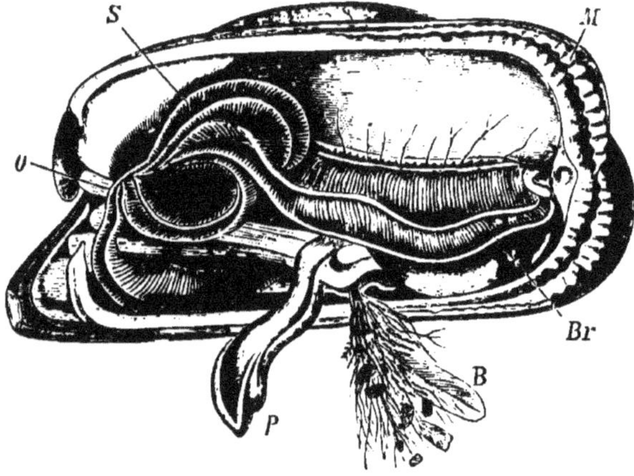

FIG. 504.—*Mytilus edulis* (règne animal). *O*, mouth; *S*, labial palps; *P*, foot; *B*, byssus secretion; *Br*, gills; *M*, thickened edge of mantle.

the oysters. *Pecten Jacobæus* L., *P. maximus* L., *P. varius* L. Mediterranean. *Spondylus gaederopus* L. *Lima squamosa* Lam.

Fam. **Aviculidæ.** With oblique unequivalved shell of laminated texture and inner mother-of-pearl layer. They possess two adductor muscles, of which, however, the anterior is very small. Mantle completely open. Foot small, secreting byssus. *Avicula hirundo* L., Gulf of Tarent. *Meleagrina margaritifera* L., pearl mussel, inhabits especially the Indian and Persian Oceans, and also the Gulf of Mexico. Secretes pearls.* The internal layer of the shell is used in commerce as mother-of-pearl. *Malleus vulgaris*, Lam., Indian Ocean.

Fam. **Mytilidæ** (fig. 504). Mussels. Shell equivalved, covered with thick epidermis, with large posterior and small anterior muscle impressions. The tongue-shaped foot fastens itself by the byssus fibres which it secretes. Mantle more or less free except a short siphonal opening fringed at the edge. *Pinna squamosa* Gm., Mediterranean. *Mytilus edulis* L., edible mussel of the North Sea and Baltic (fig. 504). *Lithodomus dactylus* Sow., in the Mediter-

* *cf.* Moebius, "Die echten Perlen, etc." Hamburg, 1857.

ranean (Temple of Serapis at Pozzuoli). *Dreyssena polymorpha* Fall., has gradually extended over many freshwater systems of Germany.

Fam. **Arcaceæ** (Archemuscheln). Shell thick, equivalved with well-developed hinge, and covered by hairy epidermis. The two adductors form two equally large anterior and posterior muscle impressions. *Arca Noæ* L., Mediterranean. *Pectunculus pilosus* L., Mediterranean.

The **Trigoniadæ (Trigoniacea)** are allied here. *Trigonia pectinata* Lam.

Fam. **Unionidæ (Najades).** Freshwater Mussels. With long equivalved but not equisided shells, which are covered externally by a strong smooth usually brown epidermis, and internally by a mother-of-pearl layer. One of the muscle impressions is divided. Foot with cutting edge; gills fused behind the foot. The outer gill plates also function as brood-pouches for the developing eggs. They live in standing or running water.

Anodonta cygnea Lam., in ponds. *A. anatina* L., more in rivers and brooks. *Unio pictorum* L., (Malermuschel). *Unio tumidus* Retz., *batavus* Lam. *Margaritana margaritifera* Retz. (Flussperlmuschel), in mountain streams of South Germany, especially in Bavaria, Saxony, and Bohemia.

II.—SIPHONIATA.

Part of the mantle edges fused, with elongated tubular siphons.

Fam. **Chamidæ (Chamacea)** (Gienmuscheln). Shell unequivalve, with strongly developed cardinal teeth and simple pallial line. The mantle edge fused, except at three points, viz., the opening for the foot, the dorsal (cloacal) and ventral (inhalent) siphons. *Chama Lazarus* Lam.

The **Tridacnidæ** are closely related to the above. *Tridacna gigas* L. *Hippopus maculatus* Lam. Indian Ocean.

Fam. **Cardiidæ (Cardiacea)** Cockles. Shell equivalve, fairly thick, heart-shaped and arched, with large incurved umbones, external ligament, and strong hinge formed of several teeth. Siphons short. Foot powerful and bent elbow-like, serves for swimming; passes out through anterior slit. *Cardium edule* L., North Sea and Mediterranean. *Hemicardium cardissa* L., East Indies.

Fam. **Lucinidæ (Lucinacea).** Shell circular, free, closed, with one or two cardinal teeth, and a second quite rudimentary lateral tooth. Pallial line simple. Mantle open in front, prolonged behind into one or two siphons. *Lucina lactea* Lam. Mediterranean.

Fam. **Cycladidæ.*** Shell equivalve, free, swollen, with external ligament and thick horny epidermis. Mantle with two (rarely one) more or less fused siphons. Live in fresh water. *Cyclas cornea* L., *Pisidium* Pf. *Corbicula* Mühlf.

Fam. **Cyprinidæ.** Shell regular, equivalve, elongated to an oval, closed, with thick and strong epidermis. One to three principal cardinal teeth, and usually a hinder lateral tooth. Pallial line simple. Mantle edges fused to form two siphonal openings. *Cyprina islandica* Lam., *Isocardia cor* L. Mediterranean.

Fam. **Veneridæ.** Shell regularly round, or oblong with three diverging cardinal teeth on each valve. Pallial line bent in. Siphons of unequal size, fused at the base. *Venus verrucosa* L., Mediterranean. *Cytherea Chione* L., edible. Mediterranean. *C. Dione* L., Atlantic Ocean.

Fam. **Mactridæ** (Fig. 499). Shell trigonal, equivalve, closed or slightly

* Fr. Leydig, "Anatomie und Entwickelung von Cyclas." *Müller's Archiv*, 1855.

gaping, with thick epidermis. Two diverging cardinal teeth. Pallial indentation short, rounded. Siphons fused, with fringed openings. *Mactra stultorum* L., Mediterranean. *Lutraria* Lam.

Fam. **Tellinidæ.** With two long, completely separated siphons; edges of mantle widely open, bearing tentacles. Triangular foot. *Tellina baltica* Gm. *T. radiata* L. *Donax trunculus* L.

Fam. **Myidæ** (Gapers). Mantle almost completely closed, with slit for the protrusion of the short or cylindrically elongated foot, and very long fleshy fused siphons. The valves gape at each end and possess a weak hinge. Bury themselves deep in mud and sand. *Solen vagina* L., razor shell. *Mya truncata* L. (Gaper).

Fam. **Gastrochænidæ (Tubicolidæ).** Shell thin, equivalve, toothless, sometimes inserted in a calcareous tube formed by an excretion of the mantle. Mantle with one small opening anteriorly and prolonged behind into two fused siphons with terminal openings. *Gastrochæna clava* L., *Clavagella bacillaris* Desh. *Aspergillum javanum* Lam., Indian Ocean.

Fam. **Pholadidæ.** Boring mussels. The valves of the two sides gaping; without cardinal teeth and ligament, but with accessory calcareous pieces which lie either on the hinge (*Pholas*) or on the siphons (*Teredo*, fig. 505). Mantle with only small opening for the passage of the thick foot. Siphons elongated. Bury themselves in mud and sand, or bore into wood and even into solid stone, calcareous rocks and corals. They form passages, from which they protrude their fused siphons. *Pholas dactylus* L. Piddock, *Ph. crassata* L. *Teredo navalis* L. (Fig. 505) Shipworm, was the cause of the famous dam-break in Holland at the beginning of last century.

FIG. 505. — *Teredo navalis*, removed from its calcareous tube, with elongated siphons (after Quatrefages).

SCAPHOPODA.

Diœcious Mollusca without head, eyes, or heart, with tri-lobed foot, and tubular calcareous shell open at the two ends.

The *Scaphopoda* are allied to the Lamellibranchs. The admirable investigations of *Lacaze-Duthiers* * first cleared up this group of Molluscs, which were for a long time known as *Cirrobranchiata* and grouped amongst the Gastropods. He showed that they are closely related to the *Acephala*, and constitute forms transitional between the latter and the *Cephalophora*.

The shell is an elongated, somewhat bent, open, conical (with the apex broken off) tube, and contains the animal, which has a similar shape and is fastened by a muscle to the thinner lower edge of the shell

* Lacaze-Duthiers, "Histoire de l'organisation et du développement du Dentale." *Ann. des Sc. Nat.*, 1856-1858.

(fig. 506). The body possesses a saccular mantle, like the shell open at both ends, and a trilobed foot; the foot is protruded through the larger of the openings of the shell from the anterior opening of the mantle, the margin of which is thickened. A separated cephalic region is not present, but there is an egg-shaped projection in the mantle cavity, at the apex of which is placed the mouth, surrounded by eight leaf-like labial appendages.

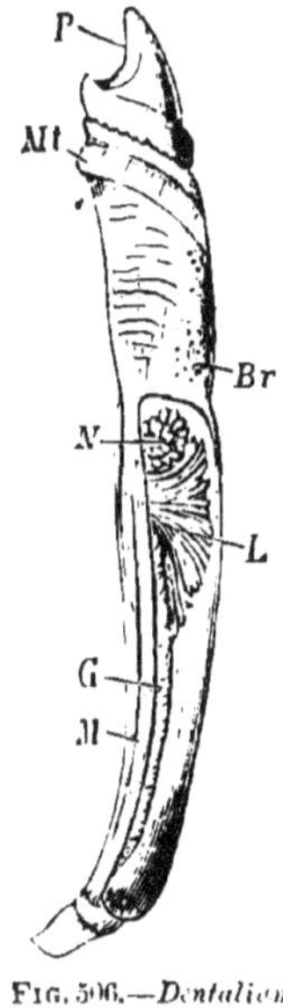

Fig. 506.—*Dentalium Tarentinum* (after Lacaze-Duthiers). Animal without shell from right side. *P*, foot; *Mt*, circular muscle of mantle; *M*, longitudinal muscle; *Br*, gills; *N*, kidney; *L*, liver; *G*, generative gland.

The **buccal armature** consists of a lateral (right and left) rudimentary jaw, and a tongue beset with five rows of plates.

The **alimentary canal** is divided into a buccal cavity, œsophagus, stomach with large liver, and an intestine, which after several coils closely pressed together, opens behind the foot into the middle of the mantle cavity.

The **circulatory organs** are reduced to two mantle vessels and a complicated system of wall-less spaces of the body cavity.

Respiration is effected by the surface of the mantle and also by the filiform tentacles, which arise from two ridges (*cervical collar*) behind the head-like buccal prolongation.

The **kidney** lies round the rectum, and opens by two openings placed on the right and left of the anus.

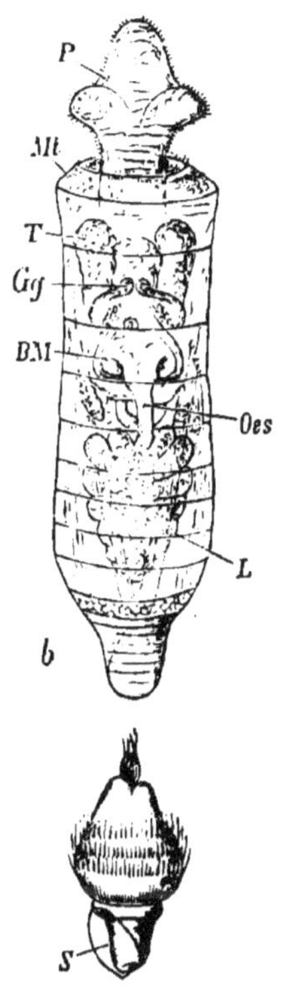

Fig. 507.—Larva of *Dentalium* (after Lacaze-Duthiers). *a*, young larva with first rudiment of shell (*S*). *b*, Older larva seen from the dorsal surface; *T*, tentacle collar; *Gg*, cerebral ganglion; *Oes*, œsophagus; *L*, liver.

The **nervous system** consists of three groups of ganglia, of which the pedal ganglion bears two otocysts. **Eyes** are absent. The numerous ciliated tentacles serve as *tactile organs*.

The *Scaphopoda* are diœcious. The **ovaries** and **testes** are unpaired finger-shaped lobed glands, which are placed behind the liver and intestine, and open to the exterior with the right kidney.

The animals live buried in mud, and creep about slowly by means of the foot. The young swim about for some time as larvæ, provided with ciliated tuft and ciliated collar; then acquire a shell, which is almost bivalve, a velum, and foot; the shell subsequently becomes tubular (fig. 507).

Order.—**Solenoconchæ.**

Fam. **Dentalidæ.** *Dentalium entalis* L., *D. elephantinum* L. Mediterranean and Indian Ocean.

Class II.—GASTROPODA.*

Mollusca with distinct head, often bearing tentacles: a ventral muscular foot and undivided mantle, which frequently secretes a simple plate-shaped or spirally twisted shell.

The anterior part of the body or head usually bears two or four tentacles and two eyes, which are placed sometimes at the apex, usually at the base of a pair of tentacles (fig. 508). The muscular foot projects from the ventral side of the body: its form and size presents numerous modifications. As a rule it has a broad and long plantar surface; but in the *Heteropoda* it has the form of a vertically extended fin. The shape of the body depends on the position and form of the mantle. The latter is placed like a cap on the dorsal surface, and consists of a more or less considerable fold of the dorsal integument; its edge is usually thickened, sometimes also prolonged into

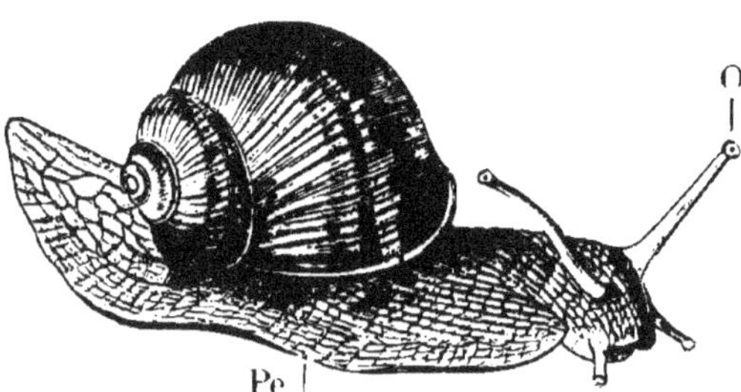

FIG. 508.—*Helix pomatia*. *O*, Eyes at the extremity of the long tentacles; *Pe*, foot.

* Martini and Chemnitz. Conchylien-Cabinet. 12 Bde. Herausgegeben von Küster. Nürnberg. 1837-1865.
Sowerby. "Thesaurus conchyliorum, or figures and descriptions of shells." London. 1832—1862.
Reeve. "Conchologia iconica, etc." London, 1842-1862.
H. and A. Adams. "The genera of the recent Mollusca," 3 vols. London. 1858.
H. Troschel. "Das Gebiss der Schnecken." Berlin, 1856-1878.
Woodward. "Manual of the Mollusca." 2nd ed., London, 1868.
Fol. "Etudes sur le développement des Mollusques." I. and II.
C. Rabl, "Ueber die Entwickelung der Tellerschnecke." Morphol. Jahrbuch. Tom. V. 1819.

lobes or drawn out into processes. The lower surface of the mantle usually serves as the roof of a cavity, which extends on to the dorsal surface and also on to the sides of the body. This cavity contains the respiratory organ, and opens to the exterior by an aperture or tubular prolongation at the mantle edge.

The body cavity is developed on the dorsal surface of the foot, usually in a visceral sac, which projects like a hernia. The visceral sac tapers gradually at its upper end, and is usually spirally twisted. The mantle and visceral sac are covered by the shell, which to a certain extent repeats the twistings of the latter and can usually completely receive and protect the head and foot when the animal is retracted. The shell is as a rule hard and calcareous, and possesses an internal nacreous layer similar to that of the mother-of-pearl layer of the Lamellibranch shell. The shell is sometimes delicate, horny, and flexible, or it may have a gelatinous (*Tiedmannia*) or cartiliginous (*Cymbulia*) consistency. More rarely the shell is so small that it only covers the mantle cavity with the respiratory organs or lies hidden completely within the mantle (*Limax*, *Pleurobranchiata*). In other cases it is thrown off at an early stage, so that the adult beast is completely without a shell (*Nudibranchiata*). The shell differs from that of the *Lamellibranchiata* in being composed of a single piece; it is either flat and cup-shaped (*Patella*) and uncoiled, or it is spirally twisted in very different ways, from a flat disc-shaped to the long drawn-out turret-shaped spiral (fig. 509). In the first case it more resembles the embryonic shell, which lies as a delicate, cap-shaped covering on the mantle. The growth of the shell keeps pace with that of the animal, the additions being made to the edge of the shell, viz., to that part which lies on the edge of the mantle. In consequence of the inequality of this growth spiral twistings arise, the diameter of which gradually and continuously increases. Inasmuch as the unsymmetrical growth of the shell is due to the unequal growth of the body, the position of the openings of the unpaired organs (anus, sexual opening) to one side of the great external lip of the shell is intelligible.

FIG. 509.—Section through the shell of *Helix pomatia*.

The following parts may be distinguished in a spirally-twisted shell; (1) the apex, as the part of the shell at which the growth began and from which the spiral twistings started; (2) the opening or *aperture*, which leads into the last and usually largest turn of the spiral; its lip (*peristoma*), swollen in the adult animal, lies on the edge of the mantle. The spiral is twisted to the right or left round an axis which is directed from the apex to the aperture, and is indicated either by a solid spindle (*columella*) or a hollow canal. When the turns of the spiral are far removed from the axis, this canal may become an almost conical space with a wide opening (*Solarium*). The turns are usually closely applied to one another; more rarely they are separated (*Scalaria pretiosa*). According to the position of the columella, a columella edge or inner lip and an outer edge or outer lip of the aperture may be distinguished. The latter may be entire (*holostomatous*), or broken by an excavation which is often prolonged into a canal (*siphonostomatous*). In many Gastropods an *operculum* is added; this is usually placed on the hind end of the foot, and closes the shell aperture when the animal is retracted. Many terrestrial Gastropods secrete before the beginning of the winter sleep an operculum, which is thrown off again in the spring.

The slimy integument consists of a superficial layer of cylindrical cells, which are frequently ciliated, and of a connective tissue dermis, which is inseparably connected with the dermal muscles. Calcareous and pigment glands are placed in the integument; they are especially numerous at the edge of the shell, where they contribute to the growth and peculiar colouring of the shell. The shell, which is a cuticular structure, is secreted by the epithelium, like other cuticular structures; it becomes hard when the calcareous salts which are mixed in the organic basis assume a hard and crystalline condition. The superficial layer of the shell often remains uncalcified as a thin delicate epidermis, while the inner surface is thickened by mother-of-pearl layers (secreted by surface of mantle). The connection of the animal to its shell is effected by a muscle, which on account, of its position on the spindle (*columella*), is called the spindle muscle. This muscle arises from the dorsal part of the foot, and is attached to the spindle at the beginning of the last turn of the spiral.

The **nervous system** presents a great resemblance to that of the *Lamellibranchiata*, but there are many differences in detail.

In the *Placophora*, whose nervous system presents close relations to that of *Neomenia* and *Chætoderma*, the ganglionic swellings are not marked (fig. 495). In all other cases the three typical groups of ganglia are present. *The cerebral ganglia (fig. 497, *C g*) are

* The subjoined account of the nervous system is slightly modified from the German.—Ed.

connected together by a transverse band, and each of them gives off a commissure to the pedal ganglia (*P g*), and a second commissure to a pair of visceral ganglia (*Plg*). The latter ganglia, which are known as the *commissural* or *pleural ganglia*, are also connected with the pedal ganglia (fig. 497). There are thus two nervous commissures round the œsophagus—the direct cerebro-pedal, and the cerebro-pedal by way of the pleural or commissural ganglia. The pleural ganglia may lie directly on the cerebral or pedal ganglia.

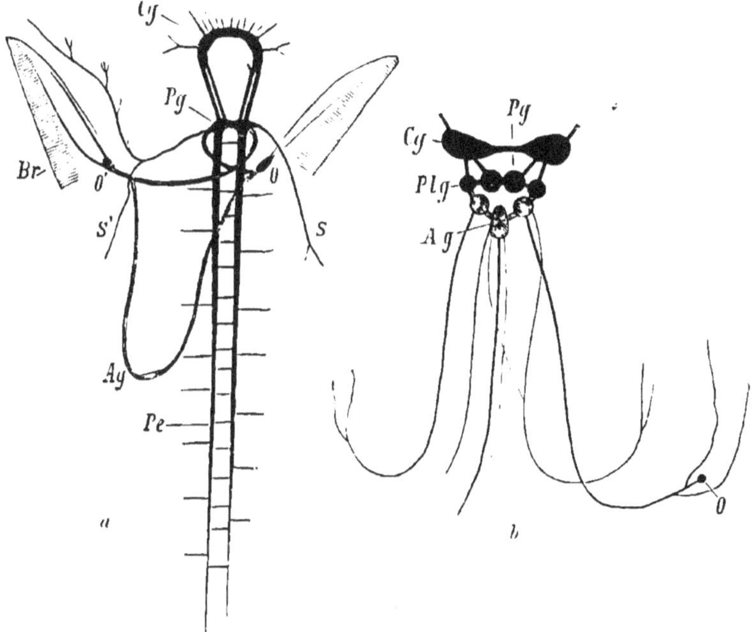

FIG. 510.—Nervous system of *Haliotis* (diagrammatic, after Spengel). *Cg*, cerebral ganglion; *Pg*, pedal ganglion; *Plg*, pleural ganglion (commissural ganglion); *Ag*, abdominal ganglion; *O* and *O'*, olfactory organs; *Pe*, pedal cord; *S* and *S'*, lateral nerves; *Br*, gills. *b*, Nervous system of *Limnæus* (after Lacaze-Duthiers).

The pleural ganglia are part of the third typical group of ganglia, viz., the visceral group. They are connected with each other by a long commissure, the *visceral commissure*, which often extends into the hinder part of the body, and contains several ganglia in its course; the latter ganglia, which also constitute part of the visceral group of ganglia, send off nerves to the sexual organs, kidney, heart, gills, olfactory organs, and mantle (fig. 497, *G s b*, *V g*, *G s p*; fig. 510 *a*, *O*, *O'*, *A g*; fig. 510 *b*, *A g*).

The visceral ganglionic system of Gastropods is therefore broken up into several ganglia, and is connected with the pedal (by the pleuro-pedal commissure) as well as with the cerebral.

In the *Prosobranchiata* the position of the visceral commissure, with its ganglia, and nerves presents a peculiar condition (*Chiastoneura*); the commissure from the right pleural ganglion passes over (dorsal to) the alimentary canal to the left side, and here forms a ganglion—the *supraintestinal ganglion* (fig. 497, *G s p*)—which supplies the left side, while the commissure from the left pleural ganglion passes under (ventral to) the alimentary canal to the right side, and there gives rise to a ganglion, the *subintestinal ganglion*, which supplies the right side (*vide* also fig. 510 *a*). The part of the visceral commissure, which connects the supra- and sub-intestinal ganglia often contains one or more ganglia (*V g*, *A g*). More rarely this crossing is less clearly marked. The cerebral ganglia always give off a pair of nerves, one on each side of the œsophagus, to the *buccal ganglia*, which give off nerves to the mouth and alimentary canal (fig. 497, *B g*).

Sense organs.—Eyes, auditory vesicles (otocysts), tactile and olfactory organs are present.

The *eyes* are paired, and are usually placed at the end of stalks, which are as a rule fused with the tentacles. The eyes are largest and most developed in the *Heteropoda*,* in which group they are fastened in special transparent capsules and admit of a movement of the bulb.

The two *otocysts* are ciliated internally, and are, except in the Heteropoda, connected with the pedal ganglion (fig. 497, *O t*), although their nerve always arises in the brain.

Tactile organs are represented by the tentacles, the edges of the lips which are often folded, and lobe-like prolongations which are found here and there on the head, mantle and foot. There are usually two tentacles; † exceptionally they are absent (*Pterotrachea*, etc.). They consist of simple contractile prolongations of the body wall, which can sometimes (*Pulmonata*) be invaginated into the interior of the body. Certain peculiar hair cells, from which tufts of hairs project in the aquatic Molluscs, are to be looked upon as the seat of a special sensation. They are scattered over the whole surface of the body, and are especially aggregated upon the parts of the body

* V. Hensen. "Ueber das Auge einiger Cephalophoren." *Zeit. für wiss. Zool.*, Tom. XV., 1865.

† W. Flemming. "Untersuchungen über Sinnesepithelien der Mollusken." *Arch. für. mik. Anatomie*, Tom. VI., 1870.

serving for the tactile sensation. The antennæ of the terrestrial Gastropods possess on their end-plates a great number of fine sense-cells (club-shaped cells with rods, Flemming), which are placed between specially-modified epithelial cells, and probably function as olfactory organs. Recently an organ, which was supposed to be a rudimentary gill and is innervated from the supraintestinal ganglion,

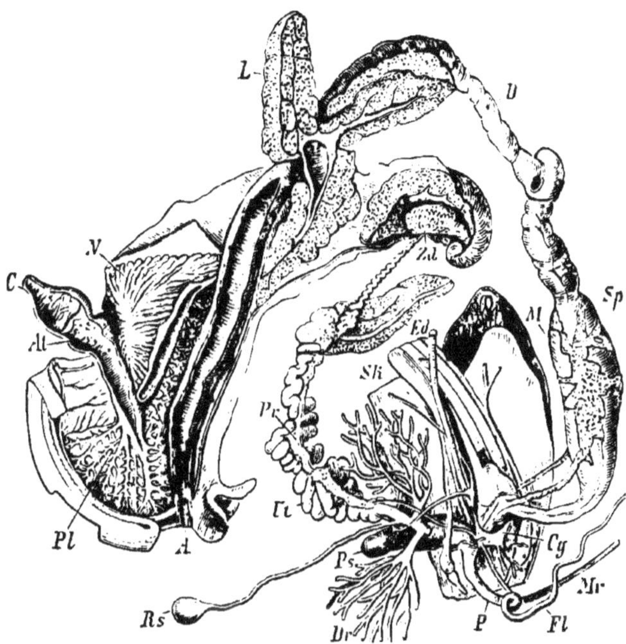

FIG. 511.—Anatomy of *Helix pomatia* (after Cuvier). The mantle cavity is opened on the left side, and the mantle is turned over to the right. The body cavity has been opened and the viscera are unravelled. *Cg*, cerebral ganglion; *Sp*, salivary gland; *M*, stomach; *D*, intestine; *L*, liver; *A*, anus; *N*, kidney; *At*, auricle; *C*, ventricle; *Pl*, lung; *Zd*, hermaphrodite gland, invested by the lobes of the liver; *Ed*, albumen gland; *Pr*, prostate; *Ut*, uterus; *Rs*, receptaculum seminis; *Dr*, finger-shaped glands; *Ps*, dart sac; *P*, penis; *Fl*, flagellum; *Mr*, retractor muscle; *Sp*, spindle muscle.

has been recognised as a sense organ and explained as an *olfactory* * organ.

In the Zeugobranchiata (*Fissurella*, *Haliotis*), two such organs are present, one on the right and the other on the left side, and are indicated by a considerable ganglion.

The **digestive organs** rarely have a straight course; they are

* J. W. Spengel, "Die Geruchsorgane und das Nervensystem der Mollusken." *Zeit. für. wiss. Zool.*, Tom. XXXV.

usually much coiled, and as a rule bend forwards to open in front on the right side in the mantle cavity. The anus, however, is sometimes on the dorsal surface behind.

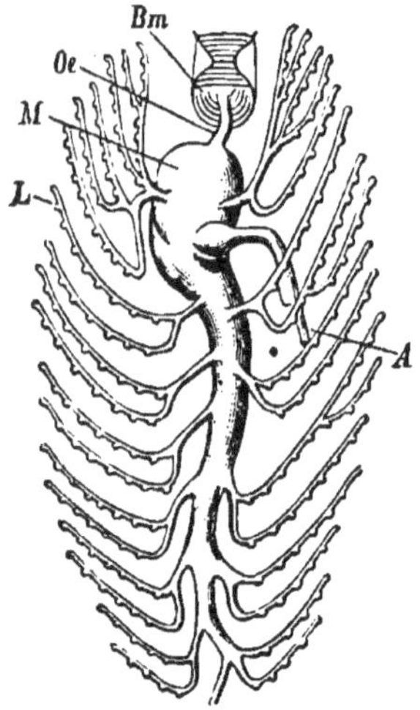

Fig. 512.—Alimentary canal of *Eolis papillosa* (after Hancock). *Bm*, buccal mass; *Oe*, œsophagus; *M*, stomach, *L*, liver sacs, which enter the dorsal appendages; *A*, anus.

Many of the higher *Gastropoda* possess an invaginable proboscis, the invagination beginning at the base; others possess one which is retractile from the point. The mouth is bounded by lips, and leads into a buccal cavity armed with hard masticating structures, and receiving the ducts of two salivary glands. The buccal cavity leads into the œsophagus, which is followed by a dilated stomach, usually provided with a cæcal appendage. The stomach opens into an intestine, which is usually long and much coiled, and surrounded by a very large, multi-lobed liver. The liver occupies nearly all the upper part (upper coils) of the visceral sac, and pours its secretion into the intestine and also into the so-called stomach (fig. 511). The arrangement of the digestive canal and of the liver presents in details many essential modifications; one of the most remarkable is that offered by the intestine with its hepatic cæca of the *Phlebenterata* (fig. 512). The terminal portion of the intestine is distinguished by its size, and may be called the rectum.

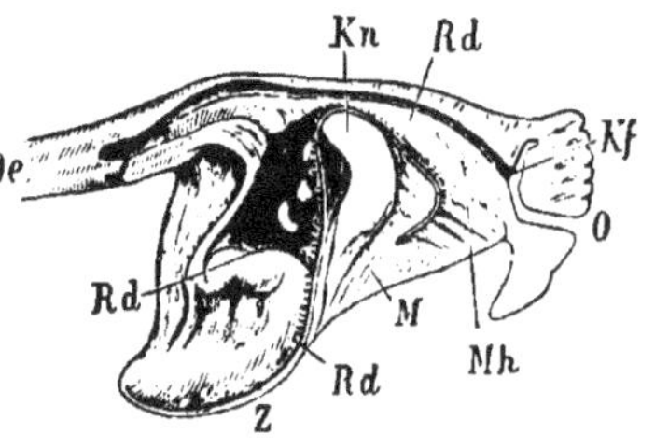

Fig. 513.—Longitudinal section through the buccal mass of *Helix* (after W. Keferstein). *O*, mouth; *Mh*, buccal cavity; *M*, muscles; *Rd*, radula; *Zn*, lingual cartilage; *Oe*, œsophagus; *Kf*, jaws; *Z*, sheath of radula.

The **armature of the buccal cavity** consists partly of jaws placed on the upper wall, partly of the so-called lingual ribbon (*radula*), placed on a tongue-like projection of the ventral surface of the buccal cavity.

The jaws consist either of a single curved horny plate, placed close behind the edge of the lip, or of two lateral pieces of very different form, between which, in some Pulmonates, there is an unpaired piece. There are no lower jaws; but on the floor of the buccal cavity there is a ridge, partly muscular and partly cartilaginous, which, from its resemblance to the tongue of the Vertebrata, has received the same name (fig. 513). The surface of this tongue is covered by a tough membrane, known as the lingual ribbon or *radula*, on which are arranged transverse rows of plates, teeth, and hooks of a characteristic form. Behind, the radula passes into a cylindrical pocket, the so-called *radula sheath* (fig. 513 *Z*), which projects in a tubular manner from the lower (ventral and posterior) end of the buccal mass. The radula is secreted in the radula sheath. The size, number, and form of the plates and teeth on the surface of the radula vary in different forms, and afford important systematic characters for genera and families.

In the transverse rows of plates—the so-called *segments* of the radula membrane—*median*, *intermediate*, and *lateral plates* may be

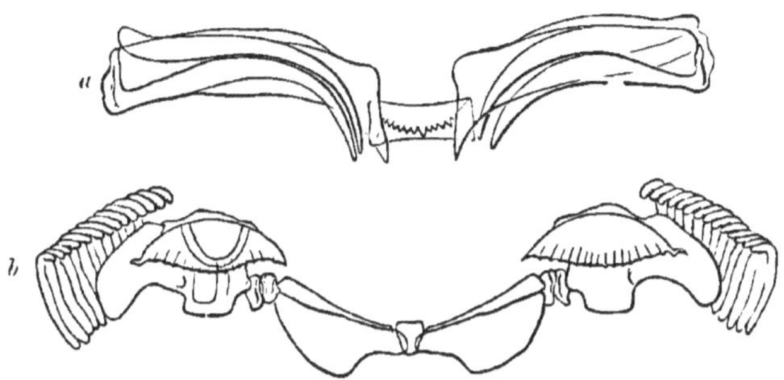

FIG. 514.—*a*, A segment of the radula of *Pterotrachea Lesueurii* (after Macdonald). *b*, ditto of *Neritina fluviatilis* (after S. Lovén).

distinguished (fig. 514 *a*, *b*). Troschel believed that natural divisions could be formed according to the special structure of the armature of the radula. But this one-sided systematic treatment requires many corrections, as has been especially shown in the case of the Tænioglossa and Rhipidoglossa.

The **vascular system** presents numerous and essential variations. The heart is enclosed in a special pericardium, and is usually placed on one side of the middle line near the respiratory organs (fig. 515). It usually consists of a conical ventricle, which gives off the aorta, and of an auricle which is turned towards the respiratory organs, and into which the blood passes by veins. In some Gastropods (Gastropods with two gills, *Haliotis*, *Turbo*, *Nerita*, *Fissurella*, etc.), the heart resembles that of the Lamellibranchs, in that there are two auricles and the ventricle is pierced by the rectum. The aorta usually divides into two arteries, of which one passes forward and gives off many branches to the head and foot; while the other passes dorsalwards to the viscera (fig. 515, *Aa*, *Ac*). The arteries terminate by opening into blood spaces of the body cavity without special walls, from which the blood passes either through the branchial (pulmonary) arteries, or directly, without traversing intermediate vessels (*Heteropoda* and many *Nudibranchiata*), to the respiratory organs, whence it is returned through branchial (pulmonary) veins to the auricle. The arrangements described as obtaining in the *Lamellibranchiata*, by which water is able to enter the blood spaces and dilute the blood, are said to occur also in *Gastropoda*.

In a small number of Gastropods only is respiration effected

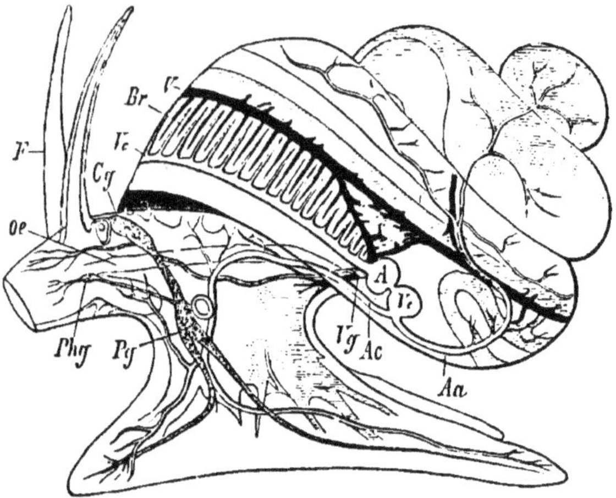

FIG. 515.—Nervous system and circulatory organs of *Paludina vivipara* (after Leydig). *F*, tentacle; *Oe*, œsophagus; *Cg*, cerebral ganglion with eye; *Pg*, pedal ganglion with adjacent otocyst; *Vg*, visceral ganglion; *Phg*, pharyngeal ganglion; *A*, auricle of heart; *Ve*, ventricle; *Aa*, abdominal aorta; *Ac*, cephalic aorta; *V*, veins; *Vc*, afferent vein; *Br*, gill.

exclusively through the general integument. By far the greater number breathe through gills, and many through lungs; a few combine branchial and pulmonary respiration. The gills are usually foliaceous or pennate cutaneous appendages, which are generally placed between the mantle and foot and enclosed by the mantle fold; in rare cases they are exposed and placed on the dorsal surface. The mantle cavity is therefore at the same time the respiratory cavity.

The primitive arrangement of the gills appears to be that found in the *Zeugobranchiata*, in which there are two, one on each side; but, usually an asymmetrical development takes place, and one gill only remains (fig. 516). The respiration of air is confined to some *Prosobranchiata* and to the *Pulmonata*. In this case also the mantle cavity serves as the respiratory cavity, but it differs from the branchial cavity by containing air, and possessing, instead of a gill, a rich network of blood-spaces and vessels on the inner surface of its roof. Both branchial and pulmonary cavities communicate by a long slit along the mantle edge or by a small round aperture, capable of being closed, with the external medium. Frequently, however, the edge of the mantle is prolonged into a long respiratory tube of variable length, which is analogous to the siphon of the *Lamellibranchiata*. This siphon corresponds, as a rule, to a notch or canal of the shell (*vide* p. 32).

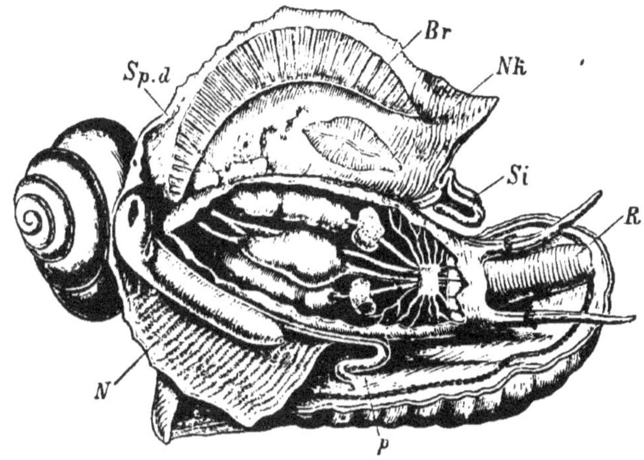

Fig. 516.—Anatomy of *Cassis cornuta* (after Quoy)). *R*, proboscis; *Si*, siphon; *Br*, gill; *Nk*, olfactory organ (formerly regarded as a rudimentary gill); *Spd*, salivary gland; *N*, kidney; *P*, penis.

The structure of the respiratory organs has become of importance for the classification of the larger groups. According to the position of the respiratory organs, with regard to the heart and its auricle, two great divisions can, as Milne Edwards has pointed out, be established: (1) the *Opisthobranchiata*, in which the auricle and gills are placed behind the ventricle; (2) the *Prosobranchiata*, in which the auricle, with the branchial vein entering from the front, lies in front of the ventricle. As far as this character is concerned, the *Heteropoda* and most *Pulmonata* are allied to the latter group: but the *Pulmonata*, in many features of their organization and in their hermaphroditism, stand closer to the *Opisthobranchiata*.

The **kidney** (fig. 516) is the most important excretory organ of the *Cephalophora*. It corresponds in position and structure to the organ of Bojanus of Lamellibranchs. It is, however, usually unpaired, and lies near the heart as an elongated triangular sac, with spongy (rarely smooth) walls of a yellowish brown colour. The secretion of the gland consists mainly of hard concrements, which arise in the lining cells, and consist of uric acid, calcareous and ammoniacal salts. It opens near the anus into the mantle cavity, either immediately by a slit capable of being closed, or by a special excretory duct running with the rectum.

The *Gastropoda* generally possess, in the roof of the respiratory cavity, a *mucous gland*, which often pours out an enormous quantity of its secretion through the mantle orifice. The *purple gland* (*Purpura*, *Murex*) lies in the roof of the mantle cavity, near the rectum. It is a long, whitish-yellow glandular mass, the colourless secretion of which, according to the investigations of Lacaze-Duthier, quickly acquires, under the influence of sunlight, a red or violet colour. The secretion of this gland was known to the ancients, and prized by them on account of its permanence. The coloured fluid, which is excreted from pores of the skin of many Opisthobranchs, *e.g.*, *Aplysia*, must not be confounded with the genuine purple.

Another gland, whose function is not accurately known, is the pedal gland of *Limax* and *Arion*. It extends through the whole length of the foot, and consists of unicellular glands, the delicate ducts of which open into the band-shaped main duct. The latter opens to the exterior between the foot and the head. In many naked Pulmonates (*Arion*) there is, in addition, a gland at the point of the tail, which secretes considerable quantities of mucus with great rapidity.

Generative Organs.—Some of the *Gastropoda* are diœcious, some are hermaphrodite. The *Pulmonata* and *Opisthobranchiata* are hermaphrodite; the *Prosobranchiata* are diœcious. Almost all Gastropods lay eggs, usually in strings. Only a few bear living young, which have developed from the fertilised eggs in the uterus.

The **female organs** consist of an ovary, oviduct, albumen gland, uterus (dilated and glandular part of the oviduct), vagina, and receptaculum seminis.

The **male organs** consist of a testis, a vas deferens with seminal vesicle, a ductus ejaculatorius, and external copulatory organs.

The hermaphrodite forms are distinguished by the close connection of the male and female generative glands and their ducts; for not only are the latter in direct communication with each other, but the ovaries and testes are, with a few exceptions (*Actæon*, *Janus*), united in one hermaphrodite gland, which is usually imbedded among the lobes of the liver. The ova and spermatozoa arise either in different but adjacent follicles of the lobed or branched hermaphrodite gland (*Nudibranchiata*), the ovarian follicles being placed peripherally to the semeniferous follicles (*Eolis*) or the epithelium of the same follicle produces in one part ova, in another part spermatozoa, not however usually at the same time, the maturity of the male element preceding that of the female (terrestrial snails). The efferent duct of the female is nearly always provided with a separated albumen gland, and a receptaculum seminis (fig. 517). In the *Helicidæ* the vagina bears two tufts of finger-shaped glandular tubes and a peculiar sac—the *dart-sac*—which produces in its interior a dart-like calcareous rod. The latter — the so-called love-dart—is attached to a papilla at the base of the sac; it is protruded during copulation, and seems to play the part of a stimulating organ. It is usually broken during use and is replaced later by a new one. The male generative opening is always in connection with a protrusible penis, and usually opens with the female into a common lateral cloaca.

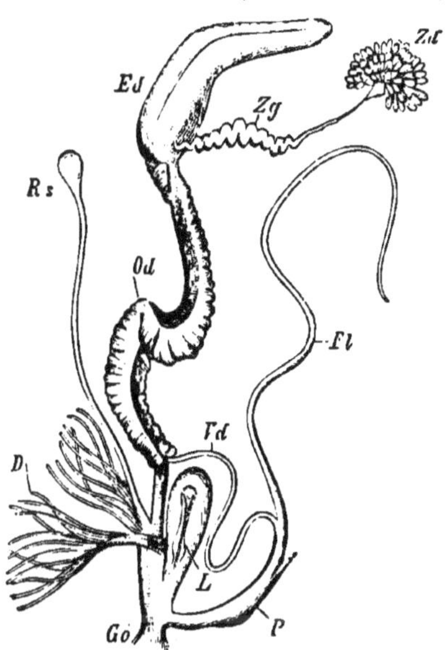

Fig. 517.—Sexual organs of the Roman Snail (Helix pomatia). *Zd*, hermaphrodite gland; *Zg*, its duct; *Ed*, albumen gland; *Od*, oviduct and seminal groove; *Vd*, vas deferens; *P*, protrusible penis; *Fl*, flagellum; *Rs*, receptaculum seminis; *D*, finger-shaped gland; *L*, Spiculum amoris; *Gö*, common genital opening. (After Baasen).

The structure of the generative organs in the diœcious Gastropods resembles that of the hermaphrodite forms. A receptaculum seminis and an albumen gland may be present in the female (*Paludina*). The ovaries and testes lie hidden among the lobes of the liver, and the sexual orifices are placed laterally. The males almost always possess a projecting penis, which is either perforated by the terminal part of the vas deferens (*Buccinum*) or traversed by a furrow, at the base of which the sexual opening is placed. When the penis is remote from the sexual opening, a ciliated furrow is present, which conducts the spermatozoa from the opening to the penis (*Murex*, *Dolium*, *Strombus*).

The embryonic* **development** begins with an unequal segmentation leading to the formation of a blastula or gastrula. Later the embryo acquires a ciliated velum, the first rudiment of the shell, foot, and primitive kidney, and rotates in the fluid albumen of the egg by the vibrations of the cilia.

The free development is either direct, the just-hatched anima possessing (excepting for the rudiments of larval organs) the form and organization of the adult (*Palmonata*), or it takes place by a metamorphosis. Almost all marine *Gastropoda* develop by meta-

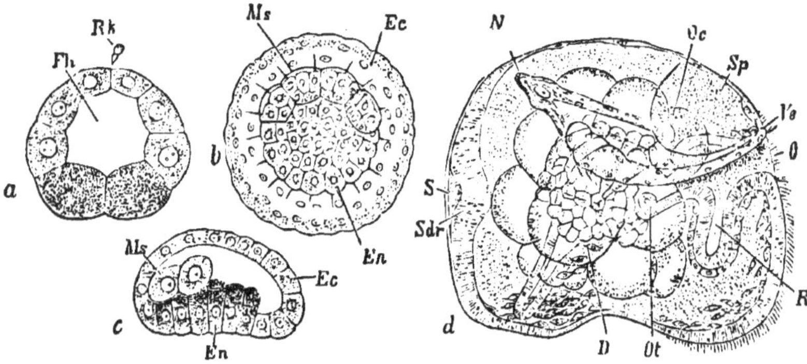

FIG. 518.—Some stages in the embryonic development of *Planorbis* (after C. Rabl). *a*, optical section through a segmenting ovum (24 segments). *Rk*, polar bodies; *Fh*, segmentation cavity. *b*, stage with four mesoderm cells, viewed from the vegetative (lower) pole. *Ms*, mesoderm cells; *En*, endoderm; *Ec*, ectoderm. *c*, Oblique optical longitudinal section through the stage with four mesoderm cells. *d*, Older embryo, in which the shell gland has shifted to the right. *Sdr*, shell gland; *S*, shell; *O*, mouth; *D*, alimentary canal; *R*, commencing radula; *Sp*, apical plate (thickening of præoral lobe); *Oc*, eyes; *Ot*, otolith; *N*, primitive kidney; *Ve*, velum.

morphosis, and the larvæ possess two large ciliated sails (velum), which serve as locomotory organs in place of the still rudimentary foot. The shell, which is already present on the dorsal surface, is still small and flat with hardly any trace of the spiral twisting, and can usually be closed by an operculum which is attached to the foot. Very often a change of shell is effected, the old embryonic shell being thrown off and a new one formed in its place.

* Cf. especially.
N. Bobretzky, "Studien über die embryonale Entwickelung der Gastropoden." *Archiv für mik. Anat.*, Tom XIII., 1876.
C. Rabl, "Ueber die Entwickelung der Tellerschnecke." *Morph. Jahrb.* Tom V.
Also Fol, Bütschli, R. Lankester, etc.

By far the majority of *Gastropoda* are marine; the *Basommatophora* and some *Prosobranchiata* (*Paludina*, *Valvata*, *Melania*, *Neretina*, etc.) inhabit fresh-water. Many *Littorina*, *Cerithia*, *Melania*, etc., live in brackish water. The *Cyclostomidæ*, and the *Stylommatophora* among the Pulmonates, are terrestrial. Further, many branchiate Gastropods are able to live for some time out of water in dry places; in such circumstances they are withdrawn into their shells, the opening of which is closed by the operculum. Almost all move by creeping; some, however, as *Strombus*, jump; others, as *Oliva* and *Ancillaria*, swim excellently by the aid of the lobes of their foot. Some marine forms, as *Magilus*, *Vermetus*, etc., are fixed by their shells; a few only are parasitic, as *Stylifer* on sea-urchins and starfishes, *Entoconcha mirabilis* in *Synapta*.

The method of nutrition differs as much as the habitat. Many, especially the *Siphonostomata*, are voracious predatory animals, and prey on living animals; some branchiate Gastropods, as *Murex* and *Natica*, with this object bore into the shells of Molluscs; several (*Strombus*, *Buccinum*) prefer dead animals. An equally large number, viz., almost all Pulmonates and holostomatous branchiate Gastropods, feed on plants.

Order 1.—Prosobranchiata.*

Diœcious branchiate Gastropods with shell, and with gills in front of the heart.

Behind the usually distinctly separated head lies the respiratory (mantle) cavity, into which the rectum, kidney, and oviduct open. In rare cases two gills are present, as a rule the right gill is absent. The branchial veins enter the heart from the front. Cerebral, pedal, pleural and visceral ganglia are present. The males are, as a rule, more slender, and are easily recognized by the large penis placed on the right side of the anterior part of the body. In the generative organs, the accessory glands are usually absent. The eggs are surrounded by albumen and laid in capsules, which are frequently fixed to foreign objects; more rarely they are attached to the foot and carried about (*Janthina*).

* Fr. Leydig, "Ueber Paludina vivipara." *Zeit. für. wiss. Zool.*, Tom II., 1850.
E. Claparède, "Anatomie und Entwickelungsgeschichte der Neritina fluviatilis." *Müller's Archiv.*, 1857.
H. Lacaze-Duthiers, "Mémoire sur le système nerv. de l'Haliotide et Mémoire sur la Pourpre." *Ann. des Sc. Nat.*, Tom XII. and XIII.
N. Bobretzky, l.c.

Sub-order 1. **Placophora.*** *Body vermiform, symmetrical, without eyes and tentacles. Ventral surface flattened: dorsal surface covered by calcareous plates placed in a segmental manner one behind the other. Gills and kidney paired.*

The Placophora are the most nearly allied of all Mollusca to certain forms of worms, to which they approximate through the genera *Neomenia* and *Chætoderma*. The symmetrical body does not possess a separated head, eyes, or tentacles. The integument presents numerous scattered spines, which are sometimes hard and chitinous, and sometimes calcified; they always arise in special follicles lined by ectoderm cells. In addition to these integumentary structures, which are also present in *Chætoderma*, there are a series of broad calcareous plates on the dorsal surface, which are only exceptionally covered by the mantle (*Cryptochiton*), and which, according to their origin, represent a multivalve Molluscan shell. The free edge of the mantle is moderately thickened, and under it on each side is placed the small mantle cavity as a furrow containing a series of leaf-like gills (fig. 519).

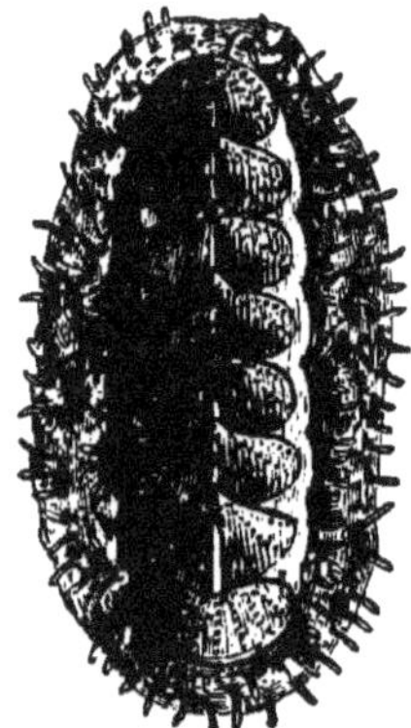

FIG. 519.—*Chiton (spiniferus) spinosus* (règne animal).

Of special interest is the simple condition of the **nervous system** (fig. 495), which greatly resembles that of the Gephyrean-like genera *Neomenia* and *Chætoderma*. Cerebral ganglionic swellings are absent, in correspondence with the want of eyes and tentacles. Four nerve trunks pass off from the double œsophageal ring, an upper lateral pair, the *pallial nerves*, and a ventral pair, the *pedal nerves*, which latter are connected by transverse commissures. Pedal and visceral ganglia are not separated as ganglionic swellings from the nerve stems. Buccal ganglia, on the contrary, are present.

The **alimentary canal** begins with the mouth, which is placed on a roundish lobe; it is much coiled, and extends through the whole length of the body, to open by the anus at the hind end. As in

* A. Th. Middendorff, "Beiträge zu einer Malacozoologica rossica. 1. Beschreibung und Anatomie neuer oder für Russland neuer Chitonen." *Mém. Acad. Imp., St. Petersburg*, 1848.

S. Lovén, "Ueber die Entwickelung der Gattung Chiton." *Archiv für Naturgesch.*, 1856.

B. Haller, "Die Organisation der Chitonen der Adria." *Arbeiten a. d. Zool Inst. in Wien.*, Tom IV., 1882.

Vide also Tullberg's and Graff's works on *Neomenia* and *Chætoderma*.

most Cephalophora (Odontophora) a large muscular mass, the **tongue**, covered by a hard chitinous plate, the *radula*, is found upon the floor of the buccal cavity. The **heart**, on the other hand, more nearly resembles in structure and position that of the Lamellibranchs, in that it consists of two auricles opening into a median ventricle, which lies over the rectum.

The **kidneys** are paired, and open right and left in the mantle furrow; [they also open, as in other Molluscs, into the pericardium]. The *Placophora* are diœcious.

Testes and **ovaries** are simple unpaired glands, which lie immediately over the liver and alimentary canal: their ducts open on each side into the mantle cavity in front of the kidneys.

The development of the egg begins with an equal segmentation; subsequently the segments of one-half of the ovum divide less rapidly. This half is invaginated, so that a gastrula arises. The larva which leaves the egg membranes resembles Lovén's worm larva in the possession of two eye-spots and a ciliated ring, and develops without a larval shell.

Fam. **Chitonidæ**. In place of the shell, eight calcareous pieces are present, which are so arranged that the hinder edge of one shell piece overlaps the anterior edge of the next following piece.

Chiton squamosus L., Mediterranean. *Cryptochiton Stelleri*, Midd.

Sub-order 2. **Cyclobranchiata**. Prosobranchiata with flat plate-shaped shell and foliaceous gills, which are arranged in a closed circle under the edge of the mantle round the broad root of the foot. The buccal lobes are little developed. The foot is powerful, and usually flat and broad. The lingual armature, like that of the *Placophora*, is formed of toothed horny plates, hence the name *Docoglossa* of Troschel. A cervical gill placed on the right side of the neck is sometimes present (*Lottia*). Two kidneys are present. External copulatory organs absent. They feed on plants.

Fam. **Patellidæ**. (Limpets). The shell is bowl-shaped, and consists of a single piece, to which the animal is attached by a horse-shoe-shaped muscle. Head with two tentacles, at the swollen base of which are placed the eyes. Tongue extraordinarily long and spirally coiled. The radula is without the median plates, while the intermediate and marginal plates are raised to hooks, and smaller lateral plates appear.

Patella L. The apex of the shell is slightly eccentric, and hardly inclined to the front. *P. cærulea* L., *P. tarentina* Lam., *P. scutellaris* Lam., Adriatic and Mediterranean. *Nacella* Schum. Circle of gills broken on the head: the apex of the pellucid shell, shining internally like mother-of-pearl, bent forwards. *N. pellucida* L.

Sub-order 3. **Zeugobranchiata**. Gills bipennate, paired and symmetrical. Anterior border of mantle deeply cleft, in correspondence with which the shell is perforated or provided with a slit on its outer lip. Kidneys paired, that of the left side rudimentary. Auricle paired; ventricle perforated by rectum. Tongue *rhipidoglossal*, in that the complicated radula bears in each transverse row, in addition to the median and intermediate plates, a great number of lateral plates which are arranged in a fan-like manner and the upper edges of which are bent into the form of hooks. They are all herbivorous, and are without a retractile proboscis or siphonal tube at the shell aperture. They often possess filiform appendages on the foot. A penis is not developed.

Fam. **Fissurellidæ.** Shell cup- or cap-shaped, with an aperture at the apex or an anterior marginal excavation for the entrance of water into the mantle cavity, which contains two symmetrical gills. Mantle edge fringed. The animals resemble the *Patellidæ*, are provided with tentacles and a large foot. *Fissurella* Brug. Shell with longish aperture through the apex, which is placed in front of the middle. *F. græca* L., Adriatic and Mediterranean. *Emarginula* Lam. An excavation at the anterior edge of the deep bowl-shaped shell. *E. elongata* Costa, Adriatic and Mediterranean. *Scutus* Montf. (*Parmophorus* Blainv.) Australia.

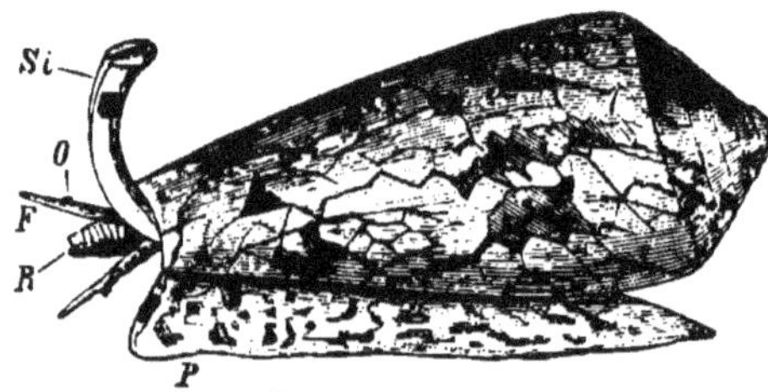

FIG. 520.—*Conus textilis* (règne animal). *R*, proboscis; *Si*, siphon; *F*, tentacle; *O*, eye.

Fam. **Haliotidæ.** Sea-ears, ormers. Shell flat, ear-shaped, internal mother-of-pearl lustre, with a row of holes on the left side. The mantle cavity is on the left side and contains two gills, of which the right is the smaller. Foot fringed, with a broad pedal surface. Head with two long tentacles and short stalked eyes. *Haliotis* L. Spiral of shell small and flat. Foot projecting slightly over the shell. *H. tuberculata* L., Adriatic and Mediterranean.

Sub-order 4. **Ctenobranchiata (Anisobranchiata**, *e.p.*). With large cervical gill of pectinate form on the left side with small olfactory organ (so-called rudimentary gill, fig. 516, *Nk*). A spiral shell is very generally present (fig. 520). The male possesses a penis on the right side. Most are carnivorous and possess a protrusible proboscis.

1. **Rhipidoglossa.** Each transverse row of the radula with numerous lateral plates arranged in a fan-like manner (fig. 514, *b*).

Fam. **Trochidæ,** (Top shells). With conical shell and spiral operculum. Foot prolonged into cirri and lobes. Eyes on short stalks. *Turbo* L. With roundish (convex) windings, round aperture, and buccal edge somewhat cut off. *T. rugosus* Lam. *Trochus* L. With angular windings, buccal edge divided above, and outer lip thin. *Tr. varius* L., Adriatic and Mediterranean.

Fam. **Neritidæ (Neritacea).** With thick, hemispherical shell and operculum. Eyes stalked, behind the two long tentacles. Proboscis short, often bilobed. Foot large, triangular. Heart perforated by rectum, with two auricles. *Nerita* L. Shell thick, hemispherical, spiral lateral; aperture semi-circular. *N. rugata* Recl.; *N.* (*Neritina*) *fluviatilis* L.; *Navicella* Lam.; *N. elliptica* Lam., Pacific Ocean.

2. **Ptenoglossa.**—Without siphon, aperture of shell entire, without excavation or canal. Tongue armed with rows of numerous small hooks and without the median plates.

Fam. **Janthinidæ.** *Janthina bicolor* Menke, Mediterranean.

Fam. **Solariidæ.** (Wentle-traps). *Scalaria communis* Lam., *Sc. pretiosa* Lam., East Indies. *Solarium perspectivum* Phil., Mediterranean.

3. **Rhachiglossa.**—With long proboscis invaginable from the base. Tongue long and narrow with at most three plates in each transverse row, a toothed median plate and an intermediate plate on each side, which are often reduced to mere hooks, and may be absent. All possess a siphon and are predatory.

Fam. **Volutidæ** (Faltenschnecken). *Voluta undulata* Lam., New Zealand; *V. vespertilio*. East Indies; *Cymbium æthiopicum* L.

Fam. **Olividæ.** *Oliva utriculus* Lam., Indian Ocean; *Harpa ventricosa* Lam., New Guinea.

Fam. **Muricidæ (Canaliferæ).** *Murex brandaris* L., Mediterranean. *Fusus australis*. Quoy Gaim. *Columbella mercatoria* L., Atlantic.

Fam. **Buccinidæ.** Whelks. *Buccinum undatum* L.; *Nassa reticulata* L. Mediterranean; *Purpura lapillus* L., North Sea.

4. **Toxoglossa.**—Tongue with two rows of long hollow hooks, which can be protruded from the mouth. All possess a siphon, and usually prey on marine animals.

Fam. **Conidæ** (fig. 520) (Kegelschnecken). *Conus litteratus* L., East Indies.

Fam. **Terebridæ** (Schraubenschnecken). *Terebra dimidiata* Lam.

Fam. **Pleurotomidæ.** *Pleurotoma nodifera* Lam.; *Cancellaria* Lam.; *C. cancellata*, Lam.

5. **Tæniogiossa.**—In each transverse row of the elongated radula there are usually seven plates. Two small jaws usually found at the mouth entrance.

Holostomatous are:—

Fam. **Littorinidæ,** (Winkles). *Littorina littorea* L.

Fam. **Cyclostomidæ.** Respire air like the *Pulmonata* by vessels of the mantle cavity. Live in damp places on land. *Cyclostoma elegans* Drap.

Fam. **Paludinidæ,** (Flusskiemenschnecken). Inhabit fresh water. *Paludina vivipara* L.; *P. impura* Lam.

Fam. **Vermetidæ,** (Wurmschnecken). *Vermetus arenarius* L.

Fam. **Cerithiidæ.** *Cerithium læve* Quoy Gaim.

Siphonostomatous are :—

Fam. **Cypræidæ**, (Cowries). *Cypræa tigris* Lam ; *C. moneta* L.

Fam. **Tritoniidæ**, (Tritonshörner). *Tritonium variegatum* Brug. ; *Ranella gigantea* Lam.

Fam. **Doliidæ.** *Cassis cornuta* Lam. ; *Dolium galea* L.. Mediterranean.

Fam. **Strombidæ (Alata)** (Flügelschnecken). *Strombus Isabella* Lam. ; *Pterocerus lambis* Lam. ; *Rostellaria rectirostris* Lam.

Fam. **Naticidæ.** *Natica ampullaria* Lam. ; *Sigaretus haliotoideus* L.. Atlantic.

Fam. **Capulidæ**, (Mützenschnecken). *Capulus hungaricus* L., Adriatic ; *Calyptræa rugosa* Desh.

Fam. **Ampullariadæ**, (Doppelathmer). With branchial and pulmonary cavity. In rivers of hot countries. *Ampullaria celebensis* Quoy. ; *A. polita* Desh.

Order 2.—HETEROPODA.*

Pelagic Gastropoda with fin-like foot, large projecting head and highly-developed moveable eyes. Diœcious.

The body (fig. 521) of the Heteropoda is usually cylindrical and elongated and prolonged into a proboscis-like projecting head, which carries large well-developed eyes and tentacles, and encloses a powerfully-armed protrusible tongue (fig. 514 *a*). The main peculiarity of the body consists in the formation of the foot, the anterior and middle portion (*pro-* and *mesopodium*) of which is modified to the form of a leaf-shaped fin, often provided with a sucker (fig. 521 *S*); while the hinder section (*metapodium*) is considerably elongated and extended far backwards, and seems to form the caudal continuation of the body. The visceral sac is either spirally twisted, and enclosed by a mantle and spiral shell (*Atlanta*), or has the form of a saccular and projecting mass, which is placed at the limit of the hinder region of the foot, and is likewise covered by the mantle and a hat-shaped shell (*Carinaria*, fig. 521); or finally the visceral sac is reduced to a very small, scarcely-projecting nucleus, which is covered on the front side by a membrane with a metallic lustre and is completely without a shell.

The **nervous system** is more highly developed than that of any other Gastropod. The two large *eyes* are placed near the tentacles in special capsules, in which they are moved by several muscles. The

* Souleyet. "Hétéropodes. Voyage autour du monde exécuté pendant les années 1836 et 1837 sur la corvette la Bonite, etc." Tom II. Paris, 1852.

R. Leuckart. "Zoologische Untersuchungen." Heft III. Giessen, 1854.

C. Gegenbaur. "Untersuchungen über Pteropoden und Heteropoden." Leipzig. 1854.

H. Fol. "Sur le Développement des Hétéropodes." *Arch. de Zool. experim.* Tom V., 1876.

large *auditory vesicles* each receive a long auditory nerve from the cerebral ganglion, and are characterised not only by the remarkable vibrations of the long tufted cilia of their epithelium, but also by the arrangement of the nerve cells (group of hair cells of the *macula acustica* round a large central cell, fig. 83). In addition numerous peculiar nerve-endings in the skin, which appear to serve the tactile sensation, and the so-called *ciliated organ* on the anterior side of the visceral sac, are present. The latter has the form of a ciliated pit, under which is placed the ganglionic swelling of a nerve which

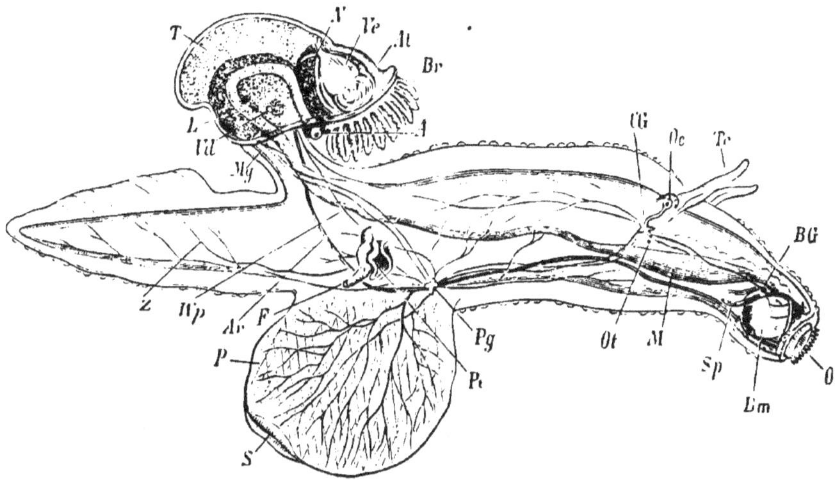

Fig. 521.—Male of *Carinaria mediterranea* (after Gegenbaur). *P*, foot; *S*, sucker; *O*, mouth; *Bm*, buccal mass; *M*, stomach; *Sp*, salivary gland; *L*, liver; *A*, anus; *CG*, cerebral ganglion; *Te*, tentacles; *Oc*, eye; *Ot*, auditory vesicle; *BG*, buccal ganglion; *Pg*, pedal ganglion; *Mg*, mantle ganglion; *N*, kidney; *Br*, gills; *At*, auricle; *Ve*, ventricle; *Ar*, anterior aorta; *Z*, posterior branch of same; *T*, testis; *Vd*, vas deferens; *Wp*, ciliated furrow; *Pe*, penis; *F*, flagellum with gland.

arises in the visceral ganglion; it has the value of an *olfactory organ.*

The males are distinguished by the possession of a large copulatory organ, which projects freely on the right side of the body: the males of *Pterotrachea* also possess a sucker on the foot. In *Atlanta* and *Carinaria* the sucker is present in both sexes. The testes and ovaries fill the posterior part of the visceral sac and are partially imbedded in the liver. The ducts, viz., vas deferens and oviduct, open on the right side of the body; the former at some distance from the organ of copulation, to which the sperm is conducted from the sexual opening in a ciliated furrow. The copulatory organ consists of two parts placed side by side, (1) the penis with the continuation of the ciliated groove; and (2) the gland rod which encloses a longish gland. The oviduct (fig. 90) is more complicated, inasmuch as a large albumen gland and a receptaculum seminis open into it; its dilated terminal part acts as a vagina.

The *Heteropoda* are exclusively pelagic animals, and they are often found in great numbers in the warmer seas. They are somewhat clumsy in their movements, which are effected with the ventral surface uppermost by oscillations of the whole body and the fin. They are all carnivorous. When the tongue is protruded, the lateral teeth fly apart from one another like the limbs of forceps, and when retracted they again fall together. By means of these prehensile movements small marine animals are seized and drawn into the mouth.

Fam. **Pterotracheidæ.** *Carinaria mediterranea* Lam., *Pterotrachea coronata* Forsk., Mediterranean.

Fam. **Atlantidæ.** *Atlanta Peronii* Less., Mediterranean.

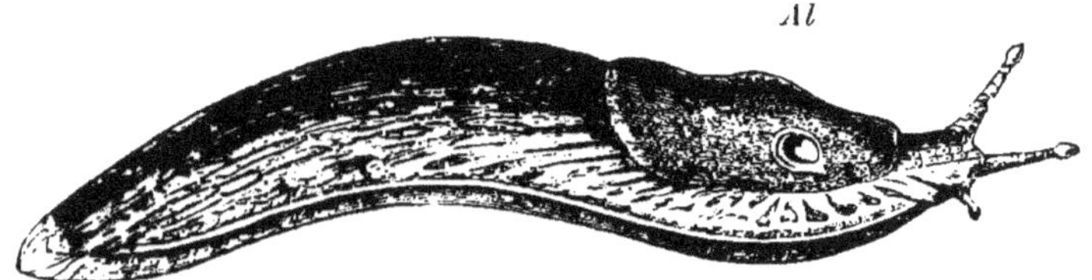

Fig. 522.—*Arion empiricorum* (règne animal). *Al*, respiratory aperture.

Order 3.—Pulmonata.*

Terrestrial and fresh-water Gastropods with lung which is placed in front of the heart. Hermaphrodite.

The roof of the mantle cavity, as in the *Cyclostomidæ*, is provided with a network of vessels for aërial respiration. The mantle (pulmonary) cavity opens to the exterior on the right side by a respiratory aperture (fig. 522.) The mantle cavity of the young of the fresh-water Pulmonates is at first filled with water, and only later with air. Some species of *Planorbis* and *Limnæus* retain, during the whole time of their life, the ability to breathe both in air and water (some *Limnæus*, with lungs full of water, have been dredged up at

* L. Pfeiffer. "Monographia Heliceorum viventium." Leipzig, 1848-1869: and "Monographia Auriculaceorum viventium." Cassel, 1856.
A. Rossmässler, "Iconographie der Land-und Süsswassermollusken Europas." Leipzig, 1835-1859.
Férussac et Deshayes. "Histoire naturelle générale et particulière des Mollusques terrestres et fluviatiles." Paris, 1829-1851.

considerable depths in Lake Constance). The anus and renal opening are placed near the respiratory aperture, sometimes in the respiratory cavity itself. The generative organs open some way in front, but on the same side. In the forms with a left-handed spiral, the respiratory orifice, anus and generative opening are on the left side. Some Pulmonates are naked, or possess only rudiments of the shell in the dorsal integument; others carry a relatively thin and usually right-handed shell. *Physa*, *Planorbis*, and *Clausilia* alone present a left-handed spiral. A true operculum is absent. On the other hand, many forms secrete temporarily a winter operculum.

While the Pulmonates (with some exceptions) resemble the Prosobranchs in the position of the heart behind the respiratory organs, in the arrangement of other organs, *e.g.*, the nervous system, they more resemble the Opisthobranchs. The dentition consists of an unpaired, horny, and usually longitudinally-ribbed upper jaw (which, however, may be absent) and of a radula, which is covered with a great number of toothed plates in longitudinal and transverse rows. All are hermaphrodite. A few, *e.g.*, species of *Clausilia* and *Pupa*, are viviparous. Most Pulmonates, however, lay eggs, either as in the fresh-water forms united in tubular or flat masses on water-plants, or as in the terrestrial forms in damp places, each one being surrounded by a protecting calcareous shell. The ovum is always contained in a large mass of albumen, which serves as nourishment to the developing embryo.

I.—Basommatophora.

The eyes lie at the base of the two tentacles. Present many resemblances to the *Tectibranchiata*.

Fam. **Limnæidæ.** *Limnæus auricularis* Drap., (Pond snail); *L. stagnalis* O. Fr. Müller; *Physa fontinalis* L., *Planorbis corneus* L., *Ancylus fluviatilis* Blainv.

Fam. **Auriculidæ.** *Auricula Judæ* Lam., *A. Midæ* Lam., *Carychium minimum*, O. Fr. Müll.

II.—Stylommatophora.

The eyes lie at the tips of two usually retractile tentacles (posterior tentacles).

Fam. **Peroniadæ** (*Amphipneusta*). Lung behind heart (Opisthopulmonate). *Peronia verruculata* Cuv., *Veronicella* Blainv., *Onchidium* Buchan.

Fam. **Limacidæ** (Naked snails). *Arion* Fér. Sexual opening beneath the respiratory orifice in front of the middle of the dorsal shield. Back without a keel, with caudal gland and mucous aperture at end of body. *A. empiricorum* Fér., *Limax* L. Respiratory aperture behind the middle of the right edge of the mantle. Sexual opening far removed from respiratory aperture behind the right tentacle. Back keeled, without caudal gland and mucous orifice. *L. agrestis* L., *L. cinereus* O. Fr. Müll.

FIG. 523.—*Doris (Acanthodoris) pilosa* (Bronn). *Br*, gills; *A*, anus; *F*, tentacle.

Fam. **Helicidæ**. *Succinea amphibia* Drap., *Pupa muscorum* L., *Clausilia bidens* Drap., *Bulimus montanus* Drap., *Helix pomatia* L. (Roman snail), *H. nemoralis* L.

Order 4.—OPISTHOBRANCHIATA.*

Hermaphrodite Gastropods, with flat foot. Branchial veins open into the auricle behind the ventricle.

The great majority of this order are without a shell. The branchial cavity never contains more than one gill. The gills are usually exposed (fig. 523) or absent. Sometimes there are dorsal processes, into which appendages of the alimentary canal enter (fig. 525). The nervous system contains cerebral, visceral and pedal ganglia (except in *Tethys*, which has a fused ganglionic mass and simple œsophageal commissure). The branchial veins, with a few exceptions (*Gastropteron*), enter the heart from behind.

Sub-order 1. **Tectibranchiata**. Gill almost always on the right side, covered by the mantle edge or placed in a dorsal branchial cavity. Shell usually present (fig. 524).

Fam. **Pleurobranchidæ**. With large gill on right side, and usually internal shell. *Pleurobranchæa Meckelii* Cuv., *Pleurobranchus aurantiacus* Cuv., *Umbrella mediterranea* Lam.

Fam. **Aplysiadæ** (Sea-hares). Shell covered by two lobes of the foot. *Aplysia depilans* L., Mediterranean.

Fam. **Bullidæ**. *Bulla ampulla* L., *Philine aperta* L., *Acera bullata* O. Fr. Müll.

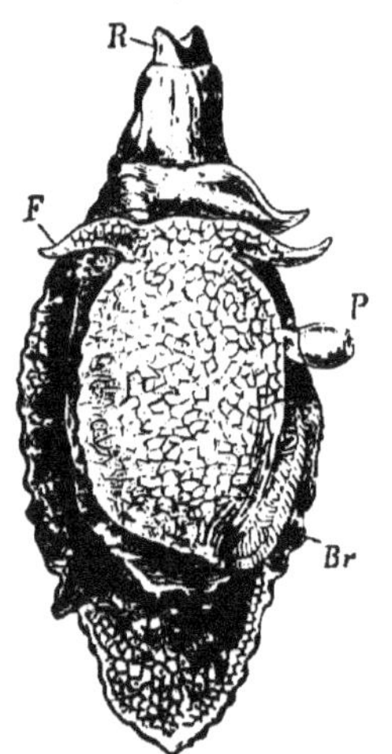

FIG. 524.—*Pleurobranchus aurantiacus* (règne animal). *Br*, gills; *P*, penis; *F*, tentacle; *R*, proboscis.

Sub-order 2. **Nudibranchiata**. Marine Gastropods, without shell

* Alder and Hancock, *l. c.* H. Müller and C. Gegenbaur, "Ueber Phyllirhoë bucephalum." *Zeit. f. wiss. Zool.*, Tom IV., 1854.

or mantle. The gills project freely on the dorsal surface, and may receive appendages of the alimentary canal.

Fam. **Tritoniadæ.** Gills in two longitudinal rows on the back. *Tritonia Hombergii* Cuv., *Scyllæa pelagica* L.

To this family is allied *Tethys fimbriata* L., with concentrated ganglionic mass, without radula and buccal mass.

Fam. **Dorididæ.** Gills in circle round anus (fig. 523). *Doris coccinea* Forb., *D. tuberculata* Cuv., Adriatic and Mediterranean.

Fam. **Æolididæ.** Numerous processes on dorsal surface, into which diverticula of the alimentary canal pass (*Phlebenterata*). *Æolis papillosa* L. (fig. 525), *Tergipes Edwardsi* Nordm. Here are allied *Phyllirhoë bucephalum* Pér., and the *Phyllidiidæ*.

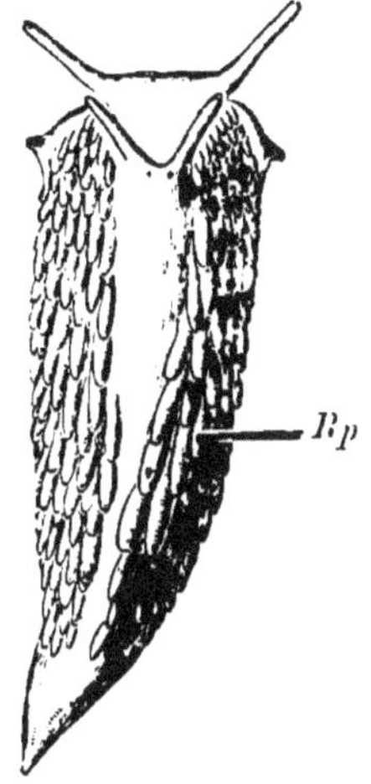

FIG. 525.—*Æolis papillosa* (Bronn). *Rp*, dorsal papillæ.

Sub-order 3. **Saccoglossa.** Gills absent, or as simple appendages of the dorsal integument. The radula with a single row of toothed plates, of which the anterior, after they are worn out, fall into a pocket developed on the floor of the buccal cavity.

Fam. **Limapontiadæ.** *Limapontia atra* Johnst.

Fam. **Elysiadæ.** *Elysia viridis* Ok.

Class III.—PTEROPODA. *

Hermaphrodite Mollusca without sharply separated head, with two large wing-like fins, often with cephalic cones.

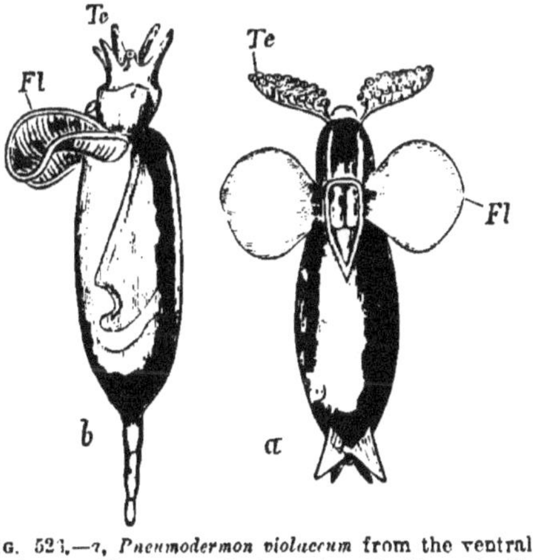

The body is sometimes elongated and straight, sometimes with its hinder part spirally rolled. The anterior region bears the

FIG. 524.—*a*, *Pneumodermon violaceum* from the ventral side. *b*, *Clione australis* from the side (Bronn). *Fl*, fins; *Te*, tentacles.

* Rang et Souleyet, "Histoire naturelle des Mollusques Ptéropodes." Paris, 1852.

C. Gegenbaur, "Untersuchungen über die Pteropoden und Heteropoden." Leipzig, 1855.

A. Krohn, Beiträge zur Entwickelungsgeschichte der Pteropoden und Heteropoden." Leipzig, 1860.

H. Fol, "Sur le développement des Ptéropodes." *Archives de Zoologie expérimentale*, etc., Tom. IV., 1875.

mouth and tentacles but is hardly separated off as a distinct head: beneath the mouth there are two large lateral fins which morphologically are to be explained as paired parts (*epipodia*) of the foot (the unpaired part is rudimentary), and which, by their wing-like flappings, cause the movement of the animal. The body is either naked (fig. 526) and without distinctly separate mantle, or there is a shell of very various shape, into which the body with the fins can usually be completely withdrawn, and which may be horny, gelatinous and cartilaginous, or calcareous, and is almost always symmetrical. In the last case (presence of shell) the mantle is usually very completely developed and encloses most of the body to the region of the fins, behind which the slit-like entrance to the mantle cavity is placed. The integument usually contains calcareous concretions, cutaneous glands, and pigment cells, which may give the body a dark brown, sometimes brownish, or even reddish colour.

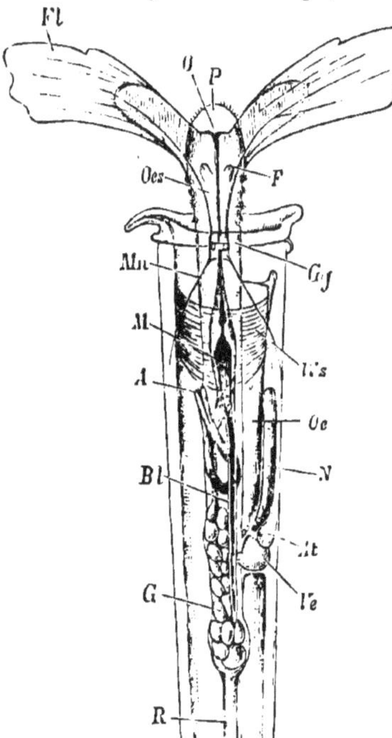

FIG. 527.—*Creseis acicula* from the dorsal side (after Gegenbaur). The hinder part of the body is omitted. *Fl*, fins; *O*, mouth; *Oes*, œsophagus; *P*, median lobe of the foot; *F*, tentacle; *Gg*, cerebral ganglion; *Mn*, mantle nerve; *Ws*, ciliated shield; *M*, stomach; *Bl*, blind sac of stomach; *A*, anus; *N*, kidney; *Oe*, opening of kidney into the mantle cavity; *At*, auricle; *Ve*, ventricle; *G*, sexual gland; *R*, retractor.

The mouth is sometimes surrounded by several arm-shaped processes (*Clio*), or by two processes beset with suckers (*Pneumodermon*), the *cephalic cones* (fig. 526). It leads into a buccal cavity, armed with jaws and toothed radula; at the bottom of the mouth the long œsophagus begins (fig. 527). The œsophagus leads into a dilated stomach, which is followed by a long, coiled intestine, which is surrounded by the liver and bends laterally and forwards. The anus is usually in the mantle cavity on the right side and near the front end.

The **circulatory organs** are reduced to arterial vessels; the main trunks arise from the spherical ventricle. The veins are replaced by a system of lacunæ of the body cavity without special walls, into which the arteries open. The blood returns from the lacunæ through the respiratory organs to the pericardial sinus, whence it enters the auricle through the venous ostium.

The **respiratory organs**, as far as they are not represented by the whole integument (*Clio*), have the form either of foliaceous branchial appendages (*Pneumodermon*) at the hind end of the body, or, in the shell-bearing forms, of internal gills placed within the mantle cavity, the entrance to which is lined with peculiar ciliated bands. The gills are always but slightly developed, and are reduced either to folded elevations of the ciliated mantle-wall, or to the mantle-wall itself.

The **kidney** is an elongated contractile sac, which communicates with the pericardial sinus by a ciliated funnel, and with the mantle cavity or directly with the exterior by a strongly ciliated opening which is capable of being closed.

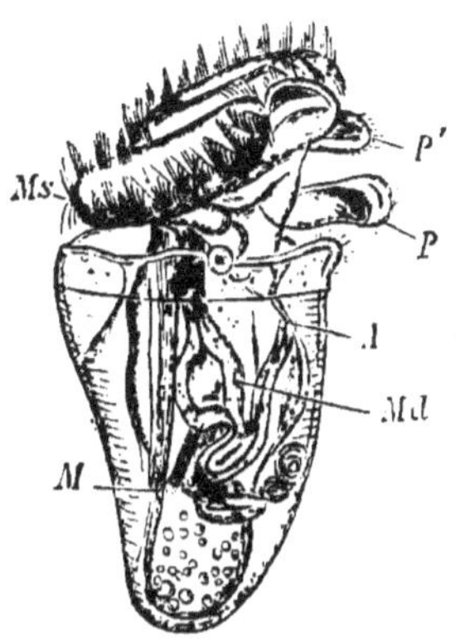

Fig. 528.—Larva of *Cavolinia tridentata* (after Fol). *Ms*, velum; *P*, foot; *P'*, the two lateral (epipodial lobes of the foot; *A*, anus; *M*, retractor muscle; *Md*, stomach.

The **nervous system** resembles that of the higher Opisthobranchs. Pleural ganglia are present. The cephalic cones receive their nerves from the brain; the two fins as parts of the foot from the pedal ganglia.

Sense organs.—A pair of auditory vesicles are always present. Eyes on the other hand are absent or very rudimentary, as red pigment spots (*Hyalea*) placed either on the visceral sac near the œsophageal ring or on the tentacles (*Clio*). Tactile organs are represented by two small tentacles (*Hyalea*, *Cymbulia*) and the larger cephalic cones which are sometimes beset with suckers (*Clio* and *Pneumodermon*).

The Pteropoda are *hermaphrodite*. The hermaphrodite gland lies near the heart behind the stomach in the visceral sac, and usually possesses a common duct which is provided not only with a seminal vesicle, but also with a kind of albumen gland and receptaculum seminis; it opens to the exterior usually on the right side in front of the anus. The penis is sometimes in the terminal part of the duct; in the *Hyaleidæ* and *Cymbuliidæ* it has the form of a rolled-up protrusible tube placed in front of the sexual opening. The eggs are surrounded by albumen and laid in long strings which float freely in the sea. The embryos acquire velar lobes and shell, and leave the egg as larvæ (fig. 528). While the velum is atrophying, the two fins gradually appear on the first-formed unpaired part of the foot, while the shell (with operculum) is usually cast off. The *Hyaleidæ* however appear to keep the larval shell and develop it further, while the *Cymbuliidæ* replace it by a new shell. The naked *Pneumodermidæ* and *Clionidæ* do not after the loss of the velum and shell grow direct into the sexual animal, but first acquire three rings of cilia and pass into a new larval phase (fig. 529). The Pteropods always live on the high sea, but may by retracting their velum sink.

Order 1.—Thecosomata.

Pteropoda with a shell. Head but little developed, often not distinct; tentacles rudimentary. The rudimentary foot remains in connection with the fins.

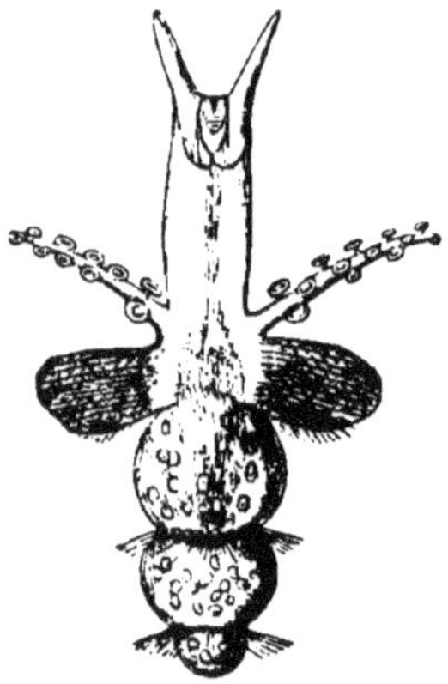

Fam. **Hyaleidæ.** Shell calcareous or horny, swollen ventrally or pyramidal, symmetrical, with pointed processes. *Hyalea tridentata* Lam., *Cleodora* Per. Les., *Creseis* Rang., *Cr. acicula* Rang., Mediterranean.

Fam. **Cymbuliidæ.** With cartilagino-gelatinous shell, boat-shaped or slipper-shaped. *Cymbulia Peronii* Cuv., *Tiedmannia neapolitana* Van Ben.

Order 2.—Gymnosomata.

Naked Pteropods, head bearing tentacles, often with external gills. Fins separated from the foot. Larvæ with rings of cilia.

Fig. 529.—Larva of *Pneumodermon* (after Gegenbaur).

Fam. **Clionidæ.** Body spindle-shaped, without gills. *Clio borealis* Pall., constitutes with *Limacina arctica* the chief food of Whales.

Fam. **Pneumodermonidæ.** Body spindle-shaped, with external gills, and two protrusible arms, which are beset with suckers and placed in front of the fins. *Pneumodermon violaceum* d'Orb.

Class IV.—CEPHALOPODA.*

With well-marked head, a circle of arms bearing suckers round the mouth and funnel-shaped perforated foot. Diœcious.

In the form of their body the Cephalopods are most nearly allied

* Férussac et d'Orbigny, "Histoire naturelle générale et particulière des Céphalopodes acétabulifères vivants et fossiles." Paris, 1835-45.

J. B. Verany, "Mollusques Méditerranéens observés, décrits, figurés et to the Pteropods. The morphological relation between these two groups was first thoroughly discussed by R. Leuckart. He showed that the cephalic cones (tentacles) of *Clio* correspond to the cephalic arms of Cephalopods, while the median lobe of the foot, represented by the cervical collar, is the equivalent of the funnel. Huxley, however, does not take this view; he holds that the arms are parts of the propodium and that the funnel, which is formed by the fusion of paired folds, is equivalent to the paired elements of the epipodium which in Pteropods form the fins.

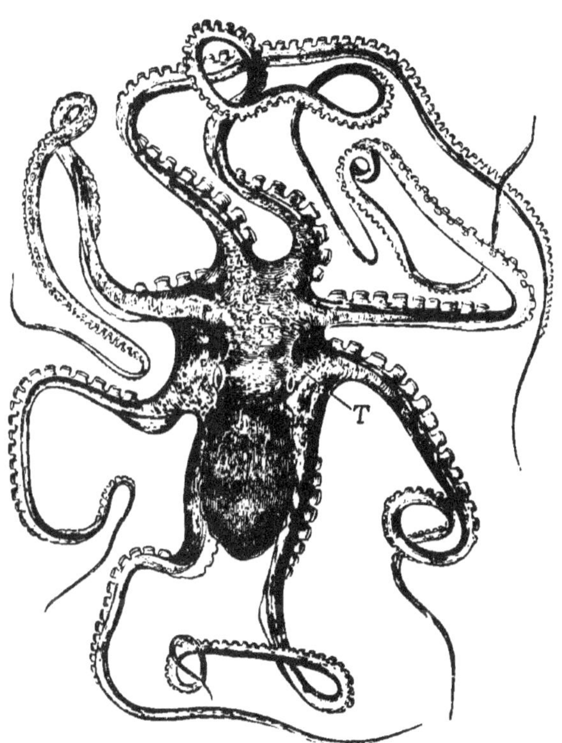

Fig. 530.—*Octopus macropus*, creeping (after Verany). *T*, funnel.

The mantle cavity is placed on the posterior surface of the body, which in the natural position is the under surface. In it are placed on each side one (*Dibranchiata*) or two (*Tetrabranchiata*) gills, the anus, the paired renal openings, and the generative opening which is sometimes single and sometimes paired. The eyes and olfactory organs are placed at the sides of the head. Anteriorly around the mouth four pairs of fleshy cephalic arms,

chromolithographiés d'après le vivant." I° Partie. Céphalopodes de la Méditerranée. Gênes, 1847-51.

H. Müller, "Ueber das Männchen von Argonauta argo und die Hectocotylen." *Zeit. für wiss. Zool.*, 1855.

Jap. Steenstrup, "Hectocotylus dannelsen hos Octopodsl., etc." *K. Danks. Vidensk. Selskabs Skrifter*, 1856. Uebersetzt im *Archiv für Naturgesch.*, 1856.

Alb. Kölliker, "Entwickelungsgesch. der Cephalopoden." Zürich, 1844.

arranged in a circle, project; they serve for creeping and swimming, as well as for the capture of prey, and usually bear rows of suckers on their oral surface. In many forms (*Octopoda*) the basal parts of the arms are united by a membrane which forms a kind of funnel in front of the mouth, the cavity of which is contracted and dilated in movement (not to be confounded with the pedal funnel, fig. 530 *T*). In others two lobe-like cutaneous appendages, the so-called fins, serve for swimming (fig. 531): these forms (*Decapoda*) possess in addition to the eight arms a pair of very long tentacles (fig. 531).

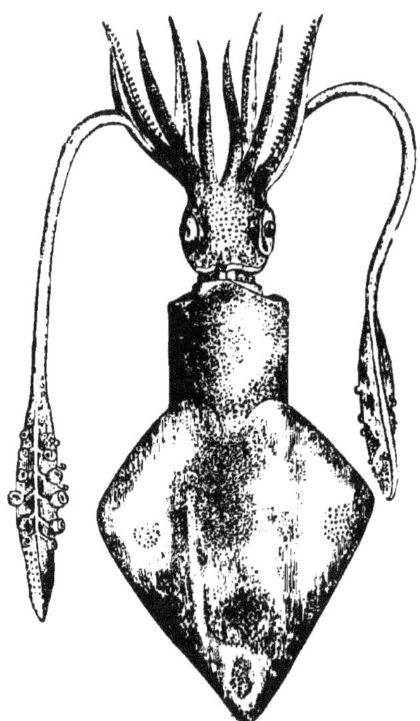
Fig. 531.—*Loligo vulgaris* (after Verany).

In *Nautilus*, the single living representative of the *Tetrabranchiata*, there is found in place of the eight arms a crown of very numerous tentacles. These, however, according to the view of Valenciennes, appear to correspond morphologically to suckers; in fact similar filaments are found on the arms of *Cirroteuthis*, as prolongations of the cylindrical nucleus of the suckers. The true arms of *Nautilus* are very short and rudimentary, forming fold-like lobes at the base of the tentacles.

The **funnel** is placed on the ventral (posterior) side and projects from the broad opening of the mantle cavity, which can be closed laterally by suckers. It has the form of a cylindrical tube, narrowed at the front (free) end, and in *Nautilus* is open along the under surface. Its broad base is placed in the mantle cavity, and it serves to conduct away to the exterior from the latter the respiratory water which has entered by the general mantle-opening, and with it the excrementitious and generative products. At the same time, acting in conjunction with the powerful mantle musculature, it serves as an organ of locomotion. The respiratory water is violently driven through the funnel by the contraction of the mantle, the general opening of the mantle being firmly closed by the sucker-like arrangement at the base of the funnel; the animal, in consequence of the reaction, is thus projected backwards.

Many *Cephalopoda* are naked (*Octopoda*), others (*Decapoda*) possess an internal rudimentary shell, a few (*Argonauta*, *Nautilus*) are provided with an external spirally-coiled shell. The internal shell rudiment of the Decapoda lies in a pocket in the dorsal mantle, and is usually a flat, lancet-shaped spongy calcareous plate (*os sepiæ*). The external shell is only exceptionally thin and simple (*Argonauta*); usually it is spirally-twisted and divided by cross partitions into a number of successive chambers. The animal lives in the anterior chamber, which is the last formed and largest. The other chambers, which diminish continuously in size backwards, are filled with air; they remain, however, connected with the large anterior chamber by a central tube (*siphon*), which perforates the partitions and contains a prolongation of the animal's body.

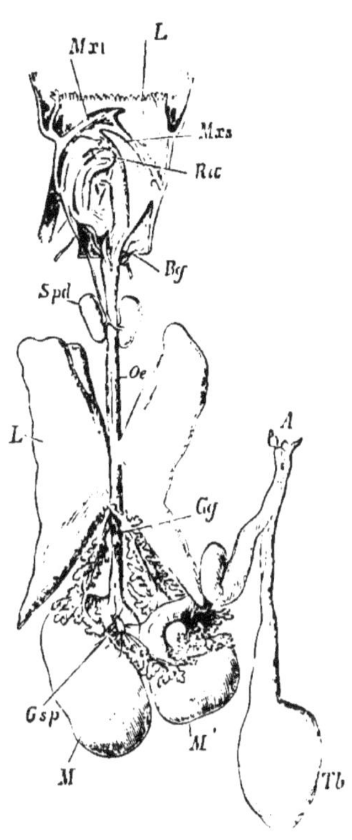

Fig. 532.—Digestive apparatus of *Sepia* (after W. Keferstein). *L*, lip; *Mxi*, *Mxs*, lower and upper jaws; *Ra*, radula; *Bg*, buccal ganglion; *Spd*, salivary gland; *Oe*, œsophagus; *L*, liver; *Gg*, bile duct; *Gsp*, splanchnic ganglion; *M*, stomach; *M'*, blind appendage of stomach; *A*, anus; *Tb*, ink sac.

The dermis of the *Cephalopoda* contains the remarkable **chromatophores**, which cause the well-known play of colours. These consist of cells filled with pigment; to their walls, which are formed of a cellular membrane, numerous radiating muscular fibres are attached. When the latter contract the cells are pulled out into a star shape; in the processes so formed the pigment is distributed. When the contraction ceases, the cell returns, in virtue of the elasticity of its walls, to its original spherical form and the pigment is again concentrated in a small space; thus the animal changes its colour. There are usually two kinds of chromatophores, as far as colour is concerned, placed above and near one another. They are connected with a special centre on the stalk of the optic ganglion and they cause a rapid interchange of blue, red, yellow and dark colours. In addition to the chromatophores, there is a deeper layer of small shining spangles which produce interference colours, and thus give rise to the peculiar iridescence and lustre of the skin.

The *Cephalopoda* possess an internal **cartilaginous skeleton,** which serves for the protection of the nerve centres and sense organs and attachment of muscles. In the Dibranchiata this skeleton constitutes a cartilaginous capsule which encloses the cerebral ganglia, œsophageal ring, and the auditory organ, while its lateral portions are hollowed out and represent the orbits. There are also (Decapods) optic cartilages, a so-called brachial cartilage and dorsal cartilage, various small cartilages for the closure of the mantle cavity, and fin cartilages for the support of the fins.

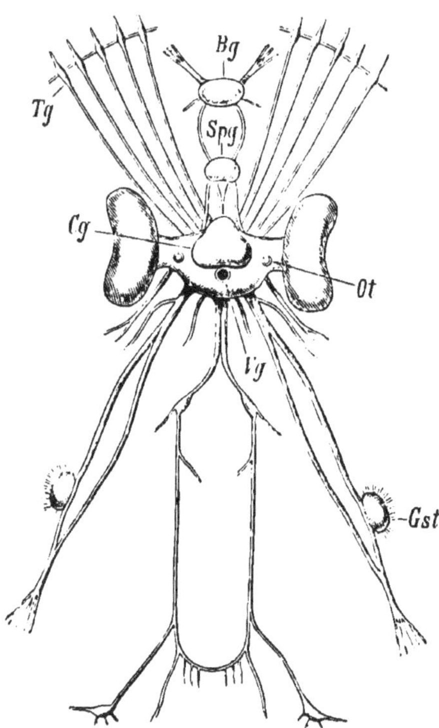

Fig. 533.—Nervous system of *Sepia officinalis* (after Chéron). *Cg*, cerebral ganglion; *Vg*, visceral ganglion; *Bg*, buccal ganglion; *Spg*, suprapharyngeal ganglion; *Tg*, ganglia of the tentacles; *Gst*, stellate ganglion; *Ot*, auditory vesicle.

Alimentary canal.—The mouth, which is placed within the circle of arms, is surrounded by a circular fold forming a kind of lip (fig. 532). It is armed with two powerful jaws, an upper and a lower, which resemble in form a reversed parrot's beak. The radula, which recalls that of the Heteropoda, bears in each row a tooth-like median plate, and on each side three long hooks, adapted for drawing in the food; in addition there may also be some flat non-toothed plates. The œsophagus usually receives two pairs of salivary glands, and either has the form of a simple narrow tube, or presents before its junction with the stomach a crop-like dilatation (Octopods, fig. 535 *Jn*).

The stomach (Fig. 532 *M*) is usually spherical; its walls are muscular and its internal lining is raised into longitudinal folds or papillæ. It possesses a large, sometimes spirally wound, cæcal appendage, which opens into it close to the point of origin of the intestine, rarely at some distance from that point. The ducts of the large liver open into this cæcum. A mass of yellow glandular lobes, which are attached to the upper part of the bile ducts, may be interpreted as pancreas (fig. 532 *Gy*). The intestine is but little convoluted and the anus always opens in the middle line of the mantle cavity.

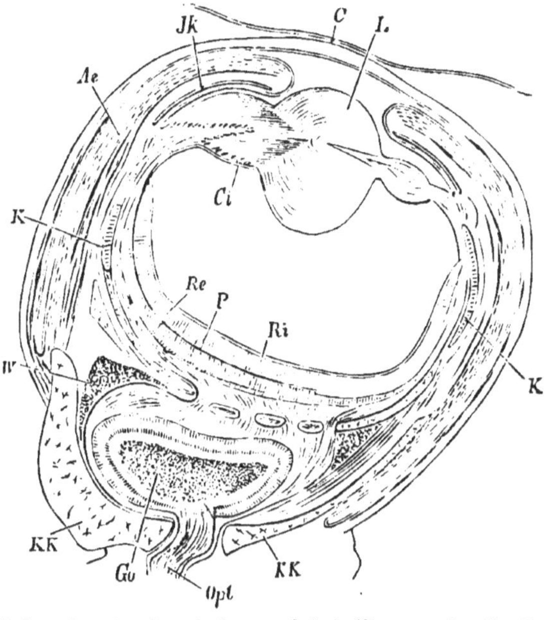

FIG. 534.—Horizontal section through the eye of *Sepia* (diagrammatic, after Hensen). *KK*, cephalic cartilage; *C*, cornea; *L*, lens; *Ci*, ciliary body; *Jk*, iris cartilage; *K*, cartilage of optic bulb; *Ae*, argentea externa; *W*, white body; *Opt*, optic nerve; *Go*, optic ganglion; *Re*, outer layer of rods, *Ri*, inner layer of rods of the retina; *P*, pigment layer of the retina.

The **nervous system** is characterised by its great concentration and high development. In the *Dibranchiata* the nerve centres constitute a large ganglionic mass which is placed in the cartilaginous cranial capsule and is perforated by the œsophagus (fig. 533). It is divided into a dorsal and a ventral portion, connected by two commissures. The former corresponds to the brain (cerebral ganglia) and sends nerves to the sense organs and to the buccal ganglia. The ventral portion consists mainly of the pedal and visceral ganglia.

The latter sends a large number of nerves to the mantle, the viscera and the gills. The large *ganglion stellatum*, which is found on each side in the mantle, a ganglion of the vena cava, two branchial ganglia, and the *ganglion splanchnicum* are all developed on the course of these nerves from the visceral ganglia.

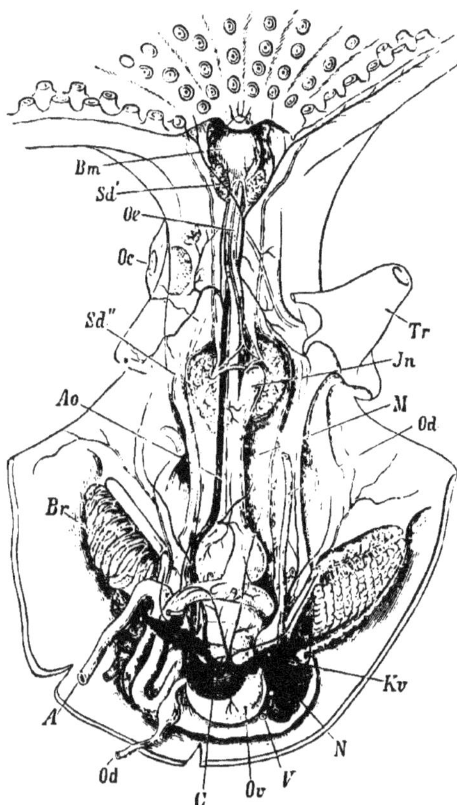

Fig. 535.—Viscera of *Octopus vulgaris* after removal of the posterior mantle wall and liver (after M. Edwards). *Bm*, buccal mass; *Sd'*, upper salivary gland; *Oe*, œsophagus, *Sd''*, lower salivary gland; *Jn*, crop; *M*, stomach; *A*, end of the rectum turned back; *Oc*, eye; *Tr*, funnel; *Br*, gills; *Ov*, ovary; *Od*, oviduct; *N*, kidney; *Kv*, auricle, receiving the branchial vein; *V*, posterior vena cava; *C*, ventricle; *Ao*, aorta.

Of the **sense organs** the large eyes, placed on the sides of the head, are the most conspicuous. Each eye-bulb is placed in a special orbit, which is partly formed by an excavation in the cephalic cartilage. It is enclosed in a strong capsule which is continued over the front of the eye as a thin and transparent membrane, the *cornea*. The cornea may, however, be entirely absent (*Nautilus*), or in other cases be pierced beneath an eyelid-like cutaneous fold by a small hole (*Oigopsidæ*), through which the water enters the anterior optic chamber, and passes into a space of various extent round the anterior surface of the bulb (fig. 534). The Cephalopod eye possesses almost exactly the parts as the Vertebrate eye. The presence of the inner layer of same retinal rods in the former may be mentioned as an essential difference between them. The eye of *Nautilus* is without the lens.

The two auditory sacs are placed in the cephalic cartilage, and in the *Dibranchiata* in special cavities of the latter, the so-called cartilaginous labyrinth. They receive from the pedal ganglion their short auditory nerves, which, however, arise in the brain.

The **respiratory organs** have the form of two (*Dibranchiata*) or four (*Tetrabranchiata*) pennate gills, which are placed at the sides of the visceral sac in the mantle cavity. They are bathed by a current of water which is continually renewed.

The **heart** lies in the hinder part of the visceral sac, more or less closely approximated to the apex of the body. It consists of a median ventricle and as many lateral auricles as there are gills (figs. 535 and 536). A large anterior aorta (*aorta cephalica*) passes off from the ventricle and gives in its course strong branches to the mantle, alimentary canal, and funnel, and breaks up in the head into vessels to the eyes, lips and arms. A posteriorly directed visceral artery also leaves the ventricle. The capillary network, which is richly developed in all the organs, passes partly into sinuses, partly into veins, which are collected through lateral veins into a large anterior and a posterior vena cava. Each of these bifurcates into two or four trunks (according to the number of gills) which carry the blood to the gills. Immediately before their entrance into the gills the walls of these so-called branchial arteries are (except in Nautilus) especially muscular and rhythmically contractile and constitute *branchial hearts*. The *Cephalopoda* also possess arrangements by which a mixture of water with the blood can be effected.

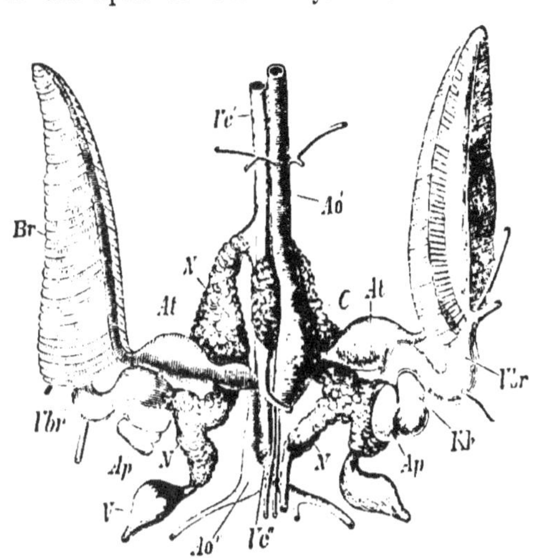

FIG. 536.—Circulatory and excretory organs of *Sepia officinalis* from the dorsal side (after Hunter) *Br*, gills; *C*, ventricle; *Ao'* and *Ao''*, the anterior and posterior aorta; *V*, lateral vein; *Vc'*, anterior vena cava; *Vc''*, posterior vena cava; *N*, renal appendages of the veins; *Vbr*, advehent branchial vessels (branchial arteries); *Kh*, branchial heart; *Ap*, appendage of the same; *At*, *At'*, auricles receiving the revehent branchial vessels (branchial veins).

Paired **kidney** sacs are always present, one on each side of the abdomen. They open into the mantle cavity, each through the apex of a papilla. The anterior walls of the sacs are pushed inwards by cæcal appendages of the venæ cavæ (branchial arteries), so as to give rise to a number of racemose lobules projecting into each renal sac (fig. 536). The renal sacs, as in other Molluscs, communicate with the body cavity, which in *Sepia* is largely developed and contains the heart, generative organs, etc., but in the *Octopoda* is reduced to a narrow tubular space (" water - vascular system " of Krohn) and only contains the sexual glands.

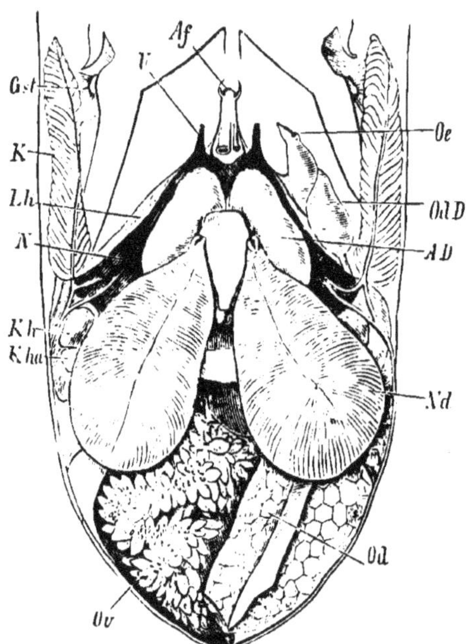

Fig. 537.—Anatomy of the body of a female *Sepia* (after C. Grobben). *Ov*, ovary in its cavity (body cavity) which is laid open; *Od*, oviduct; *Oe*, opening of the same; *OdD*, oviducal gland; *Nd*, nidamental gland; *AD*, accessory nidamental gland; *N*, kidney; *U*, ureter; *Lh*, canal of the body cavity (water canal); *Kh*, branchial heart; *Kha*, pericardial gland (appendage of branchial heart); *K*, gills; *Af*, anus; *Gst*, stellate ganglion.

An excretory organ very generally present is the **ink-sac.** It is a piriform sac, whose duct opens to the exterior with the anus, and empties an intensely black fluid, which surrounds the body of the animal as in a black cloud, and so protects it from the pursuit of larger marine animals.

The *Cephalopoda* are diœcious. Males and females present external sexual differences which principally concern a particular arm. According to the discovery of Steenstrup, one of the arms in the male always becomes modified, *hectocotylized* as it is called, as an intromittent organ. The two sexes of *Argonauta* differ considerably, inasmuch as the small male has no shell.

The **sexual glands** lie freely in the body cavity. Their products are dehisced into the body cavity, from which they are taken up and conveyed to the exterior by special ducts. The ovary is unpaired and racemose, and the oviduct is a double (*Octopoda*) or unpaired (usually left) duct opening into the mantle cavity; it receives in its course a round gland, and its terminal portion possesses glandular walls. In addition, the so-called nidamental glands (fig. 537) are present in the *Decapoda* and *Nautilus*; they open into the mantle cavity near the generative opening and secrete a cementing substance which surrounds and unites together the eggs. The eggs are surrounded—either singly (*Argonauta*, *Octopus*) or in great number (*Sepia*) —by capsules with long stalks, which are united together in racemose masses (so-called sea-grapes), and fastened to foreign objects in the sea. In other cases the eggs are aggregated in gelatinous tubes (*Loligo*, *Sepiola*).

The male generative apparatus presents a similar arrangement (fig. 538, *a*). The testis (*T*) consists of an unpaired gland formed of long cylindrical tubes. The duct of the testis is placed on the left side and is long, coiled and complicated. The following parts may be distinguished in it: (1) a much coiled vas deferens (*Vd*), which opens into the body cavity, (2) a long dilated vesicula seminalis (*Vs*) with two prostatic glands (*Pr*) opening into its terminal portion, (3) a spacious sac, known as Needham's sac, in which the spermatophores are formed, and which opens into the mantle cavity at the apex of a papilla placed on the left side.

Fig. 538*a*.—Male sexual organs of *Sepia officinalis* (after Duvernoy), modified from C. Grobben. *T*, testis, with a piece of peritoneum; *To*, opening of the testis into the body cavity; *Vd*, vas deferens; *O*, opening of the vas deferens into the body cavity; *Vs*, vesicula seminalis; *Pr*, prostate; *Sp*, spermatophore reservoir; *Oe*, sexual opening.

Fig. 538 *b*.—Spermatophore of *Sepia* (after M. Edwards).

In copulation the large spermatophores (fig. 538, *b*) are introduced by means of the hectocotylised arm into the female sexual opening. In some Cephalopoda (*Tremoctopus violaceus*, *Philonexis Carenæ*, and

Argonauta argo) the hectocotylised arm of the male appears as an individualized intromittent organ which is filled with spermatophores, then separates from the body of the male, moves about for a time independently, and finally conveys the semen into the mantle cavity of the female (fig. 539).

The **development*** of the egg is introduced by a discoidal (partial) segmentation which takes place at the pointed pole of the egg. As in the bird's egg, the segmented portion of the ovum (formative yolk) gives rise to a *germinal disc* which in the subsequent growth is raised more and more from the lower part of the blastoderm which forms the yolk sac. Soon several projections appear on the embryonic rudiment (fig. 540); first in the centre of the germ a flattened ridge is formed around a central depression (*M*) which it soon grows over. This is the mantle [the depression is the so-called shell gland]; on each side of it the two parts of the funnel appear (*Tr*), and between these and the mantle the gills (*Br*). Also laterally but external to the folds of the funnel the first traces of the head appear as two pairs of elongated lobes, of which the external anterior pair bears the eyes. On the outer edge of the disc papilliform structures are formed, the first rudiments of the arms. In the later growth of this absolutely symmetrical embryo the Cephalopod form becomes more and more apparent: the mantle projects considerably, and grows over the gills and two parts of the funnel, which fuse to form the definitive funnel. The cephalic lobes grow together between the mouth and funnel, and on their oral sides become more sharply constricted off from the yolk, which with a few exceptions persists for some time as a yolk sac (fig. 541).

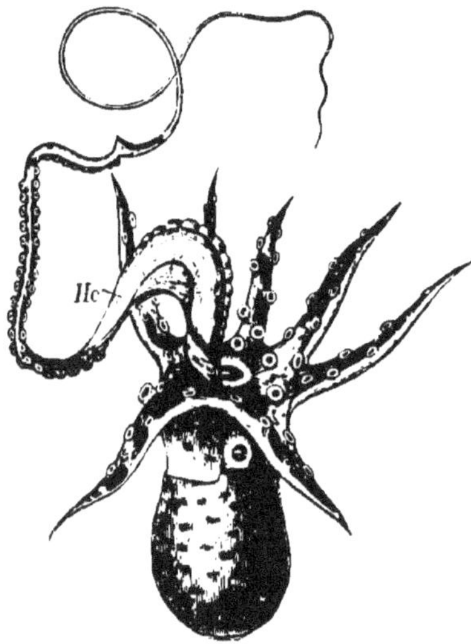

FIG. 539.—Male of *Argonauta argo* (after H. Müller). *Hc*, hectocotylised arm.

The Cephalopods are marine animals, some frequenting the coast and others the high seas. They feed on the flesh of other animals,

* Cf. besides van Beneden and Kölliker; Ussow, "Zoologisch-embryologische Untersuchungen." *Archiv für Naturgesch.*, 1874.

especially *Crustacea*. Some of them attain a great size. The flesh is eaten, and the colouring matter of the ink-sac (sepia) and the dorsal shell (*os sepiæ*) are used by man. The remains of Cephalopods occur in all formations from the oldest Silurian and constitute important characteristic fossils (*Belemnites*, *Ammonites*).

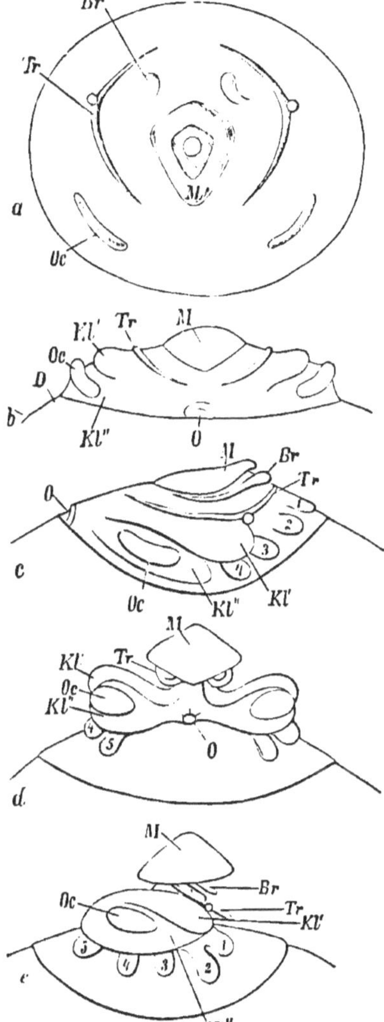

FIG. 540.—Embryonic development of *Sepia officinalis* (after Kölliker). *a*, View of germinal disc from above, commencing embryo lying on the yolk. *Br*, gills; *Tr*, folds of the funnel; *Oc*, eye; *M*, mantle. *b*, Somewhat older stage, seen from the front. *D*, yolk; *Kl'*, anterior; *Kl''*, posterior cephalic lobe; *O*, mouth. *c*, later stage from the side. 1-4, first rudiments of the arms. *d*, older stage from the front. 5, fifth pair of arms. *e*, Still later stage in lateral view The halves of the funnel have united.

Order 1.—TETRABRANCHIATA.*

Cephalopoda with four gills in the mantle cavity and numerous retractile tentacles on the head, with split funnel and many-chambered shell.

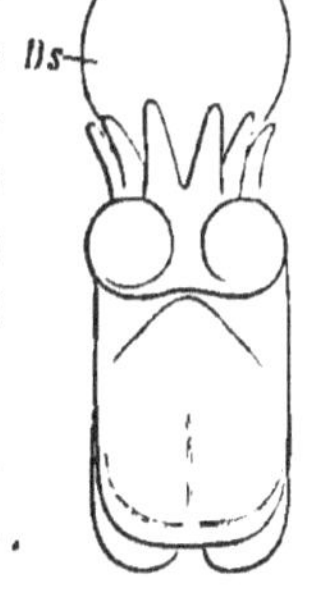

FIG. 541.—Almost ripe embryo of *Sepia officinalis* from the dorsal (anterior) face (after Kölliker). *Ds*, yolk sac.

The appendages of the head are peculiar. In place of the arms there are a number of filiform tentacles round the mouth. In *Nautilus* there are on each side of the body (*a*) nineteen external tentacles, of which the dorsal pair constitutes a kind of hood which can close the orifice of the shell; (*b*) two ocular tentacles on each side near the

* Van der Hoeven. "Beiträge zur Kenntniss von Nautilus" (in Dutch), Amsterdam, 1856.

W. Keferstein in Bronn, Classen und Ordnungen des Thierreichs. Dritter Band, Cephalopoda. 1865.

eye and (c) twelve internal tentacles, the four ventral of which on the left side are in the male modified to form the *spadix*, an organ analogous to the hectocotylised arm. Finally, in the female there are on each side, within the latter, fourteen or fifteen ventrally-placed labial tentacles. (Fig. 542.)

The cephalic cartilage, instead of forming a complete ring, consists of two horse-shoe-shaped limbs on which the central parts of the nervous system lie. The eyes are stalked, and are without a lens or other refractile media. The funnel has the form of a lamina rolled upon itself, but the edges are free and not fused. There is no ink-sac. The branchiæ are four in number as are also the branchial vessels and the kidneys.

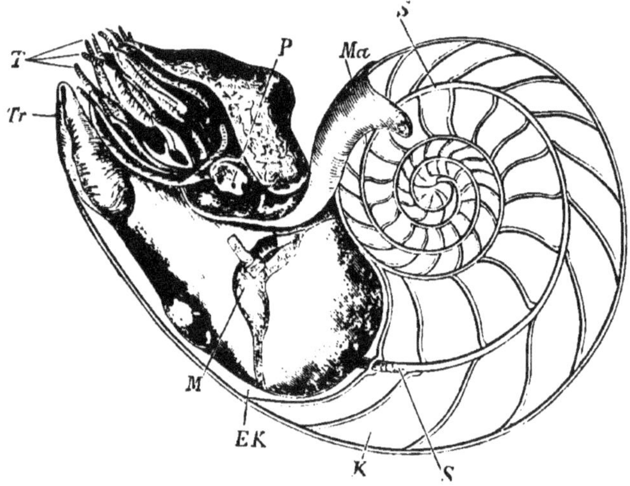

Fig. 542.—*Nautilus* (règne animal). *T*, tentacles; *P*, pupil of the eye; *Ek*, terminal chamber; *Tr*, funnel; *K*, chambers of the shell; *S*, siphon; *Ma*, mantle; *M*, muscle.

The hinder part of the thick external shell of the Tetrabranchiata is divided by cross partitions into numerous chambers, which are filled with air and are traversed by a siphon. The shell consists of an external, frequently coloured calcareous layer, and an internal mother-of-pearl layer. The similar structure of many fossil shells allows us to infer a similar organisation for their unknown inhabitants. The position and structure of the siphon, as well as the form of the septa, and the lines of fusion of the latter with the shell, are important characters for the classification of the fossil Tetrabranchiata. The small number of living species of the genus *Nautilus* are found in the Indian and Pacific Oceans.

Fam. **Nautilidæ.** The septa are simply bent and concave towards the anterior chambers. Line of suture simple, with a few large wavy curves or a lateral lobe. Siphon usually central; shell orifice simple. *Orthoceras*, shell straight. *O. regularis* v. Schl., calcareous strata of the North German plain. *Nautilus*, shell coiled. *N. pompilius* L., Indian Ocean.

Fam. **Ammonitidæ.** The septa much folded at the sides, always with lobe on the outer side, in the middle usually convex forwards. Siphon on the outer side. Contains only fossil forms. *Goniatites retrorsus* v. Buch., *Ceratites nodosus* Bosc., *Ammonites capricornus* v. Schl.

Order 2.—DIBRANCHIATA.*

Cephalopoda with two gills in the mantle cavity, eight arms bearing suckers or hooks, complete funnel and ink-sac.

The *Dibranchiata* possess round the mouth eight arms provided

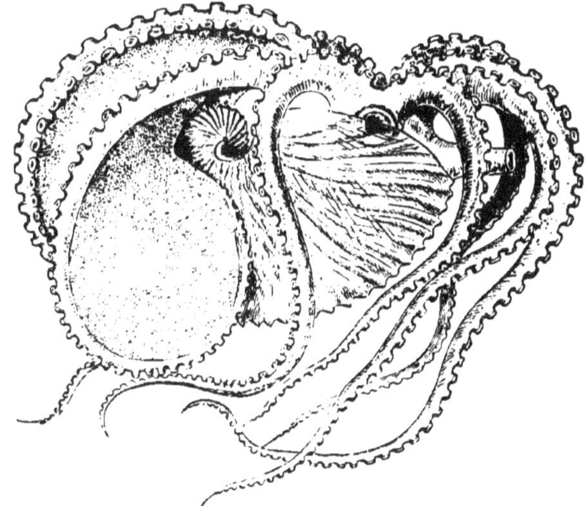

FIG. 543.—*Argonauta argo* (female), swimming.

with suckers or hooks; in the *Decapoda* there are, in addition, two long tentacles placed between the ventral arms and the mouth. The cephalic cartilage constitutes a completely closed ring surrounding the central parts of the nervous system; its slightly arched lateral parts serve for the support of the sessile eyes. There are only two gills in the mantle cavity and the same number of branchial vessels and kidneys. The funnel is closed. An ink-sac is usually present. The shell is in many forms completely absent; in others it is reduced to a horny or calcareous dorsal lamella. A spirally-coiled shell is

* Chief works: Férussac et d'Orbigny *l.c.*, also Verany *l.c.*

rarely present. In the female *Argonauta* (fig. 543) there is a single-chambered spiral shell with thin walls; in *Spirula* (fig. 544) there is a multilocular spiral shell, the chambers of which are traversed by a siphon.

Sub-order 1. **Decapoda.** In addition to the eight arms, there are two long tentacles between the third and fourth pairs of arms (ventral). The suckers are stalked and provided with a horny rim. The eyes are without a sphincter-like lid. The mantle bears two lateral fins, and at the mantle edge a well-developed apparatus for closing the mantle opening. An internal shell is present.

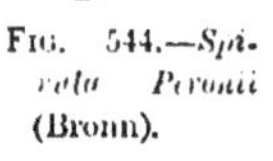

Fig. 544.—*Spirula Peronii* (Bronn).

Fam. **Spirulidæ.** *Spirula Peronii* Lam., Pacific Ocean.

Fam. **Belemnitidæ.** *Belemnites digitalis* Voltz, Upper Lias.

Fam. **Myopsidæ.** With closed cornea and covered lens. *Sepia officinalis* Lam., *Loligo vulgaris* Lam., Mediterranean (fig. 531). *Sepiola vulgaris* Grant., Mediterranean, *Rossia macrosoma* Fér. d'Orb., Mediterranean.

Fam. **Oigopsidæ.** Eyes with widely-opened cornea, so that the crystalline lens is exposed and bathed by the sea-water. *Onychoteuthis Lichtensteini* Fér., *Ommastrephes todarus* d'Orb.

Sub-order 2. **Octopoda.** The two tentacles are not present. The eight arms bear sessile suckers without a horny ring, and are connected at their base by a membrane. Eyes relatively small, with sphincter-like lid. The short, rounded body is without the internal shell, and usually also the fin-like appendages. Mantle without cartilaginous apparatus for closing mantle opening, and attached to the head by a broad cervical band. Funnel without valve; oviduct paired.

Fam. **Octopodæ** (fig. 530). *Octopus vulgaris* Lam., *O. macropus* (fig. 535), *Eledone moschata* Lam.

Fam. **Philonexidæ.** *Philonexis Carenæ* Ver., *Tremoctopus violaceus* Dell. Ch., *Argonauta argo* L. The small male is without a shell (fig. 539). The large female possesses fin-like expansions of the dorsal arms, and bears a boat-shaped, delicate shell, round the sides of which the arm-fins are spread (fig. 543).

CHAPTER II.

MOLLUSCOIDEA.

Attached bilateral unsegmented animals, with crown of ciliated tentacles or spirally rolled buccal arms; enclosed by a cell or by a bivalve shell, the valves of which are dorsal and ventral: with a simple ganglion or with several ganglia connected by a pharyngeal ring.

The two groups, *Bryozoa* and *Brachiopoda*, which are included in the Molluscoidea were formerly placed amongst the Molluscs, to which they do indeed present affinities. With the increase in our knowledge of their developmental history, it appears more and more probable, not only that the two groups are descended from an ancestral form common to them and the Annelids, but also that in spite of the considerable differences between them in the adult state, they are in reality closely related, a supposition which agrees with the great resemblance of their larvæ. Should this view of the close relationship of the Brachiopoda, which are always solitary, with the Bryozoa, which almost always form colonies, turn out to be well grounded, then the tentacular crown and the simple ganglion of the latter would be homologous with the spiral arms and subœsophageal ganglion of the former respectively.

Class 1.—BRYOZOA * = POLYZOA.

Small animals usually united together to form colonies: with ciliated tentacular crown, horse-shoe-shaped alimentary canal and simple ganglion.

The *Bryozoa* owe their name to the moss-like dendritic appearance of their colonies, on which the small individual zooids are arranged in a regular manner. The colonies may, however, have a foliaceous or polyparium-like form, or they may form crusts on the surface of foreign objects. Solitary Bryozoa are rare exceptions (*Loxosoma*). As a rule the colonies possess a horny or parchment-like, frequently

* F. A. Smitt, "Kritisk förteckning öfver Skandinaviens Hafs-Bryozoer Öfvers." *Kongl. Vetensk. Akad. Förhandl.*, 1865, 1866, 1867.
H. Nitsche, "Beiträge zur Kenntniss der Bryozoen." *Zeit. für wiss. Zool.*, 1869 and 1871.
J. Barrois, "Recherches sur l'embryologie des Bryozoaires." Paris, 1877.

also calcareous, rarely gelatinous exoskeleton, which arises from the hardening of the cuticle around the individual zooids. Each zooid (*zoœcium*) (fig. 545) is accordingly surrounded by a very regular and symmetrical case—the **ectocyst** or **cell**; through the opening of which the anterior part of the soft body of the contained zooid with its tentacular crown can be protruded.

The form of the cells, and the manner in which they are connected together, are very different in the different groups, and give rise to a great variety in the form of the colonies composed of them. The cells are usually completely shut off from each other. With regard to their connection, they sometimes project obliquely or at a right angle; sometimes they are spread out horizontally on the same plane; sometimes arranged in rows on a branched axis. Their openings are usually turned towards one side or towards two opposite sides. The soft body wall, or **endocyst** (fig 545, *En*) is closely applied to the inner wall of the ectocyst: it consists of an external layer of cells (matrix of the ectocyst) and of a network of crossing muscular fibres (the external fibres are transversely, the internal longitudinally arranged), which are separated from the first layer by a homogeneous membrane. On the inner side of the muscular layer there is, at least in the fresh-water Bryozoa, a delicate layer of ciliated cells which line the body cavity. At the opening of the cell the soft endocyst is invaginated inwards, and passes thence on to the anterior and extrusible part of the body, of which it forms the only investment. In most fresh-water Bryozoa this reduplicature of the endocyst is always present even when the zooid is protruded (fig. 545). The greater part of the anterior region of the body, with its crown of tentacles, can, however, always be protruded from the cell and retracted into it again by special muscles traversing the body cavity (fig. 545).

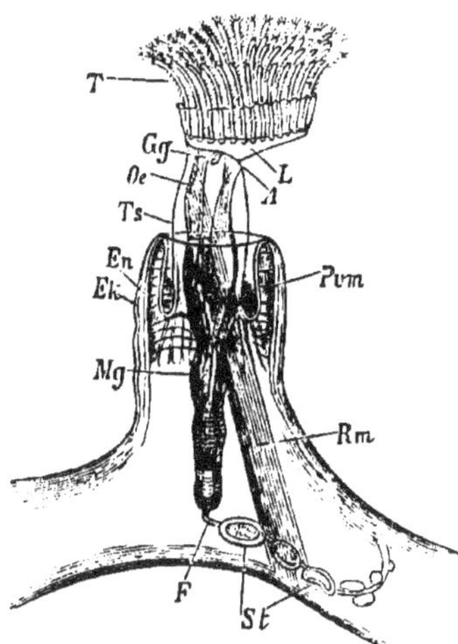

FIG. 545.—*Plumatella repens* (after Allman). *T*, Tentacles; *L*, lophophore; *Oe*, œsophagus; *Mg*, stomach; *A*, anus; *F*, funiculus; *St*, statoblasts; *Ts*, tentacular sheath; *Ek*, ectocyst; *En*, endocyst; *Gg*, ganglion; *Pvm*, parieto-vaginal muscles; *Rm*, retractor muscle.

The disc on which the mouth is placed is known as the **lophophore.** The lophophore is either circular (*Stelmatopoda*), or it is drawn out into two lobes so as to have a horse-shoe shape (*Lophopoda*, fig. 545), and its margins are produced into a number of richly ciliated tentacles. The tentacles are simply hollow processes of the body wall; they are provided with longitudinal muscles, and their cavity communicates with the body cavity, from which they are filled with blood. They serve both for procuring food (setting up by means of their cilia whirlpools in the water) and for respiration.

The **digestive organs** lie freely in the body cavity, and are attached to the integument by the so-called *funiculus* and by bundles of muscles. The body and tentacular apparatus has been incorrectly regarded as a kind of individual, and opposed to the cell or *Cystid*, in which it is placed, as the *Polypid*. The mouth is placed in the centre of the circular or horse-shoe shaped lophopore, and a moveable epiglottis-like process, known as the **epistome**, often projects over it. The alimentary canal is bent on itself, and consists of (1) an elongated ciliated œsophagus often dilated to a muscular pharynx; (2) a spacious stomach, with a blind backward prolongation, the hind end of which is attached to the body-wall by a cord (**funiculus**), and (3) a narrow intestine, which is bent up nearly parallel with the pharynx and is directed forwards. The intestine opens by the dorsally-placed anus, near but usually outside the buccal disc (*Ectoprocta*, fig. 545). In a few forms the anus is within the circle of tentacles (*Endoprocta*), *e.g.*, *Pedicellina* and *Loxosoma* (fig. 546).

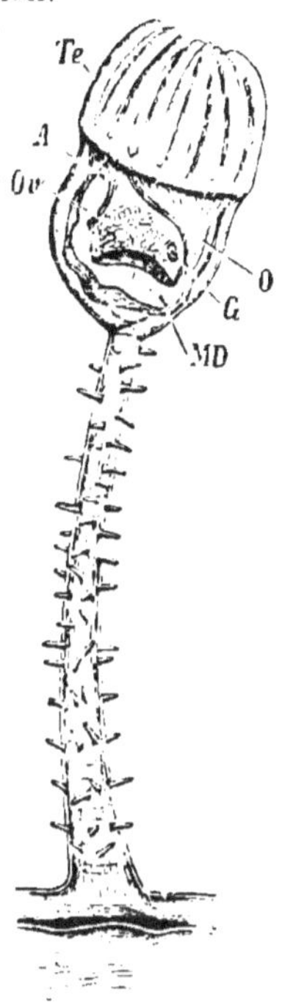

FIG. 546.—*Pedicellina echinata*. *Te*, tentacular crown; *O*, mouth; *MD*, alimentary canal; *A*, anus; *G*, ganglion; *Ov*, ovary.

Heart and **vascular system** are absent. The blood fills the whole body cavity, through which it is circulated chiefly by the cilia of the body-wall. The whole surface of the anterior protrusible part of the body, and especially of the tentacles, serves as a **respiratory organ**. The ciliated canal of the *Endoprocta* is to be regarded as a **kidney**.

The **nervous system** consists of a ganglion placed on the œsophagus between the mouth and the anus. This ganglion in the *Lophopoda* is contained in the cavity of the lophophore, and is attached to the œsophagus by a delicate circum-œsophageal ring; it sends off numerous nerves to the tentacles and œsophagus. According to Fr. Müller there is in *Serialaria* a so-called *colonial nervous system* which connects the individual zooids of one colony and enables them to co-ordinate their activities. Claparède * describes the same for *Vesicularia*, also for *Scrupocellaria scruposa* and *Bugula* (*avicularia*). Special organs of sense have not been recognised.

Many forms of Bryozoa present examples of a well-marked polymorphism. In *Serialaria* and its allies the joints of the stalk represent a special form of individual; they have a considerable size and a simplified organization, and serve as the ramified substratum on which the nutritive individuals are placed. In addition, there are here and there joints of the roots which, under the form of tendril- and stolon-like processes, serve to attach the colony. The peculiar appendages known as *avicularia* and *vibracula*, which are modified individuals and seem to have the function of food-procuring organs, are found in many marine Bryozoa. The avicularia (fig. 547, *Av*) resemble birds' heads and consist of two-armed pincers, which are attached to the colony near the openings of the cells and occasionally snap. They may seize small organisms, *e.g.*, worms, and hold them till they are dead; the decomposing organic remains are swept into the mouth by the currents caused by the cilia of the tentacles. The vibracula have a similar arrangement, but present in place of the snapping beak a long and extremely moveable flagelliform filament (fig. 548). Finally there are the *ovicells* (oœcia), each of which is filled with an egg; they have the form of helmet or dome-shaped appendages and are sessile on the zoœcium (fig. 547 *Ovz*).

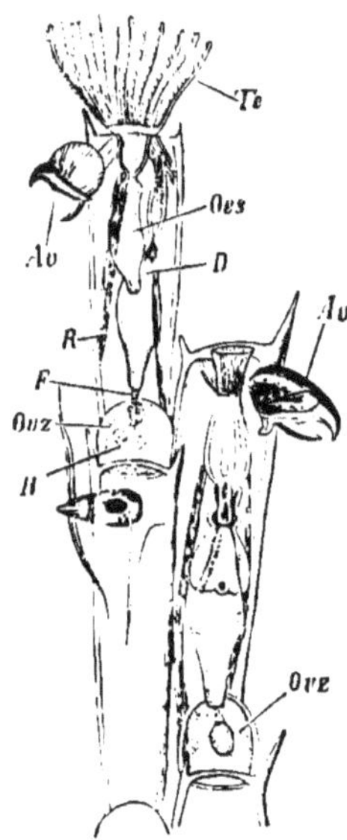

Fig. 547.—*Bugula avicularia* (after Busk). *Te*, Tentacular crown; *R*, retractor muscle; *D*, alimentary canal; *F*, funiculus; *Av*, avicularia; *Oes*, œsophagus; *Ovz*, Ovicells.

The **reproduction** is partly sexual and partly asexual; in the latter case it may be effected by the so-called *statoblasts* or by budding. The male and female sexual organs are reduced to groups of cells producing either spermatozoa or ova, which usually arise in

* Ed. Claparède, "Beiträge zur Anatomie und Entwickelungsgeschichte der Seebryozoen." *Zeit. f. wiss. Zool.*, Tom. XXI., 1871.

the same animals, more rarely in different individuals. The ovaries which are filled with many ova are placed on the inner surface of the anterior part of the body wall; while the testes with their seminal capsules are developed either on the upper part of the funiculus or near the point of attachment of the latter to the body wall. Both kinds of generative products are dehisced into the body cavity where fertilization takes place. From the body cavity the fertilized egg passes either into a bud of the body wall (*Alcyonella*), or, as in marine Bryozoa, into an external appendage, — the oœcium.

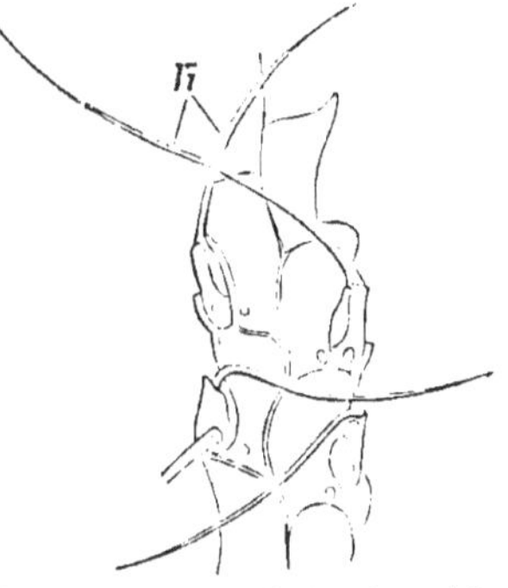

FIG. 548.—*Scrupocellaria ferox* (after Allman). *Vi*, Vibracula.

The name **statoblast** (fig. 549) was given by Allman to certain peculiar reproductive bodies, which were formerly regarded as hard-shelled winter eggs, but by him were recognised to be germs which are not fertilised. The statoblasts are found only in the fresh-water forms. They arise from masses of cells which appear principally towards the end of summer on the funiculus (fig. 545). They usually possess a lens-like, biconvex form, and are covered by two watchglass-shaped, hard chitinous shells, the edges of which are often enclosed by a flat ring formed of cells containing air (float), and sometimes (*Cristatella*) provided with a crown of projecting spines (fig. 549).

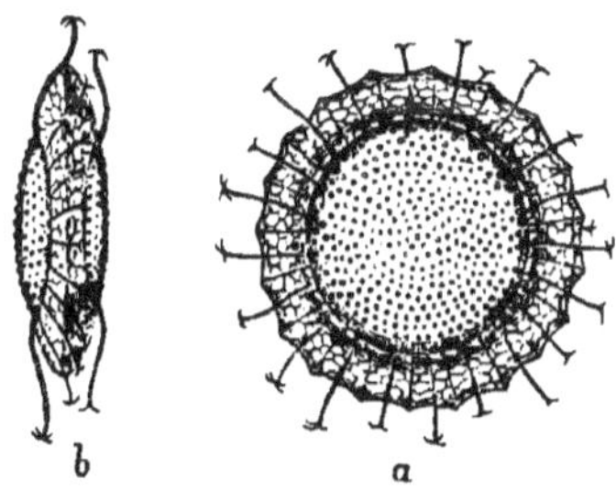

FIG. 549.—Statoblasts of *Cristatella mucedo* (after Allman). *a*, From the surface; *b*, from the side.

A very important part of the reproduction is effected by buds which remain permanently attached. The process of budding begins very early in life, before the development of the embryo is completed, and gives rise to the formation of colonies. Parts separated off from the colony are rarely able to produce new colonies (*Cristatella*, *Lophopus*).

The **development** is always connected with a metamorphosis. The budding always begins in the embryo. In the fresh-water forms, after the alimentary tract and tentacular apparatus have made their appearance, a second alimentary canal and tentacular apparatus arise, so that the ciliated embryo still enclosed in the egg membranes represents a small colony of two individuals. In the marine chilostomatous Bryozoa the fertilized egg passes into the ovicell, which consists of a helmet-shaped capsule and a vesicular operculum. Here the egg segments and develops into an embryo, which passes out as a ciliated larva, and swims about freely in the sea. The irregularly globular larva possesses a ring of cilia (fig. 550, *a*, *b*, *c*). After some time the larva attaches itself and develops the tentacular crown. The primary zoœcium soon produces new zoœcia by budding; avicularia are developed, and finally, but not until after the death of the older zoœcia, root filaments.

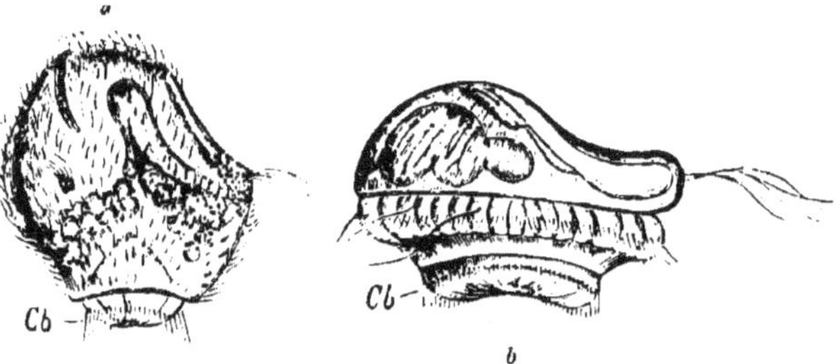

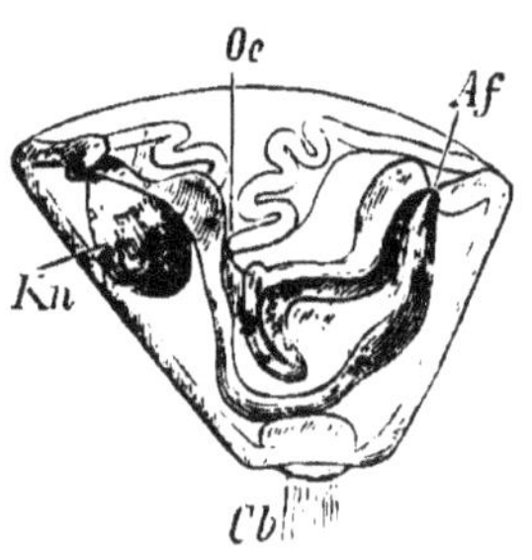

Fig. 550.—*a*, Larva of *Canda reptans* (after Barrois). *b*, Larva of *Lepralia* (after Barrois). *c*, *Cyphonautes* (diagrammatic after Hatschek). *Oc*, mouth; *Af*, anus; *Cb*, tuft of cilia; *Kn*, bud.

In the *Endoprocta* the egg develops in a brood-pouch placed on the oral side of the animal. The segmentation is complete, and leads to the formation of a blastosphere; the endoderm arises by invagination, and gives rise to the lining of the midgut; the œsophagus and rectum being formed from the ectoderm (fig. 551). The mesoderm arises from two cells. The larvæ of the Endoprocta possess an alimentary canal bent into the form of a horse-shoe, and a ciliated collar which is protruded at the front end; further, they contain a bud (fig. 551 *c*, *Kn*), as the first rudiment of a second individual, and a cement gland at the hind end (*Dr*).

Other larval forms, which are apparently of a very different structure, are reducible to the same type—*e.g.*, *Cyphonautes* (fig. 500, *c*), a larva which is found in all seas, and is, according to Schneider, the larva of *Membranipora pilosa*.

After the winter the contents of the statoblasts give rise to simple, non-ciliated animals, which possess, when they are hatched, all the parts of the adult animal, at once become attached, and produce new colonies by budding.

The Bryozoa are for the most part marine, and they attach themselves to stones, Lamellibranch shells, corals and plants. Some fresh-water forms belonging to the genus *Cristatella* have the power of moving about.

The Bryozoa were widely distributed in the earlier periods, as their numerous fossil remains, which increase in number from the Jurassic period onwards, prove.

Order 1.—Endoprocta.*

Bryozoa with anus within the circle of tentacles.

In the structure of their bodies and the formation of their colonies

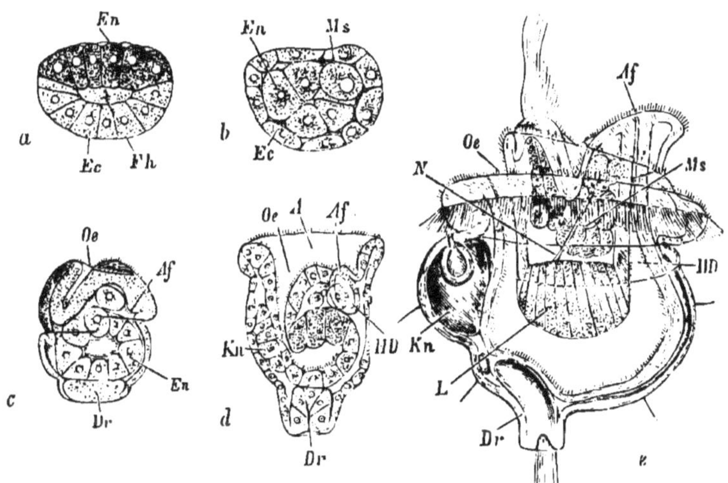

Fig. 551.—Development of *Pedicellina echinata* (after B. Hatschek). *a*, Blastosphere with flattened side of endoderm. *Ec*, Ectoderm; *En*, endoderm; *Fh*, segmentation cavity. *b*, Later stage in optical median section. One of the two first mesoderm cells (*Ms*) which lie to the right and left of the middle line is indicated. *c*, Later stage in optical median section. *Dr*, Cement gland; *Oe*, œsophagus; *Af*, first rudiment of the rectum. *d*, Young larva in optical median section. *A*. Atrium; *HD*, rectum; *Kn*, bud. *e*, Free-swimming larva, extended. *N*, Excretory canal; *L*, liver cells; *Ms*, mesoderm cells.

the Endoprocta present simpler, more primitive conditions since they retain essentially the organization of the Bryozoan larva. The tentacular apparatus of the adult is from its origin directly reducible to the ciliated crown of the larva. Mouth and anus both open within the tentacular circlet into a kind of atrium, which forms

* Besides Nitsche, cf., B. Hatschek. "Embryonal Entwickelung und Knospung der Pedicellina echinata." *Zeit. für wiss. Zool.*, Tom. XXVIII.

a brood-pouch in which the testes and ovaries open and the embryos are developed. A pair of ciliated excretory canals is present.

Fam. **Pedicellinidæ.** Stocks with stolons, on which the long-stalked individuals project. *Pedicellina echinata*, Sars. (fig. 552).

Fam. **Loxosomidæ.** Long-stalked solitary animals. *Loxosoma singulare* Kef., *L. neapolitanum* Kow.

Order 2.—Ectoprocta.

Bryozoa with anus opening outside the tentacular circlet.

This group includes by far the greater number of the Bryozoa; their structure has been especially referred to in the precedent description of the class. The anus always opens outside the ring of tentacles, which are either arranged in a closed circle or on a two-armed horseshoe-shaped lophophore.

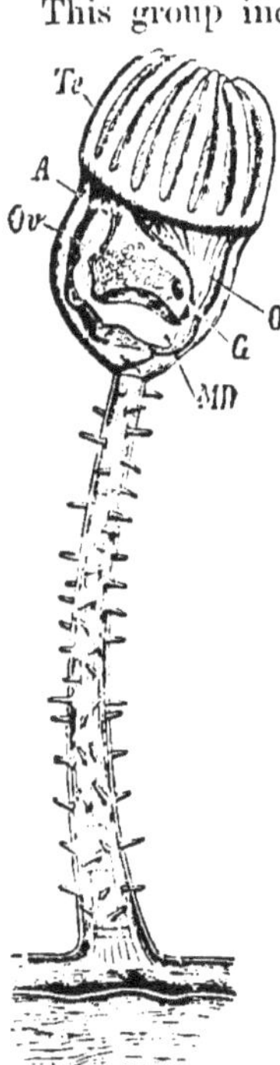

Fig. 552.—*Pedicellina echinata.* *Te*, Tentacular crown; *O*, mouth; *MD*, alimentary canal (*mesenteron*); *A*, anus; *Ov*, ovary; *G*, ganglion.

Sub-order 1. Lophopoda * (Phylactolæmata Allm.).

Fresh-water Bryozoa (excepting the marine Rhabdopleura) with horseshoe-shaped lophophore and epistome.

The Lophopoda are mainly distinguished by the bilateral arrangement of the numerous tentacles on the two-armed lophophore (fig. 553). There is always present above the mouth a moveable, tongue-shaped process, the *epistome*, whence the name *Phylactolæmata* given by Allman to this sub-order. The zooids are usually of considerable size, and, as opposed to the marine Bryozoa, they are all alike (*i.e.*, there is no polymorphism). The cells frequently communicate with each other and give rise to ramified, or more spongy massive stocks of always transparent, sometimes horny, sometimes softer (either leathery or gelatinous) consistency. Statoblasts are very generally present.

Fam. **Cristatellidæ.** Free-moving colonies on the upper surface of which the individual zooids are arranged in concentric circles. *Cristatella mucedo* Cuv.

Fam. **Plumatellidæ.** Attached, massive or ramified colonies of fleshy or coriaceous consistence. *Lophopus crystallinus* Pall., *Alcyonella fungosa* Pall., *Plumatella repens* L. (figs. 545, 553).

* G. J. Allman, "Monograph of Fresh-water Polyzoa." Ray Soc., 1856.

Sub-order 2. **Stelmatopoda (Gymnolæmata).**

Bryozoa with discoidal lophophore, tentacles in a closed circle; mouth without epistome.

The Stelmatopoda are, with the exception of the *Paludicellidæ*, all marine forms. They are always without the epiglottis-like epistome, and possess a complete circle of less numerous tentacles, which arise from a round buccal disc (fig. 517). In many forms, as in *Alcyonidium gelatinosum*, *Membranipora pilosa*, a flask-shaped ciliated canal in the body cavity has been observed; it opens to the exterior near the tentacles, and probably corresponds to the nephridia of segmented worms. Statoblasts are only rarely present. The eggs usually give rise to ciliated larvæ. The colonies are for the most part polymorphic, being often composed of root- and stem-cells, with vibracula and avicularia. The ectocysts are sometimes horny, sometimes incrusted with calcareous matter, and present great variety of form.

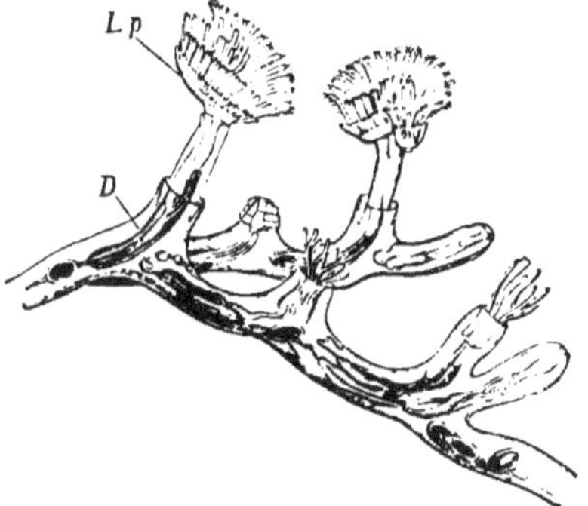

FIG. 553.—*Plumatella repens*, slightly magnified (after Allman). *Lp*, Lophophore; *D*, alimentary canal.

Tribe 1. **Cyclostomata.** The orifices of the cells wide and terminal, without movable appendages. Most of the species are fossil. The living species inhabit the Northern Seas.

Fam. **Crisiadæ.** Colonies erect and jointed. *Crisia cornuta* Lam., Mediterranean and North Sea; *C. eburnea* L.

Fam. **Tubuliporidæ.** The zoœcia disposed in continuous rows. *Idmonea atlantica* Forb., *Phalangella palmata* Wood, Arctic Ocean.

Tribe 2. **Ctenostomata.** Apertures of the cells terminal; when the tentacular sheath is retracted they are closed by a circle of spines as by an operculum. Stem-cells and root-filaments frequently occur

Fam. **Alcyonidiidæ.** Zoœcia united to form gelatinous stocks of irregular form. *Alcyonidium gelatinosum* L., Northern Seas.

Fam. **Vesicularidæ.** The zoœcia project as free tubes on the branched, creeping or erect colonies. *Vesicularia uva* L., *Farella pedicellata* Ald., Norway, *Serialaria Coutinhii*, Fr. Müll.

Fam. **Paludicellidæ.** Fresh-water forms. *Paludicella Ehrenbergii*, Van Ben.

Tribe 3. **Chilostomata.** The apertures of the horny or calcareous cells can be closed by a movable operculum or by a sphincter muscle. Avicularia, vibracula, and ovicells are often present.

Fam. **Cellulariidæ.** Dichotomously branched colonies; zoœcia in two or several rows. *Cellularia* Pallas, *C. Peachii* Busk, *Scrupocellaria* Van Ben. *S scruposa* L.

Fam. **Bicellariidæ.** Zoœcia conical or quadrangular, bent. Lateral face on which the aperture is placed is elliptical, and placed obliquely to the median plane of the axis. *Bugula* Oken, *B. avicularia* L. (fig. 547).

Fam. **Membraniporidæ.** Zoœcia more calcified and united to form an incrusting colony. *Membranipora* Blainv., *M. pilosa* L., Adriatic; *Lepralia pertusa* Esp., Adriatic; *Flustra membranacea* L.

Fam. **Reteporidæ.** Zoœcia oval-cylindrical, united to a reticulated colony. *Retepora* Lam., *R. cellulosa* L., Mediterranean to the Arctic Ocean.

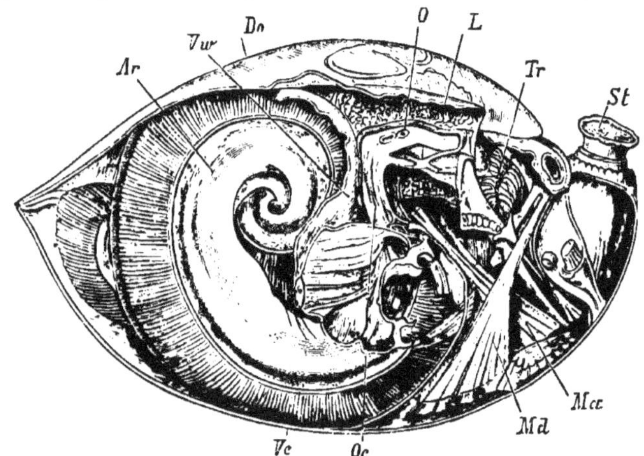

FIG. 554.—Anatomy of *Waldheimia australis*, seen from the side (after Hancock) *Do*, Dorsal side; *Ve*, ventral side of the mantle; *St*, peduncle; *Ma*, adductor; *Md*, divaricator *Ar*, arms; *Vw*, anterior body wall; *Oc*, œsophagus; *D*, intestine ending blindly; *O*, point of opening of the liver (*L*); *Tr*, funnel of the oviduct.

Class II.—BRACHIOPODA.*

Fixed Molluscoidea, with anterior (dorsal) and posterior (ventral) shell-valves, with two spirally-coiled buccal arms.

The more recent researches into the development have shown that

* R. Owen. "On the anatomy of the Brachiopoda." *Transact. Zool. Soc.*, London, 1835.

A. Hancock, "On the organisation of the Brachiopoda." *Phil. Trans.*, London, 1858.

Davidson, "Monograph of the British fossil Brachiopoda." 1858.

Lacaze-Duthiers, "Histoire naturelle des Brachiopodes vivants de la Mediterranée." *Ann. des. Sc. Nat.*, 1871, Tom. XV.

Kowalevski, "Russische Abhandlung über Brachiopoden-Entwickelung." Moskau, 1874.

W. K. Brooks, "The development of Lingula, and the systematic position of the Brachiopoda." *Chesapeake Zool. Lab. Scient. Results.* 1878.

the Brachiopoda, which have hitherto been regarded as Molluscs, are closely related to the *Bryozoa.*

The Brachiopoda possess a large body, enclosed in a bivalve shell, of which one valve is anterior (dorsal valve), the other posterior (ventral valve) (fig. 554). Both valves lie upon corresponding folds of the integument (mantle lobes), and are often connected on the back by a kind of hinge, above which the usually more arched ventral valve projects like a beak. This ventral valve is either directly fused with foreign bodies, or the animal is attached by a peduncle projecting through the opening of the beak (fig. 554 *St*). The peduncle may, however, pass out between the two valves (*Lingula*). The valves of the shell are cuticular structures secreted by the skin and impregnated with calcareous salts; they are not opened by a ligament, but by special groups of muscles (fig. 554 *Md*); they are closed also by muscles which are placed near the hinge, and pass transversely from the dorsal to the ventral surface through the body cavity (fig. 554 *Ma*).

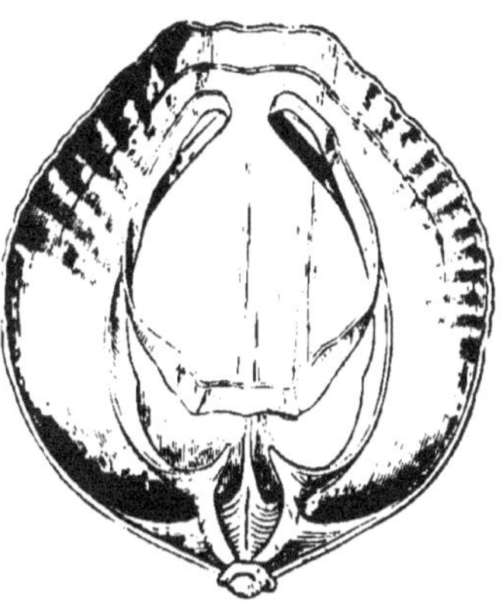

Fig. 555.—Dorsal valve of shell of *Waldheimia australis* with the brachial skeleton (after Hancock).

The body is bilateral and enclosed by the shell; it possesses two large reduplications of the integument, the two mantle lobes, which are applied to the inner surface of the shell. The edges of the mantle lobes are thickened, and carry very regularly-arranged setæ. The mantle may also produce within its own substance calcareous spicules or a continuous calcareous network.

The mouth is placed between the bases of the two spiral arms and leads into the œsophagus; the latter passes into the intestine, which is attached by ligaments and surrounded by large hepatic lobes. The intestine either describes a single bend, or is of considerable length and coiled (*Discina, Lingula*). In the latter case it opens into the mantle cavity by an anus placed on one side of the middle line; while in the hinged Brachiopoda (*Terebralula, Waldheimia*) there is no anus, and the intestine ends blindly in the body cavity (fig. 554). Sometimes the end of the intestine is continued into a string-like organ (*Thecidium*).

The two buccal arms are supported by a hard framework, consisting of calcareous processes of the dorsal valve of the shell (fig. 555). They have the form of long appendages rolled up in a conical spiral on the anterior side of the body; and they are traversed, as are the labial palps of many Lamellibranchs, by a groove. The edges of the groove give rise to close-set and long fringes composed of stiff

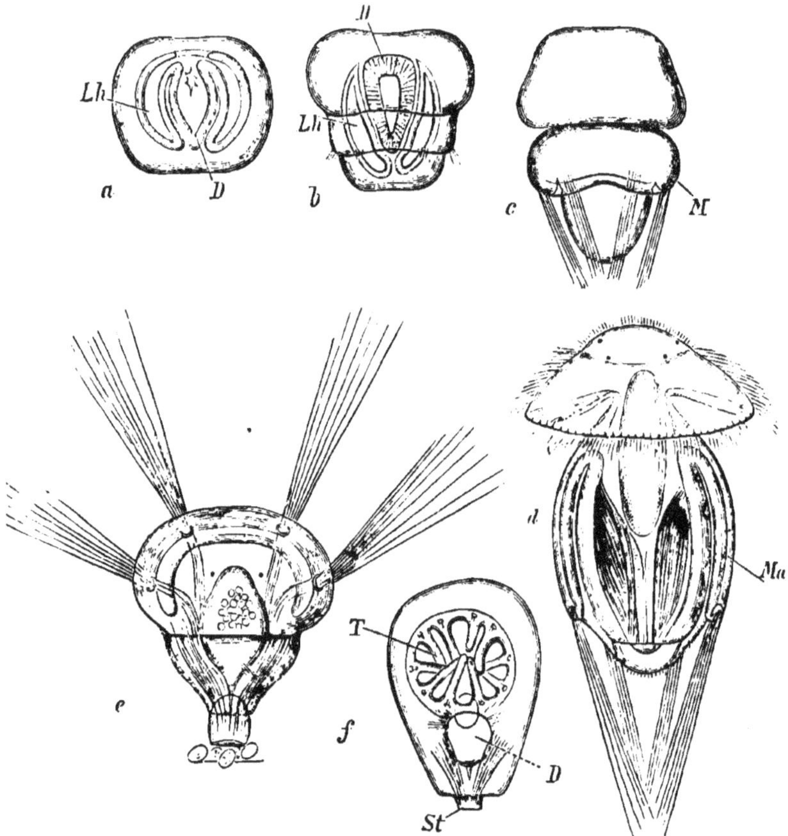

FIG. 556.—Development of *Argiope* (after Kowalevski). *a*, Larva, the gastric cavity of which has given rise to the diverticula of the body cavity (*Lh*); *D*, gut. *b*, Larva with three regions or segments. *c*, Larva with four bundles of setæ in the mantle-lobes of the middle segment. *d*, Later stage. *e*, Attached larva with mantle lobes bent anteriorly. *f*, The tentacles (*T*) are developed; *St*, peduncle.

and movable filaments, the ciliated covering of which produces a strong current which leads small particles of food to the mouth opening.

The **heart** is placed on the dorsal side of the anterior part of the intestine (stomach). It receives the blood through a venous trunk running on the œsophagus, and gives off several lateral arterial trunks. The vascular system is not closed, but is in connection with a blood sinus surrounding the alimentary canal, with the lacunæ of the viscera and with a well-developed system of lacunæ in the mantle and arms. In the latter the blood is brought into close osmotic relation with the water, over a large surface: the inner surface of the mantle and the spiral arms are, therefore, correctly regarded as **respiratory organs.**

Excretory organs.—Two, rarely four, canals, which are provided with glandular walls and open on each side of the intestine with a funnel-shaped aperture (fig. 554 *Tr*) into the body cavity, and on either side of the mouth to the exterior, are to be regarded as kidneys (corresponding to the segmental organs of Annelida.) They function at the same time as generative ducts, and were called *oviducts* by Hancock.

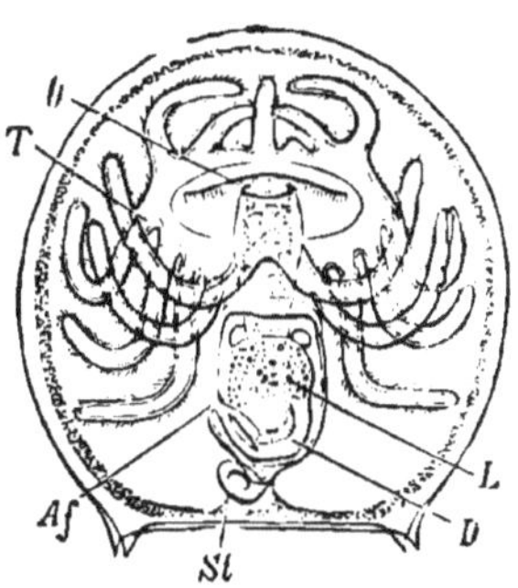

Fig. 557 *a*.—Larva of *Lingula* (after Brooks). *T*, Tentacles; *O*, mouth; *D*, alimentary canal; *Af*, anus; *L*, liver; *St*, rudiment of peduncle.

The **nervous system** consists of a circumœsophageal ring on which two small supraœsophageal ganglia are inserted. The subœsophageal ganglionic swelling of the ring is, however, much larger, and from it nerves pass out to the dorsal mantle lobe, the arms and adductor muscles, and to two small ganglia which supply the ventral mantle lobe and the peduncular muscle with nerves. Sense organs are not known with certainty.

Generative organs.—In all probability most Brachiopoda, as *Discina*, *Thecidium* and *Terebratulina* are diœcious. The sexual organs consist of thick yellow bands and ridges which have a paired arrangement and project from the body cavity into the lacunæ of the mantle, and are there considerably ramified. The eggs pass from the glands into the body cavity, and are conducted to the exterior by the oviducts (excretory organs) whose funnel-shaped internal openings have already been mentioned.

Development (fig. 556).—After a total segmentation a kind of gastrula is formed, usually by invagination, and the archenteron (*Argiope*) becomes divided as in *Sagitta* into a median cavity, and two lateral diverticula which are constricted off and give rise to the body cavity (fig. 556 *a*, *b*). The oval larva then elongates and becomes divided by constrictions into three segments (fig. 556 *b*, *c*), of which the anterior becomes umbrella-shaped, and develops cilia and eye-spots; subsequently it atrophies and gives rise to the upper lip. A fold is formed on the middle segment; this gives rise to the two mantle lobes, which soon cover the body and a part of the caudal segment (fig. 556, *d*). Four bundles of long setæ, which, as in the Worms, can be drawn in and protruded, make their appearance on the ventral lobe of the mantle of the developing larva. Later the larva becomes attached and the metamorphosis begins. The fixed posterior segment becomes the peduncle; the mantle lobes bend forward and produce the shell. The bundles of setæ are thrown off; the deposition of calcareous matter in the shell begins, and the tentacular filaments (which are at first arranged in a circle) of the later arms make their appearance. In *Thecidium* the inner layer (mesoderm and endoderm) arises from masses of cells which are budded off into the segmentation cavity. The subsequent metamorphosis of the larva when provided with tentacles has been most accurately investigated by Brooks for *Lingula*, the larvæ of which are still free-swimming when the tentacles are being developed (fig. 557, *a*, *b*).

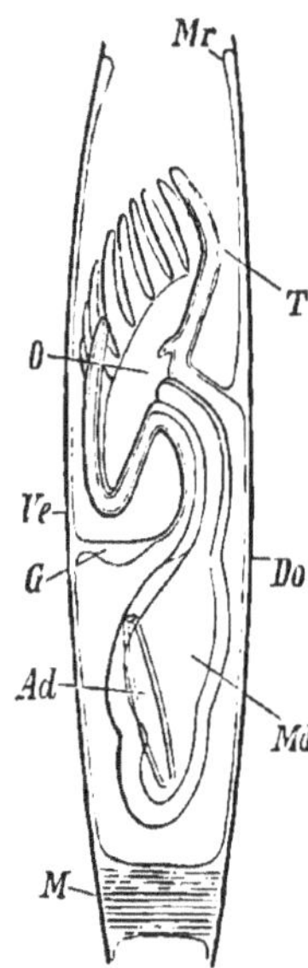

FIG. 557 *b*.—Longitudinal section of an older larva (after Brooks). *Do*, Dorsal; *Ve*, ventral valve of the shell; *Mr*, thickened mantle edge; *T*, tentacles; *O*, mouth; *Md*, stomach; *Ad*, intestine; *M*, posterior muscle; *G*, ganglion.

At the present day but few Brachiopods are found in the different seas, as compared with the much larger number in the earlier formations; certain species of these fossil Brachiopods have great importance as characteristic fossils. The oldest fossils also belong to the Brachiopoda and certain genera which first appeared in the Silurian have persisted to the present day (*Lingula*).

Order 1.—ECARDINES (INARTICULATA.).

Shell without hinge and brachial skeleton. Alimentary canal with laterally-placed anus. Edges of the mantle lobes completely separated.

Fam. **Lingulidæ.** *Lingula anatina* Lam., Indian Ocean,

Fam. **Discinidæ.** *Discina lamellosa* Brod., South America.

Fam. **Craniadæ.** *Crania anomala* Müll., North Sea; *Cr. rostrata* Hoev., Mediterranean; *Cr. antiqua* Defr., fossil from the Chalk.

Order 2.—Testicardines.

The shell is calcareous, with hinge and brachial skeleton. The intestine ends blindly.

The exclusively fossil families of the **Orthidæ** and **Productidæ** (*Productus* Sav.), the edge of the shells which have no hinge, form the transition between the two orders.

Fam. **Rhynchonellidæ**. *Rhynchonella psittacea* Lam., fossil species found in the Silurian. *Pentamerus* Sow., contains only fossil species from the Silurian and Devonian formations. The fossil *Spiriferidæ* are allied here (*Spirifer* Sow.).

Fam. **Terebratulidæ**. *Thecidium mediterraneum* Riss., *Waldheimia* King; *Terebratula vitrea* Lam., Mediterranean; *Terebratulina caput serpentis* L., North Sea; *Argiope* Dp.

CHAPTER III.

Tunicata*

Bilateral, saccular, or barrel-shaped animals; the respiratory cavity with two wide openings, between which is placed a simple nerve ganglion. Heart and branchiæ are present.

The *Tunicata* owe their name to the presence of a gelatinous or cartilaginous envelope or mantle (the *tunica externa* or *testa*), which completely surrounds the body. The body is saccular (*Ascidians*) or barrel-shaped (*Salpæ*). In all cases there is at the anterior end a wide opening (figs. 558, 559, *O*), which can be closed by means of muscles, and often also by valves. Through this opening water and nutrient matters pass into the pharyngeal cavity, which also serves as a respiratory organ. At some distance (*Ascidians*) from this first opening, or at the opposite end of the body (*Salpæ*), there is a second opening (figs. 558, 559 *A*), which can also be closed; this serves as the exhalent opening of the cloacal cavity (*Kl*), which communicates with the pharyngeal cavity.

* J. C. Savigny, "Mémoires sur les animaux sans vertèbres, II." Paris, 1815.

Chamisso, "De animalibus quibusdam e classe Vermium." Berlin, 1819.

Milne Edwards, "Observations sur les Ascidies composées de côtes de la Manche." *Mém. Acad. Sc. Paris*, 1839.

A. Kowalevski, "Weitere Studien über die Entwickelung der einfachen Ascidien." *Arch. für mikr. Anatomie*, Taf. VI., 1870.

The integument is sometimes gelatinous and sometimes of leathery or cartilaginous consistency, and is often clear as crystal or transparent, but sometimes opaque and variously coloured. The outer surface is smooth or warty, sometimes spiny or felted. This external integument, which completely envelopes the body, is called the external mantle (*tunica*), and was formerly regarded as a sort of shell and compared to the bivalve shell of the Lamellibranchs. This view seemed to be supported by the interesting discovery of Lacaze-Duthiers* that there are Ascidians in which the stiff cartilaginous tunic is split into two separate valves which can be closed by special muscles (*Chevreulius*). As a matter of fact, this is simply an external analogy, for the mantle space corresponds to an atrial cavity, and the branchial sac to the pharyngeal sac. The substance of the mantle arises as a cuticular excretion; it consists of a matrix containing *cellulose* and cells, and therefore with respect to its structure is a kind of connective tissue. In the colonial Tunicates the external mantles of all the individuals may fuse together to form a common mass.

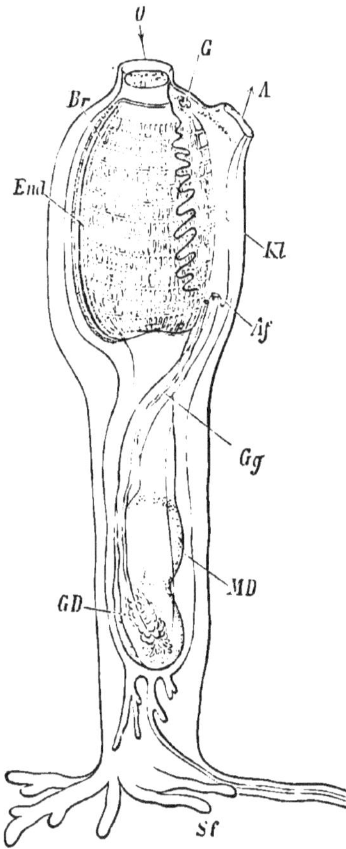

FIG. 553.—*Clavellina lepadiformis* (règne animal), somewhat diagrammatic. *O*, Mouth; *Br*, gills; *End*, endostyle; *Oe*, œsophagus; *G*, nervous centre; *MD*, stomach; *Kl*, cloacal space; *A*, exhalent pore; *Af*, anus; *GD*, genital gland; *Gg*, genital duct; *Sf*, stolons.

Beneath the saccular mantle lies the body wall of the animal, the outer cellular layer of which is applied to the mantle and represents the ectodermal epithelium which has produced the mantle and also the subjacent so-called internal mantle layer. Within the latter all the organs of the

* Lacaze-Duthiers, "Sur un nouveau d'Ascidien." *Ann. des Sc. Nat.* Vᵉ. Ser., Tom. IV., 1865.

body,—the muscles, nervous system, the digestive apparatus, the generative and circulatory organs,—lie embedded in a kind of body cavity.

The **nervous system** is confined to a simple ganglion, the position of which near the inhalent aperture marks the dorsal surface. The nerves which radiate from the ganglion branch and pass, some to the muscles and viscera, some to the sense organs—such as eyes, auditory and tactile organs—which are found principally in the free-swimming Tunicates.

The **muscular system** is chiefly developed around the respiratory cavity, and serves for the dilatation and contraction of this space as well as for closing the inhalent and exhalent pores. In the Ascidians there are three layers of muscles, an external and internal longi-

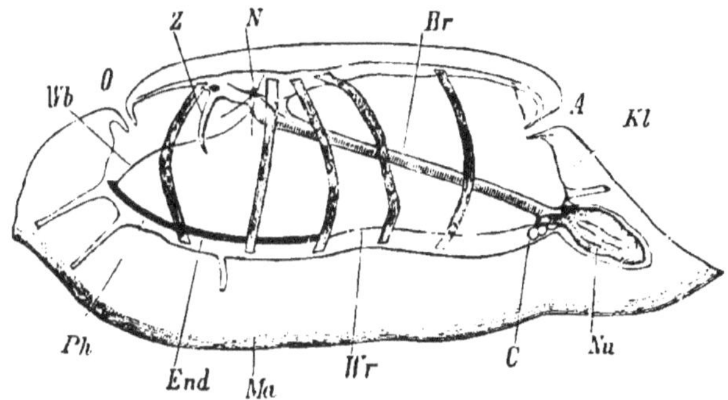

FIG. 559.—*Salpa democratica* from the side, somewhat diagrammatic. *O*, Mouth; *Ph*, pharyngeal cavity; *Kl*, cloaca; *A*, exhalent opening; *Br*, gills; *N*, nervous centre; *Ma*, mantle; *M*, muscular rings; *Z*, languet; *Wb*, ciliated arc; *End*, endostyle; *Wr*, ciliated groove *Nu*, nucleus; *C*, heart.

tudinal and an internal circular layer, while in the Salps there are band-like rings of muscles embedded in the substance of the body-wall, and effecting not only the renewal of the water used in respiration, but also the movements of the free-swimming barrel-shaped body. A special organ of locomotion is present in the small *Appendicularia* and the free-swimming Ascidian larvæ; it is placed on the ventral surface as indicated by the position of the heart, and consists of a vibratile whip-like caudal appendage supported by a notochordal rod (*urochord*).

The **alimentary canal** begins in all cases with a wide pharyngeal cavity, which functions as a respiratory organ. The anterior mantle opening, which must be looked upon as the mouth, leads into this cavity. The œsophageal opening is placed at a distance from the mouth inside this respiratory cavity, which in the Ascidians has the form of a latticed branchial sac. A ciliated groove bounded by two folds extends along the middle ventral line of the pharyngeal cavity, between the mouth and the opening of the œsophagus. The glandular walls of this ventral groove are distinguished as the **endostyle** (figs. 558 and 559, *End*). It begins with two lateral ciliated arcs, which unite to form a complete ring near the inhalent aperture (mouth), and somewhat in front of the ganglion pass over a small cone projecting into the pharyngeal cavity.

The digestive canal which follows the pharyngeal cavity consists of a ciliated œsophagus, which is usually narrowed into the form of a funnel; of a stomach, usually provided with a liver; and of a small intestine, which bends round, forming a loop, and opens into the cloacal cavity.

There is always a **heart**, which is placed on the ventral side of the intestine and is surrounded by a delicate pericardium. The contractions, which are active and regular, pass from one end of the heart to the other.

The sudden change in the direction of the contractions (discovered in the Salps by Hasselt), by which after a momentary period of rest the direction of the blood stream in the heart is reversed, is worthy of note. The vascular trunks (lacunæ) passing from the heart lead into a system of spaces in the body wall through which the blood passes. In the Ascidians there are also vascular loops in the mantle, in that diverticula of the body wall, containing blood and covered with epidermis, project into the mantle. Two principal channels for the blood are placed in the middle line—one on the dorsal side, and the other on the ventral beneath the ventral groove; they are connected by transverse channels placed in the wall of the branchial cavity. The latter communicate with the blood spaces of the variously-shaped branchia, which is formed by the walls of the pharynx, and over the surface of which the water is continually renewed by means of the vibratile cilia which cover it. In the Ascidians almost the entire wall of the pharynx takes part in the formation of the gill. In these animals the pharynx has the form of a sac with net-like walls—*i.e.*, its walls are perforated by a number of slits, which lead from the pharynx into a chamber which is developed round it. This chamber is derived from the cloacal cavity, and is known as the *peribranchial* chamber. The branchial sac or pharynx is fixed to the walls of the peribranchial cavity along the whole length of the endostyle, and by numerous short trabeculæ which pass from the bars of the branchial network to the outer wall of the peribranchial chamber. In other cases, the number of gill-slits is considerably reduced, and the gill is confined to the dorsal part of the pharyngeal wall (*Doliolum*, *Salpa*).

Generative Organs.—The Tunicates are hermaphrodite; the male and female generative products, however, often attain maturity at different times. The Salps especially, at the time of their birth, have only the female organs, and it is not until later when they are pregnant that the male organs attain maturity. In *Perophora* the testes become mature first, in the *Botryllidæ* the ova. The testes and ovaries lie, as a rule, among the viscera in the hind part of the body. The ovaries have the form of racemose glands, the testes of blind tubes united in tufts. The generative ducts of both sexes open into the cloacal chamber, in which (rarely in the place where the germs originate) the fertilization of the ovum and the development of the embryo takes place. The embryo either leaves the cloacal chamber through the exhalent aperture while still enveloped by the egg-membranes, or is nourished by a sort of placenta and born at a more advanced stage of development (*Salpa*).

In addition to the sexual reproduction, the asexual reproduction by means of budding is very general, and frequently leads to the formation of colonies with very characteristic grouping of the individuals. The budding sometimes takes place on different parts of the body, sometimes is confined to definite places or to a germ-stock (*stolo prolifer*). The colonies thus produced do not by any means always remain fixed; but, as *e.g.*, *Pyrosoma*, may possess the power of moving from one place to another, or, as in the *Salp-chains*, they can swim tolerably rapidly.

The **embryonic development** of the Ascidians presents a great resemblance to that of the lower Vertebrates, and more especially to that of *Amphioxus*. After the completion of the total segmentation a two-layered gastrula is formed, from the ectoderm of which the neural tube is developed. At the same time an axial skeletal structure, like the chorda dorsalis, arises from a double row of endoderm cells. The relative positions of the alimentary canal, the nervous system and the notochord are analogous to those of the Vertebrates.

The **post-embryonic development** of the Ascidians is complicated. The embryos leave the egg-membranes as movable larvæ (Ascidian tadpoles) provided with a swimming organ (tail) and an eye-spot. They swim about freely for some time, and in many cases produce a small colony by budding before becoming fixed. In the *Salps* and

Doliolum there is an alternation of generations, which was discovered in the case of *Doliolum* by Chamisso long before Steenstrup. The solitary Salp, developed from the fertilized ovum of the viviparous sexual form remains asexual all its life, but from its *stolo prolifer* chains of Salps are produced, the individuals of which differ considerably in form from the asexual animals and are sexual. In *Doliolum* the alternation of generations is much more complicated, inasmuch as several generations succeed one another in the cycle of development.

All the *Tunicata* are marine animals and feed on *Algæ*, *Diatoms*, and small *Crustacea*. Many, and especially the transparent *Pyrosomidæ* and *Salpidæ*, are phosphorescent, emitting a beautiful and intense light.

CLASS I.—TETHYODEA* (Ascidians).

For the most part fixed Tunicata with saccular bodies. The inhalent and exhalent pores are placed close together, and the branchial sac is large. Development by means of tailed larvæ.

The body of these animals, as the name Ascidia implies, has the form of a more or less elongated tube or sac with two openings, which are usually close to one another; of these openings the anterior is the mouth and the posterior the cloacal opening. More rarely, as in the *Botryllidæ* and the free-swimming *Pyrosomidæ*, the two openings are placed at a considerable distance from one another at the opposite ends of the body. The mouth can be closed by a sphincter muscle, and in many cases by four, six, or eight marginal lobes (fig. 560). The edge of the exhalent opening, which can also be closed, and which is placed behind the mouth on the neural (dorsal) side, is often similarly divided into four to six lobes. The spacious pharynx which, as a rule, has the form of a latticed branchial sac, contains at some

* Besides the already quoted works of M. Edwards and Savigny, Cf. J. C. Savigny, "Tableau systematique des Ascidies, etc." Paris, 1810.

Eschricht, "Anatomisk Beskrivelse af Chelyosoma Mac-Leyanum." Kjövenhavn, 1842.

Van Beneden, "Recherches sur l'Embryogénie, l'Anatomie et la Physiologie des Ascidies simples." *Mém. de l'Acad. roy. de Belgique*, Tom. XX., 1846.

A. Krohn, "Ueber die Entwickelung von Phallusia mammillata." *Müller's Archiv*, 1852.

A. Krohn, "Ueber die Fortpflanzungsverhältnisse bei den Botrylliden und über die früheste Bildung der Botryllusstöcke." *Archiv für Naturgeschichte*, Tom. XXXV., 1869.

Th. Huxley, "Anatomy and development of Pyrosoma." *Trans. Lin. Soc.*, Vol. XXIII., 1859.

distance from the mouth a circle of usually simple tentacles. On the neural side of the branchial sac is the cloacal cavity which receives not only the water flowing out through the branchial slits, but also the fæces and the generative products. The digestive canal, together with the other viscera, is sometimes placed as in all the simple Ascidians rather to the side of the branchial sac or, as in the elongated forms of the compound Ascidians, simply behind the same, and in the latter case often occasions a constriction of the body, so that Milne Edwards was able to distinguish a thorax and abdomen, or even a thorax, abdomen and post-abdomen.

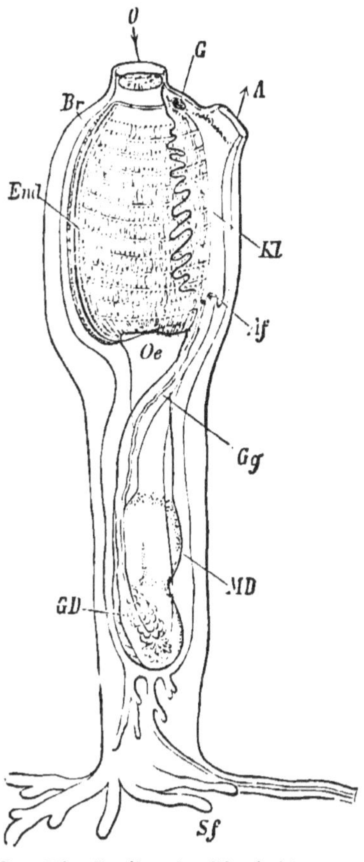

Fig. 560.—*Clavellina lepadiformis* (règne animal), somewhat diagrammatic. *O*, Mouth; *Br*, branchiæ; *End*, endostyle; *Oe*, œsophagus; *G*, nervous centre; *MD*, stomach; *Kl*, cloacal chamber; *A*, exhalent pore; *Af*, anus; *GD*, genital gland; *Gg*, duct of genital gland; *Sf*, stolons.

The Ascidians either remain solitary, and then usually attain a considerable size (*A. solitariæ*), or by budding and throwing out root-processes they produce branched colonies, the individuals of which are connected together by their body walls, and are not embedded in a common mantle covering (*A. sociales*). In other cases (*Synascidiæ*) numerous individuals live in a common mantle; they often have a characteristic arrangement around a common central opening (*A. compositæ*), so that each group has its central cavity, into which the exhalent (*i.e.* atrial) openings lead as into a common cloacal cavity (fig. 561). There are solitary (*Appendicularia*) as well as compound Ascidians (*Pyrosoma*) which can move freely. The solitary *Appendicülariæ* execute the most perfect swimming movements. In their external form they resemble the free-swimming Ascidian larvæ, and like these they have a whip-like swimming tail, which by its undulating movements propels the body forward.

In order to understand the structure of the Ascidians, it will be well to start from these simply organised forms. The most striking character of the *Appendicularia*, next to the possession of the ventrally-placed swimming tail with its notochord-like skeletal axis (*urochord*), consists in the absence of a cloacal chamber for the reception of the excreta. The anus is placed in the middle line of the ventral surface; further, there are two funnel-shaped atrial canals which begin on either side with a strongly-ciliated opening into the pharyngeal sac, and open to the exterior right and left, usually rather in front of the anus. These branchial passages arise as invaginations of the ectoderm, which come into connection with corresponding evaginations of the pharyngeal sac. The introduction of nourishment is regulated by two ciliated arcs, which begin at the front end of a short endostyle, surround the entrance of the pharyngeal sac, and run obliquely towards the dorsal surface, where they unite to form a median row of cilia (composed of two rows of ciliated cells). The latter passes back to the opening of the œsophagus, opposite a narrow ventral ciliated band, which begins at the hind end of the endostyle (fig. 562).

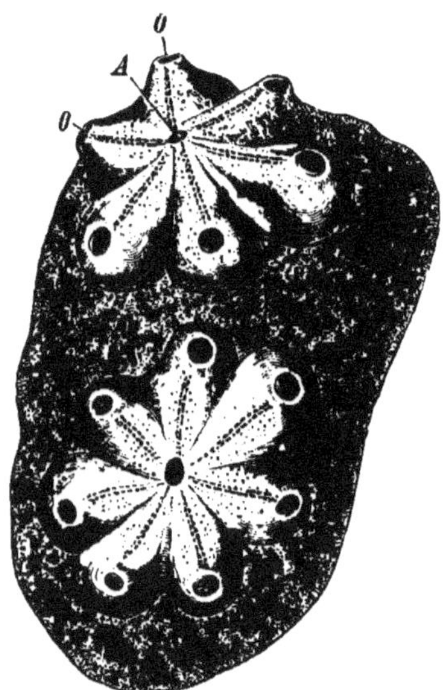

Fig. 561.—*Botryllus violaceus* (after M. Edwards). *O*, Mouth; *A*, common cloacal opening of a group of individuals.

The Ascidian larvæ (*Phallusia*) also have, as Krohn long ago discovered, two branchial slits with corresponding atrial passages. The latter, according to Kowalevski, arise as invaginations of the ectoderm, and later on unite on the dorsal side, and then open by a common cloacal orifice. The ectodermal lining of the atrial cavity, which grows round the sides of the pharyngeal sac, consists, therefore, of a branchial layer which is applied to the pharynx, and a parietal layer which forms the internal lining of the outer wall of the peribranchial or atrial cavity. The atrial cavity extends round the pharynx as far as the sides of the endostyle. The wall of the pharynx becomes perforated by an ever-increasing number of slits, and thus gives rise to the branchial basket-work.

The special form of the branchial basket-work presents numerous modifications of systematic value. Not only is the external surface of the branchial sac attached to the body-wall by trabeculæ and bands,

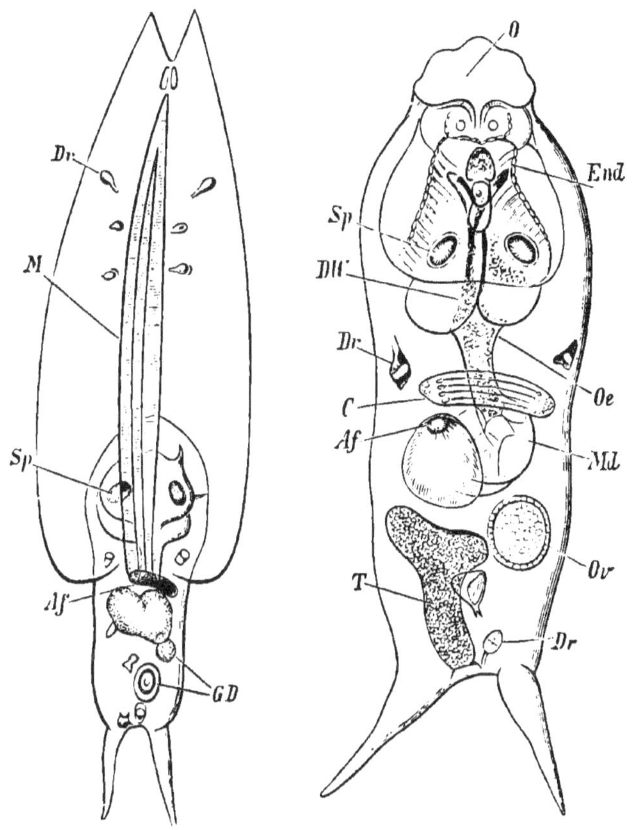

FIG. 562.—*Appendicularia (Fritillaria) furcata.* *a*, From the ventral side with the tail bent forwards. *GD*, Genital glands; *M*, muscles of the tail. *b*, From the ventral side after the caudal appendage has been removed. *O*, Mouth; *End*, endostyle; *Sp*, the two ciliated passages of the pharyngeal cavity; *DW*, the dorsal row of cilia; *Oe*, œsophagus; *Md*, stomach; *Af*, anus; *Dr*, glands; *C*, heart; *Ov*, ovary; *T*, testis.

through which the blood passes, but the internal surface also often presents folds and projections of varying form. Similarly the branchial openings with which the wall of the pharynx is pierced differ in size, number, and form; they may be rounded, elliptical, or even spirally coiled.

The ciliated arrangements in the perforated branchial sac of the

Ascidians correspond to those of the *Appendiculariæ*, and consist of the so-called endostyle with the ventral groove and the two ciliated arches.

The ciliated œsophagus is short and funnel-shaped, and leads into a dilated portion distinguished as stomach, whose walls have a layer of large entodermal cells and are complicated by the presence of fold-like projections. Glands, which are sometimes follicular, sometimes composed of bundles of tubes, or of tubes united in a network, lie upon and open into the stomach; they are generally known as * *liver*, but would be better called *hepatopancreas*. The small intestine which follows the stomach is of considerable length, is usually bent on itself (hæmal curvature), and is continued into a short rectum (piriform in *Appendicularia*), which opens into the cloacal chamber. Besides the glands already mentioned, a gland-like organ has been found in many Ascidians: as there is no opening to this gland, the concretions found in its lumen are probably not in general removed. It may, perhaps, be regarded as a kidney, since Kupffer † has shown that uric acid is present in the concretions.

The **heart** is placed on the ventral side of the intestinal canal. It is a contractile tube, each end of which is prolonged into a vessel. In the *Appendiculariæ* (*Copelata*) the heart is placed transversely, and is pierced by only two slits. The so-called vascular system of the Ascidians consists of a rich net-like system of lacunæ, which cannot, however, be said to have special walls.

The **nervous system** is reduced to an elongated ganglion (cerebral ganglion) placed on the dorsal side of the branchial cavity. From this ganglion nerves are given off, especially forwards towards the entrance of the pharyngeal sac; but unpaired sense nerves and lateral and posterior nerves also arise from it. In the *Copelata* and Ascidian larvæ the cerebral ganglion is more complicated. In these animals it has the form of a cord, primitively containing a cavity, and divided later by constrictions into three regions, and is connected with ganglia in the tail (fig. 563). The anterior conical part of the brain gives off paired sensory nerves to the region of entrance into the branchial sac; on the median globular part are placed the auditory vesicle and a stalked ciliated organ; while the attenuated posterior part gives off two lateral nerves to the atrial canals, and is prolonged

* Th. Chandelon. "Recherches sur une annexe du tube digestive des Tuniciers." *Bull. de l'acad. roy. de Belgique*, Tom. XXXIX., 1875.

† Cf. Besides Kowalevski l.c. Kupffer. "Zur Entwickelung der einfachen Ascidien." *Arch. für mikr. Anat.*, Tom. VIII. 1872.

Lacaze-Duthiers, *Arch. de Zool. expérim.*, 1874.

into a long nerve, which at the base of the tail dilates to a ganglion, and in its further course forms a number of smaller ganglia (fig. 563). The reduction of the central nervous system to the simple ganglion of the Ascidian begins after loss of the tail, and after development of the branchial basket.

Of **sense organs** the processes of the integument which serve for tactile purposes (the lobes of the oral and atrial apertures and the tentacles), and peripheral nerves, ending in epithelial cells, are most widely distributed. The large ciliated cells on the edge of the mouth of the *Copelata* must be placed in the same category. The ciliated pit is to be regarded as an **olfactory** organ. It consists of a depression in the wall of the pharynx lined with ciliated cells, and is situated in front of the ganglion. According to Julin, it is, together with a gland situated beneath the ganglion, to be regarded as the equivalent of the hypophysis. In the *Copelata* the ciliated pit is elongated and lies on the right side of the ganglion.

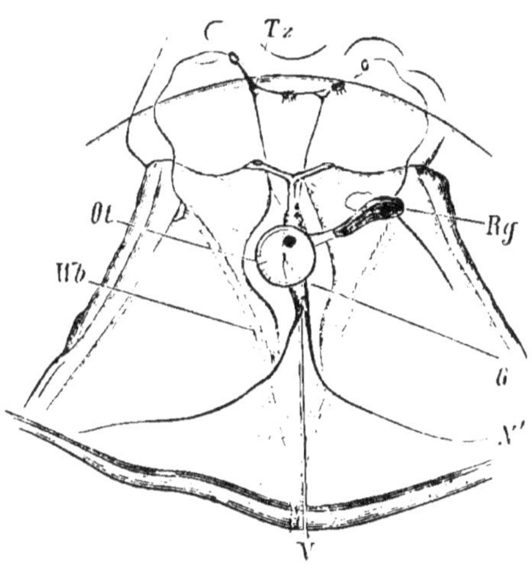

FIG. 563.—Nervous system of *Appendicularia* (*Fritillaria furcata*) (after Fol.). *G*, Ganglion; *N*, body nerve; *N'*, lateral nerve; *Ot*, otolith vesicle; *Rg*, olfactory pit; *Tz*, tactile cells with their nerve; *Wb*, arch of cilia.

There is an **auditory vesicle** on the left side of the ganglion in the *Copelata*. This structure, which is developed from a cell of the wall of the ganglion, is found in the Ascidian larvæ, but degenerates soon after the attachment of the larva. Paired auditory vesicles appear in *Pyrosoma* where they are connected with the ganglion by a short stalk.

Masses of pigment which are present with great regularity on the lips of the large openings of the body in the simple and compound Ascidians may be interpreted as **eye spots**. The eye of the Ascidian larvæ, which lies on the ganglion and originates from a part of the neural canal, has a more complicated structure. Later it degenerates, but in *Pyrosoma* it is retained in the adult condition and possesses a lens-like structure.

The **generative organs** are always united in the same animal. The formation of villi on the surface of the egg-membrane by the follicular cells surrounding the ovum is remarkable. The origin of

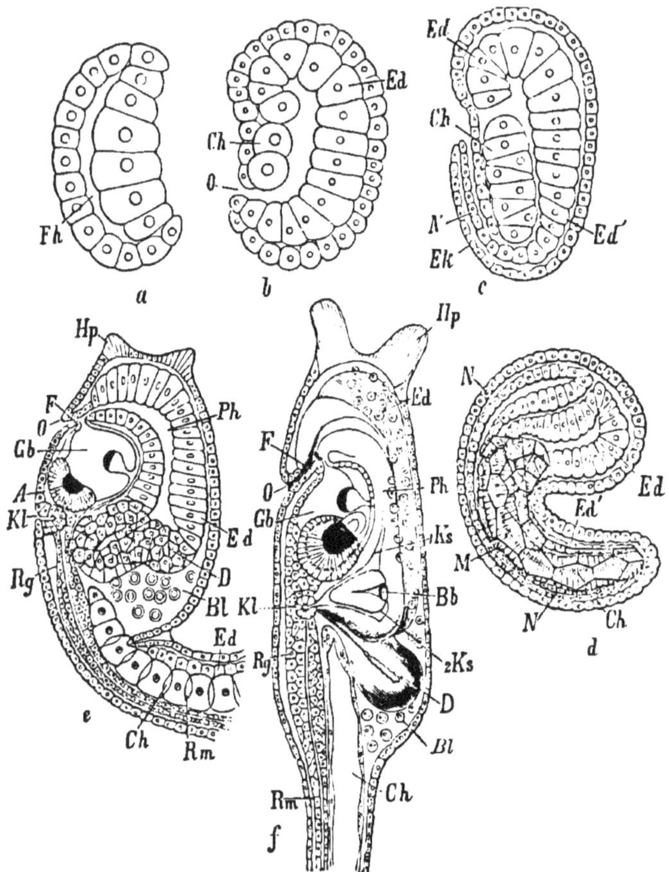

FIG. 564.—Development of *Phallusia mammillata* (after Kowalevski). *a*, Blastosphere beginning to invaginate; *Fh*, segmentation cavity. *b*. Gastrula with blastopore (*O*); *Ed*, endoderm; *Ch*, commencing notochord (urochord). *c*, Later stage. *Ek*, Ectoderm; *N*, rudiment of the still open neural canal. *d*, Stage with body and tail; *Ed'*, endodermal layer in the tail; *M*, muscular cells in tail. *e*, Just hatched larva; *Rg*, anterior swelling of the spinal division of the neural tube; *Rm*, posterior part of neural tube; *Gb*, dilated anterior part of neural tube (cerebral vesicle), with otolith projecting into it; *F*, opening of *Gb*; *A*, eye; *O*, invagination of mouth; *Ph*, pharyngeal cavity; *Ed*, endostyle; *D*, commencing intestine; *Kl*, atrial opening; *Bl*, blood corpuscles; *Hp*. papilla for attachment. *f*, Two days' larva (only the beginning of the tail is represented); $_1Ks$, $_2Ks$, branchial stigmata; *Bb*, branchial vessel between them; *B*, intestine.

the so-called test-cells (follicle-cells which have migrated inwards) over the substance of the yolk on the inside of the egg-membrane, is also worthy of note.

Development. —The segmentation is complete, and leads, according to Kowalevski, to the formation of a blastosphere as in Amphioxus (fig. 564). The wall of the blastosphere then begins to invaginate. After the completion of the invagination the blastosphere becomes a gastrula, with the remains of the segmentation cavity between the ectoderm and entoderm (fig. 564, *Fh*). The mouth of the gastrula is at first wide, but soon becomes narrower and narrower, until finally it becomes transformed into a small opening placed on the dorsal surface at the hind end of the body. A flat median groove on the ectoderm appears along the dorsal side of the already bilaterally symmetrical embryo extending from the blastopore forwards. This groove, into the hind end of which the blastopore opens, is the first rudiment of the central nervous system. It is known as the medullary groove. Its edges project and form the medullary folds which grow round and close the narrow blastopore, and gradually fuse with one another from this point forwards in such a manner as to convert the groove into a canal, the walls of which separate from the external ectoderm and give rise to the central nervous system. This canal is known as the medullary canal: behind it is shut off from the exterior, but communicates with the cavity of the gastrula (archenteron) by way of the blastopore (fig. 564 *c*), which is now known as the neurenteric canal; while in front it remains open for some time. Before these processes are completed two rows of the endoderm cells of that part of the gastric wall which immediately underlies the neural tube become different from the remaining endoderm cells and give rise to the first rudiment of the notochord. The anterior part of the archenteron only gives rise to the pharynx and intestine (fig. 564, *e*), while the posterior part furnishes the cell material not only for the notochord, but also for the muscular system and the blood corpuscles. It may accordingly be asserted that the mesodermal organs in the Ascidians arise from the entoderm, which is as good as saying that the hinder half of the gastral sac has the value of mesoderm.

In the further course of development the somewhat elongated spheroidal body grows out at the posterior and inferior end, opposite to the blastopore and rather to the right,* into a tail-like prolongation, the axis of which is formed by the cells of the notochord (at this period arranged in a simple row). The neural canal is prolonged into

* In *A. mammillata*, according to Kowalevski, on the contrary, this growth takes place at the other end towards the left, and therefore agrees with that of *Amphioxus*.

the tail dorsal to the notochord. The tail, thus developed, becomes bent and applies itself to the side of the body opposite to that on which the nervous system is placed (Fig. 564*e*). Subsequently the skin begins to thicken at the anterior end and gives rise to three papillæ, the future papillæ for attachment. The rudiment of the nervous system, on which two pigment spots provided with refractive organs make their appearance (eye and auditory organ, fig. 564*e*, *f*), is converted at its anterior extremity into a vesicle and is continued above the chorda into the tail (as a cord with a central canal) (*A. canina*).

The branchial sac, still closed and formed of columnar epithelium, lies close to the nervous system: it is separated from the ventral wall of the body by roundish uncoloured cells, which are probably the formative elements of the blood and of the wall of the heart. It has at this period the position and relative size of the future pharynx and its posterior dorsal extremity grows out to form the, at first cæcal, rudiment of the digestive canal (fig. 564 *e*, *D*). The mouth is formed from an invagination of ectoderm on the dorsal surface immediately in front of the anterior end of the cerebral vesicle (fig. 564 *e*, *O*). The cloaca first appears as a pair of dorsally-placed epiblastic involutions (fig. 564 *e*, *Kl*): these ingrowths meet and fuse with the wall of the branchial sac so that two perforations are formed. The embryo surrounded by the mantle (formed of gelatinous substance with amœboid test-cells which have wandered into it) now breaks through the villous egg-membrane and passes into the stage of the free-swimming larva, which presents on the right side of the endostyle the first rudiments of the heart, and possesses all the organs of the later Ascidian except the vessels and the generative glands: in its subsequent development, however, it has to go through a decidedly retrogressive metamorphosis. After the larva has attached itself by means of its papillæ, the tail aborts, the muscles and notochordal sheath degenerate, and the axial string of the notochord contracts. The nervous system with the pigment organs degenerates, and the cavity in it disappears; the branchial sac, on the contrary, increases in size, and the œsophagus, stomach, and intestine proper become more sharply distinct. The mantle then becomes firm, the mouth opening perforates the gelatinous covering and becomes the entrance to the branchial sac: behind the mouth the arch of cilia appears at the anterior end of the ventral furrow, which was formed at an earlier stage and gives rise to the endostyle. The opening to the œsophagus becomes funnel-shaped and more distinct. The first branchial slits soon become visible. The blood with its amœboid corpuscles is already moving in the body cavity beneath the skin, and indeed on the branchial sac, through definite channels in the connective tissue, which connects the walls of the branchial sac with the skin. The water which flows through the slits of the branchial sac is collected in the peribranchial space the opening of which coincides with the cloacal opening.

Asexual Reproduction.—In addition to the sexual reproduction, multiplication by means of budding plays an important part, particularly in the *Synascidians*. According to Krohn, Metschnikoff, and Kowalevski, an entodermal layer (arising in *Botryllus* from the covering of the atrium) and mesodermal cells as well as the ectoderm take part in the formation of the buds. Many Ascidians, as *Perophora* and *Clavellina*, produce stolons by budding, and from these new individuals are developed, but the latter are not united together into a compact system. Complex systems of buds are developed in the Synascidians, the individuals of which are embedded in a common cellulose mantle. In some cases the larva may form buds while it is still in the tailed stage (*Didemnum*). In *Botryllus*, a genus which is distinguished by the star-like grouping of the individuals round a common cloaca (fig. 561), and by the rich branching of the blood canals, the larva is simple, and does not, as Sars believed, form a colony. Metschnikoff and Krohn, whose accounts agree, have both shown that the eight knob-like buds of the larva are only processes of the ectoderm and contain diverticula of blood spaces. The young *Botryllus* produces only one bud (first generation), and before the latter is mature perishes without attaining sexual maturity. The bud of the first generation produces two buds (second generation), and dies without reaching sexual maturity. The buds of the second generation each produce two buds, which arrange themselves in a circle, and after the death of their producers form the first system with a common cloaca. In a similar manner new buds are formed, and the older generation dies; the new systems are, however, as transitory and are replaced by others, so that as the stock increases the old generations are continually being replaced by new. In this continuous process of renewal the first-formed generations have only the provisional value of establishing the colony. The later generations alone become sexually mature, and the female maturity is attained before the male. The ova of the still young hermaphrodite generations are fertilized by the sperm of the older; and it is not until after the death of the latter that the testes of the former become fully developed. The young generations, therefore, now have the double task of caring for their own already fertilized eggs and of fertilizing those of the succeeding generations.

Order 1.—COPELATÆ* (Ascidians with larval tail).

Small free-swimming Ascidians of long oval form, with swimming tail; they resemble in the whole of their organization the larvæ of other Ascidians (fig. 562). The anus opens directly to the exterior on the ventral side. The pharyngeal sac is pierced by only two branchial slits. The heart has two slits and no vessels. The ovaries and testes lie in the hind part of the body, close to one another, and are without ducts. The elongated cerebral ganglion is divided by constrictions into three parts; it is connected with a ciliated pit and an otolithic vesicle, and is prolonged into a nerve-cord of considerable size. The latter is continued into the tail, at the base of which it swells out to a ganglion: in its further course it forms several small ganglia, whence lateral nerves pass out. In consequence of a torsion of the axis of the tail, the originally dorsally-placed caudal nerve comes to have a lateral position. The segmentation of the nerve-cord in the tail (as shown by the ganglionic swellings) corresponds to the segmental divisions of the muscles, which recall the myotomes of *Amphioxus*. The large chorda (urochord), which extends along the whole length of the tail, constitutes another point of resemblance to Amphioxus.

Some species have a pellucid gelatinous covering, comparable to a shell. The development of these small animals, which were formerly erroneously held to be larvæ, has been insufficiently investigated.

Fam. **Appendicularidæ.** *Oikopleura* Mertens (*Appendicularia* Cham.). *Oi. cophocerca* Gegbr., *Oi. furcata* Gegbr., *Fritillaria* Fol. The integument forms a hood-like reduplicature in front: the tail one and a-half times as long as the elongated body; the endostyle is curved. *Fr. furcata* C. Vogt., *Fr. formica* Fol. *Kowalevskia* Fol. Without heart and endostyle. Rectum absent. *K. tenuis* Fol, Messina.

Order 2.—ASCIDIÆ SIMPLICES.†

This order comprises solitary forms as well as branched stocks.

* Cf. C. Gegenbaur, "Bemerkungen über die Organisation der Appendicularien." *Zeitsch. für wiss. Zool.*. Tom. VI.. 1855.

H. Fol, "Études sur les Appendiculaires du détroit de Messine." *Mém. Soc. de Phys. et d'Hist. Nat. de Génève.* Tom. XXI.. 1872.

† Cf. besides Lacaze-Duthiers l.c., C. Heller. "Untersuchungen über die Tunicaten des Adriatischen Meeres. I., II., III.," *Denkschr. der k. Akad. der Wissensch. Wien.* 1874-1877.

The latter, or social Ascidians, are placed on branched root-processes, and have for a time, or permanently, a common circulation. The mantle-parenchyma is usually of transparent hyaline consistency. The body of the solitary Ascidians is far larger and is surrounded by a hard cartilaginous, very thick and usually completely opaque mantle, the surface of which often has wart-like protuberances and incrustations of various kinds (fig. 560).

Fam. **Clavellinidæ.** Social Ascidians, the stalked individuals of which arise from a common branched stolon, or on a common stem. The body is sometimes (*Clavellina*) divided into th ee regions, like that of the *Polyclinidæ*. *Clavellina* Sav.: *Cl. lepadiformis* Sav., North Sea: *Perophora Listeri* Wiegm., North Sea.

Fam. **Ascidiadæ.** Solitary Ascidians, usually of considerable size. The individuals reproduce themselves, as it seems, only occasionally by budding, and are connected, when they are aggregated together, neither by a common mantle covering nor by bloodvessels. *Ascidia* L. (*Phallusia* Sav.): *A. mammillata* Cuv., Mediterranean; *A.* (*Ciona*) *intestinalis* L., etc.; *Cynthia* Sav., *C. papillosa* Sav., *C. microcosmus* Cuv., *Chevreulius* Lac-Duth., Mediterranean.

The deep-sea Ascidians are very remarkable aberrant forms. *Hypobythius calycodes* Mos., and *Octacnemus bythius* Mos.

Order 3.—Ascidiæ compositæ.*

Numerous individuals lie in a common mantle layer, and form soft, brightly-coloured colonies, which have a spongy or lobed form, and not unfrequently form crusts round foreign objects.

In almost all cases the individuals are grouped in a definite number round a common cloaca (*Botryllidæ*), so that round or star-shaped systems with central openings are formed in the colony (fig. 561). The body is sometimes simple and short, sometimes long and divided into two or three regions, and sends out branched processes containing blood into the common mantle mass, so that the latter is permeated by vascular canals.

Fam. **Botryllidæ.** The viscera of the simple body, which is not divided into thorax and abdomen, lie by the side of the respiratory cavity; without lobes round the inhalent opening. *Botryllus stellatus*, Pall.: *B. violaceus* Edw.

Fam. **Didemnidæ.** The viscera are for the most part placed behind the respiratory cavity, and the body is divided into two parts, the thorax and abdomen. *Didemnum* Sav.: *D. candidum* Sav.: *D. styliferum* Kow.

Fam. **Polyclinidæ.** The body of the individual is much elongated, and is

* Besides Savigny, cf. M. Edwards, "Observations sur les Ascidies composées des côtes de la Manche." *Mém. Acad. sc.*, Tom. XVIII., Paris, 1842.

A. Giard, "Recherches sur les Synascidies." *Arch. de Zool. expér.*, Tom. I., Paris, 1872.

Kowalevski, "Ueber die Knospung der Ascidien." *Arch. für mikr. Anat.*, Taf. X., 1874.

divided into thorax, abdomen, and postabdomen. The heart lies at the hind end of the body. *Amaroecium* Edw.; *A. proliferum* Edw.

Order 4.—Ascidiæ salpæformes.*

Free-swimming colonies, which float on the surface of the sea, and have in general the form of a fir cone, hollowed out like a thimble. They are composed of numerous individuals arranged in the common gelatino-cartilaginous mass in a direction at right angles to the long axis of the colony. The inhalent openings lie in irregular circles on the external surface: the exhalent openings open opposite to them into the space which serves as a common cloaca. The branchial sac is wide and latticed as in the Ascidians. The intestine and ovary are compressed together, and lie in a rounded prominence like a nucleus

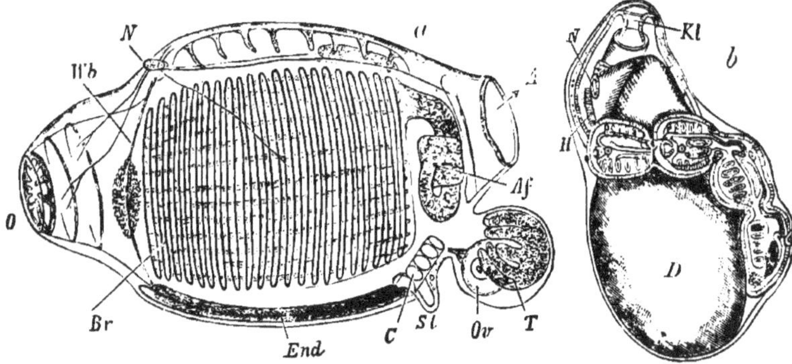

Fig. 565.—An individual of *Pyrosoma*, (after Keferstein). *O*, Mouth; *A*, atrial aperture; *Af*, anus; *Ov*, ovary; *T*, testis; *N*, ganglion; *End*, endostyle; *Br*, branchial sac; *Wb*, arch of cilia; *C*, heart; *St*, stolo prolifer. *b*, Cyathozoid of *Pyrosoma* (after Kowalevski). *H*, Heart; *Kl*, cloaca; *D*, yolk around which are the four individuals (ascidiozoids).

(fig. 565 *a*); near them is the heart. The ovary brings only one ovum to maturity, which is surrounded by a saccular follicle with a long stalk. The stalk constitutes the oviduct, and opens into the cloacal cavity. The eye lies on the ganglion. By the presence of the eye, as well as by the position of the two respiratory openings and of the viscera, by the method of reproduction and the free locomotion, the *Pyrosomidæ* are allied to the Salps.

Budding takes place by means of a stolon, which begins at the hind end of the endostyle, and contains an endodermal process of

* Th. Huxley, "Anatomy and development of Pyrosoma." *Trans. Linn. Soc.*, 1860.
W. Keferstein und Ehlers, "Zoologische Beiträge." Leipzig, 1861.
Kowalevski, "Ueber die Entwickelungsgeschichte der Pyrosomen." *Arch. für mikr. Anatomie.* Tom. XI., 1875.

the latter (endostylic cone). Sexual reproduction and gemmation take place in the same individual.

The egg develops within an ovarian sac into an embryo, which has the form of a stunted Ascidian-like individual (*cyathozooid*), and produces, by budding from a stolon, a group of four individuals (*ascidiozooids*). The peculiar mode of origin of these individuals has been minutely described by Huxley and Kowalevski (fig. 565 *b*). The process of budding, by which the colony is increased, is no less complicated: it takes place on a germ-stock (*stolo prolifer*, fig. 565 *a*, *St*) placed behind the endostyle. Each commencing bud receives a prolongation of the foundation of the ovary,* as well as of the endoderm.

The *Pyrosomidæ* derive their name from the bright light which they emit. According to Panceri, this light proceeds from a paired group of cells in the region of the mouth.

Fam. **Pyrosomidæ.** These animals were discovered by Péron in the Atlantic Ocean, and were at first regarded as solitary individuals. *Pyrosoma* Pér.: *P. atlanticum* Pér.; *P. elegans* and *giganteum* Les., from the Mediterranean.

Class II.—THALIACEA.†

Free-swimming transparent Tunicata with cylindrical or cask-shaped body. The mantle apertures are terminal, and at opposite ends of the body. The branchiæ are band-shaped or lamellar, and the viscera are compressed together into a nucleus.

The Thaliacea (fig. 566 *a*, *b*) are transparent, cylindrical or cask-shaped animals, of gelatino-cartilaginous consistency; they are either solitary, or the individuals are united in chains (usually in double rows). They move on the surface of the sea by the rhythmically alternating contraction and dilatation of the branchial cavity. The two openings are placed at opposite ends of the body; the mouth at the anterior end, the atrial at the posterior end, near the dorsal

* [Generative blastema, or indifferent tissue from which the reproductive organs of the parent were developed (Huxley).]

† Compare Th. Huxley, "Observations upon the anatomy and physiology of Salpa and Pyrosoma, together with remarks upon Doliolum and Appendicularia." *Phil. Trans.*, London, 1851.

R. Leuckart, "Zoologische Untersuchungen." Heft II., Giessen, 1854.

C. Gegenbaur, "Ueber den Entwickelungscyklus von Doliolum nebst Bemerkungen über die Larven dieser Thiere." *Zeitschr. für wiss Zool.*, Tom. VII.

C. Grobben, "Doliolum und sein Generationswechsel, etc." *Arbeiten aus dem Zool. Institute in Wien*, Tom. IV., 1882.

surface. The mouth has usually the form of a broad transverse slit, bounded by movable lips, and leads into the large respiratory cavity, which consists of the pharyngeal cavity and the cloaca, and contains the lamellar or band-shaped gill, extended from the dorsal surface obliquely backwards and ventralwards. In *Doliolum* the gill has the form of an oblique partition, which is pierced by two lateral rows of large transverse slits, through which the water flows

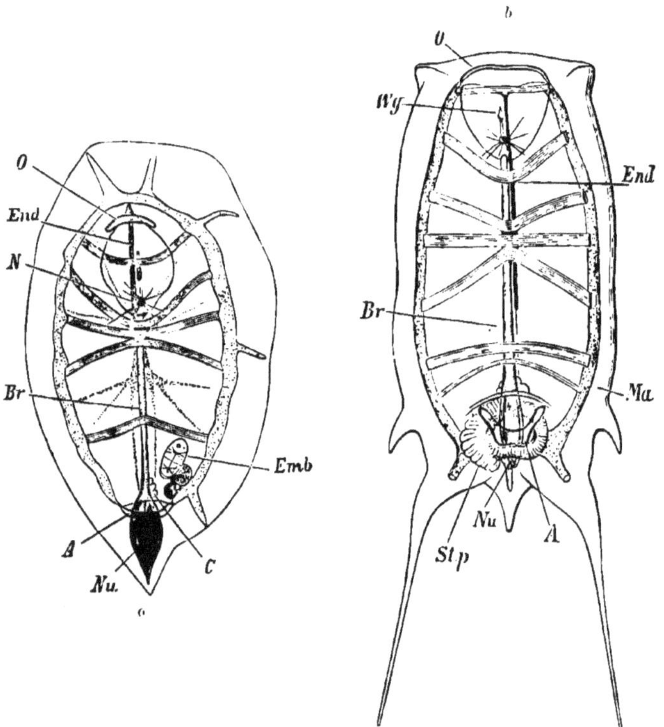

FIG 556.—*a*. *Salpa mucronata*. *b*, *S. democratica*. *O*, Mouth ; *A*, cloacal aperture ; *N*, ganglion ; *Br*, gill *End*, endostyle ; *Wg*, ciliated pit ; *Ma*, mantle ; *Nu*, nucleus ; *C*, heart ; *Emb*, embryo ; *Stp*, stolo prolifer.

from the pharyngeal cavity into the cloacal chamber. In *Salpa* the transverse slits are represented by one very large gill-slit on each side, so that the branchial wall is reduced to a median band (the median part of the gill of *Doliolum*). The two arches of cilia which bound the entrance to the respiratory cavity, and the ventral endostyle (mucous gland) from which a ciliated groove leads to the œsophagus, are placed in the wall of the pharyngeal cavity.

The **digestive canal**, together with the other viscera, the heart and the generative organs, are closely packed together in a brightly-coloured mass, the **nucleus**, at the ventral side of the hind end of the body. The mantle is often thickened round the nucleus so as to form a globular swelling.

The **nervous system**, the **sense organs**, and the **organs of locomotion**, in correspondence with the power of free locomotion, present a higher grade of development than in the Ascidians. The ganglion, with its numerous nerves, lies above the point of attachment of the branchial band, and attains a considerable size. On the ganglion there is usually (*Salpa*) a piriform or spherical process, with a horse-shoe-shaped brownish-red pigment spot and numerous rod-shaped structures, which prove beyond all doubt that this structure is an eye. In other cases (*Doliolum*) there is on the left side of the body an auditory vesicle connected with the ganglion by a long nerve. The median ciliated pit, too, is placed in the respiratory cavity in front of the ganglion. Peculiar sense organs, probably tactile in function, have been observed in *Doliolum* in the lobes of the two mantle apertures and also on other parts of the external skin. These have the form of groups of roundish cells into which nerves enter.

Locomotion is effected by means of broad bands of muscles, which span the respiratory cavity like hoops, and by their contraction narrow it. Part of the water is thus driven out of the cloacal aperture, and the body is propelled in the opposite direction.

The **reproduction** of the Salps is alternately sexual and asexual. The solitary Salps are produced sexually, the chains of Salps asexually. The individuals of the chains of Salps are sexual animals, which form no stolon; the solitary Salps only reproduce themselves asexually by budding on a ventrally-placed stolon. Since these two forms, which differ both in size and shape, as well as in the course of their muscular bands, and in certain features of the gills and viscera, alternate regularly in the developmental cycle of the species, the development represents an alternation of generations, which may even be still further complicated (*Doliolum*). This alternation of solitary Salps and chains of Salps was discovered long before Steenstrup by the poet Chamisso.

The Salps which form the chain are hermaphrodite, but the two kinds of sexual organs are neither developed nor ready to discharge their functions at the same time. Soon after birth the female organs attain maturity, while the testicular cæca are not developed till later, and produce the sperm still later. In *Salpa* the female parts are almost always reduced to a capsule enclosing a single egg and surrounded by blood. This capsule opens into the respiratory cavity on the right side, some distance from the nucleus, by a narrow stalk-like duct (fig. 567 *b*). After fertilization the stalk becomes shorter, so that the egg, which is increasing in size, approaches closer and closer to the lining of the respiratory cavity, and forms with its capsule a projecting cone in which, as in a brood pouch, the embryonic development takes place.*

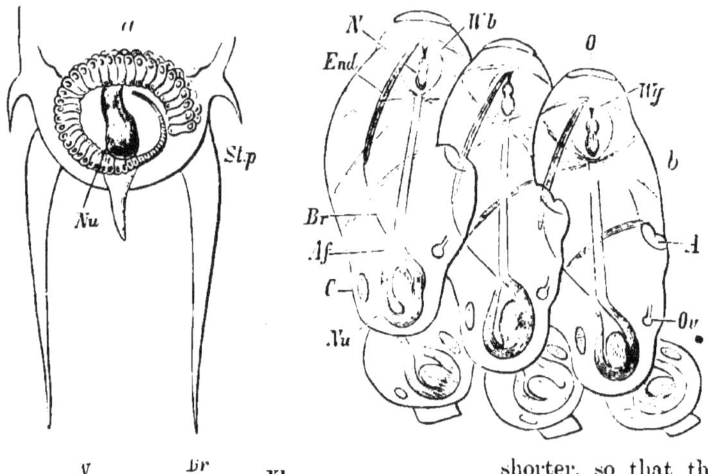

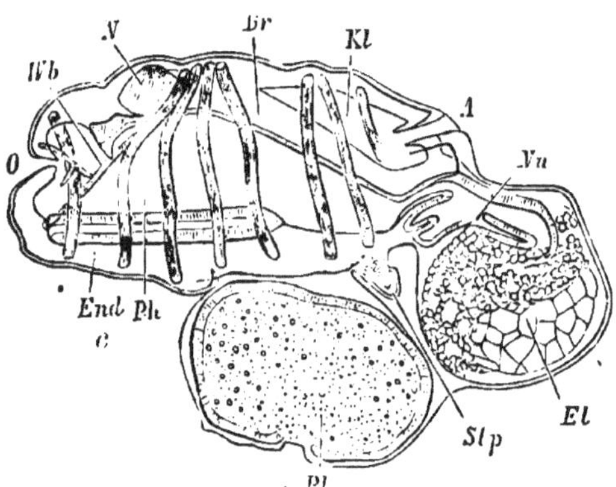

FIG. 567.—*a*, Posterior end of *Salpa democratica*, seen from the ventral side. *Stp*, Stolo prolifer; *Nu*, nucleus. *b*, Terminal portion of stolon = young chain, strongly magnified; *O*, mouth; *A*, cloacal aperture; *N*, nervous centre (ganglion); *Wg*, ciliated pit; *Wb*, arch of cilia; *End*, endostyle; *Af*, anus; *Br*, gill; *Nu*, nucleus; *Ov*, ovary; *C*, heart. *c*. Embryo of *Salpa democratica* (after C. Grobben). *El*, Elæoblast; *Pl*, placenta; *Ph*, pharyngeal cavity; *Kl*, cloacal cavity.

In the course of development a placenta is formed between the embryo

* Besides R. Leuckart l.c. compare Kowalevski, "Beitrag zur Entwickelungsgeschichte der Tunicaten." "Entwickelungsgeschichte der Salpen." *Nachr. von der königl. Gesellsch. der Wissensch.*, Nr. 19, Göttingen, 1868.

W. Salensky, "Ueber die embryonale Entwickelungsgeschichte der Salpen." *Zeitschr. für wiss. Zool.*, Tom XXVII., 1876.

W. Salensky, "Ueber die Knospung der Salpen." *Morph. Jahrb.*, Tom. III., 1877.

and the mother, and this structure plays an important part in the nourishment and growth of the embryo. In the further development of the organs, which agrees in its general features with that of the Ascidians, the placenta becomes more sharply marked off from the body of the embryo, at the posterior end of which a structure known as the *elæoblast* the equivalent of the notochord —makes its appearance (fig. 567 *c*).

It is only after a relatively long period that the embryo is born as a small fully-developed *Salpa*, which, however, still possesses the remains of the placenta and the *elæoblast*.

This solitary *Salpa*, which has been produced sexually, grows considerably during its free life, but always remains asexual, while by budding on its stolon it produces a number of individuals united together in chains. This stolon or germ-stock is a process of the body containing the rudiments of the most important organs. Its central cavity is traversed by a stream of blood, and on its walls the buds sprout out. In *Salpa*, as in the Ascidians, the stolon lies on the ventral side, and later enters into a special, open excavation of the body covering (fig. 567 *a*).

On account of the extraordinary fertility of the stolon several groups of buds of different ages are always present one behind the other; they separate successively as independent chains.

In *Doliolum* the reproductive processes are much more complicated, for not only do the sexually produced young undergo a metamorphosis but a new series of generations is introduced into the life history. The eggs are laid, and the larvæ which issue from them are provided with tails and resemble Ascidian larvæ (fig. 568, *e*). They develop into asexual forms which differ from the sexual forms, and are provided with a *dorsal* stolon (fig. 568 *b*, *Std*); the ventral stolon (stolon of Salpa) is rudimentary (*Stv*) (rosette-shaped organ). Two different kinds of buds are formed on this dorsal stolon, viz., *median buds* and *lateral buds* (Gegenbaur). The lateral buds have a slipper-like form, and are without the cloacal cavity; they do not reproduce themselves, but are concerned with the nourishment of the asexual form. The latter as it increases in size loses its gills and alimentary canal, while its muscular system becomes powerfully developed. The median buds develop into individuals, which resemble the sexual animals except that they are without genital organs; they therefore represent a second generation of asexual forms, which become free and produce the sexual generation from a *ventral* stolon.

Order 1.—DESMOMYARIA (SALPÆ).

Thaliacea of cylindrical, usually dorso-ventrally flattened form, with band-shaped muscular hoops and thick mantle (fig. 566). The

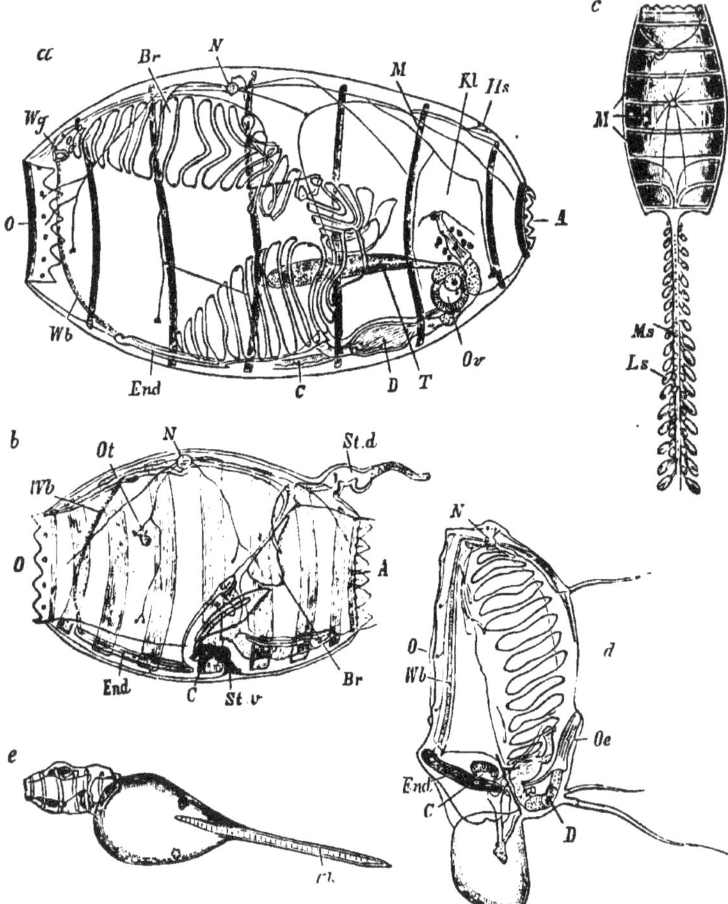

FIG. 568.—The forms of *Doliolum denticulatum* (*a*, *b*, *d*, *e*, after C. Grobben; *c*, after Gegenbaur). *a*, Sexual animal. *O*, Mouth; *A*, cloacal aperture; *Kl*, cloacal space; *N*, nerve centre; *Hs*, cutaneous sense organ; *Wb*, arch of cilia; *Wg*, ciliated pit; *End*, endostyle; *Br*, gills; *C*, heart; *D*, intestine; *T*, testis; *Ov*, ovary; *M*, muscular hoops. *b*, First asexual generation. *Stv*, Ventral stolon; *Std*, dorsal stolon; *Ot*, auditory organ. *c*, A later stage of *b*, with developed dorsal stolon and aborted intestine and gills, less highly magnified. *Ms*, Median buds; *Ls*, lateral buds. *d*, Nutritive animal produced from the lateral buds with large mouth and without cloaca; *Oe*, œsophagus. *e*, Doliolum larva with larval tail; *Ch*, chorda (*urochord*).

anterior opening is furnished with a valve-like lip which can be shut down. The gill reaches from the ganglion to the oral region, and in consequence of the development of two large lateral branchial slits is reduced to a median band. The viscera are compressed together at the end of the ventral side and form the so-called nucleus. Solitary generations which reproduce themselves by means of stolons regularly alternate with sexual animals which are budded from the stolon and united in chains. The maturity of the female sexual organs precedes that of the male organs. The single egg develops into an embryo which is nourished within the brood-pouch of the viviparous mother by means of a placenta, and becomes a solitary *Salpa* (asexual form) (fig. 567 *c*).

Fam. **Salpidæ.** *Salpa* Forsk., *S. pinnata* Forsk., *S. democratica* Forsk., *S. mucronata* Forsk. (chain form). Adriatic and Mediterranean. *S. africana* Forsk., *S. maxima* Forsk. (chain form), Adriatic and Mediterranean. *S. chordiformis* Quoy. Gaim., *S. zonaria* Pall. (chain form).

Order 2.—CYCLOMYARIA.

Body cask-shaped, mouth and atrial opening surrounded by lobes, with delicate mantle (fig. 568). The muscles are in the form of closed rings. The dorsal wall of the pharyngeal cavity is formed by a branchial lamella which is pierced by numerous slits, is placed obliquely or is bent and stretched far forwards (*D. denticulatum*). The digestive canal is not compressed into a nucleus. The ovaries contain several eggs. The testes attain maturity simultaneously with the ovaries. In the first asexual generation there is a large auditory vesicle on the left side. The development takes place by means of a complicated alternation of generations.

Fam. **Doliolidæ.** *D. denticulatum* Quoy, Gaim., the gill is bent and is pierced by about forty-five slits. *D. Mülleri* Krohn. The gill is straight, with ten to twelve slits on either side. Mediterranean.

CHAPTER IV.

VERTEBRATA.*

Bilaterally symmetrical animals with an internal skeleton (vertebral column), of which dorsal processes (upper vertebral arches) enclose the nervous centres (brain and spinal cord), and ventral processes (ribs) the cavity in which the vegetative organs are enclosed. There are at most two pairs of limbs.

The various animals included in this group were first put together

* Besides the works of Cuvier, F. Meckel and J. Müller, compare R. Owen, "On the anatomy of Vertebrates." Vol. I., II., III., London, 1866-68.
C. Gegenbaur, "Grundzüge der vergleichenden Anatomie," 2 Aufl. Leipzig, 1878.
Th. H. Huxley, "A Manual of the Anatomy of vertebrated animals." London, 1871.

by Aristotle, who called them "animals with blood" (vol. I., p. 132); he also put forward the possession of a bony or cartilaginous skeletal axis as a common characteristic. But it was Lamark who first recognised the presence of a vertebral column as the most important character, and introduced before Cuvier the name of *Vertebrata* into the science. This designation, however, in its strict significance, is only an expression for a definite grade of development of the skeleton, which may persist in its first unsegmented condition as the notochord (*Amphioxus, Myxine*). The most important characteristics therefore of the Vertebrata do not depend upon the presence of internal vertebræ and of a vertebral column, but upon a *combination of characters which have to do with the general relations of position, the mutual arrangement of the organs and the mode of embryonic development.* We may accordingly define the *Vertebrata* as laterally symmetrical organisms with an axial skeleton, on the dorsal side of which is placed the central nervous system (brain and spinal cord), while on its ventral side lie the alimentary canal with its two openings (oral and anal) and the rest of the viscera and the heart; the latter being placed on the ventral side of the alimentary canal.

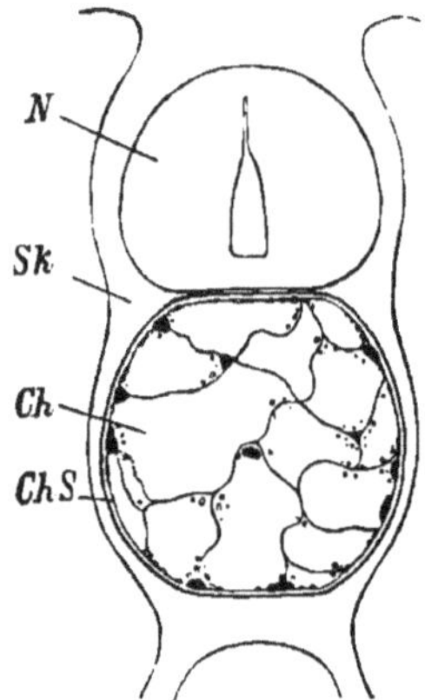

FIG. 569.—Transverse section through the chorda dorsalis (*Ch*) of the larva of *Bombinator igneus* (after Götte). *ChS*, Notochordal sheath; *Sk*, skeletogenous layer; *N*, spinal cord.

The Skeleton.—The presence of an internal skeleton is a character of great importance. While in the Invertebrates the firm supporting structures are almost always produced by the hardening and segmentation of the external skin, in the Vertebrates the relation of the hard to the soft parts of the body is reversed. The hard parts are placed in the axis of the body, and send out processes towards the dorsal and ventral surfaces, which constitute respectively a dorsal canal for the reception of the central nervous system (brain and spinal cord) and a ventral arch over the vascular trunks and the viscera. In the simplest and lowest Vertebrates the axial skeleton remains as an elastic cord—the *notochord* (*chorda dorsalis*), which in the higher Vertebrates is present in embryonic life and constitutes the first rudiment of the vertebral column (fig. 569). When the internal skeleton acquires a firmer consistency it, like the external skeleton of the Invertebrates, becomes segmented. This modification is introduced by alterations in the notochordal sheath as well as in the surrounding skeletogenous sheath (fig. 570, *a*). The latter gives rise to cartilaginous or bony rings, which represent the first rudiments of the vertebral bodies. These rings constrict the notochord till they assume the form of biconcave cartilaginous or bony discs, and become connected with cartilaginous or bony arches which are developed round the spinal cord and the perivisceral cavity (fig. 570 *a*, *b*). Each **vertebra** therefore consists of a principal median portion, the **body of the vertebra** or **centrum**, which frequently retains the remains of the notochord in its axis; of a dorsal or neural arch, and a ventral or hæmal arch. The two limbs of the dorsal arch are called **neurapophyses**, those of the ventral arch **hæmapophyses**, and the unpaired median prolongation of each arch is known as the **spinous process** (fig 570, *D*, *D'*). The transverse processes (**pleurapophyses**) which arise from different parts of the vertebræ, either from the neural arches or from the centra, are not independent structures but merely processes. The **ribs**, on the other hand, are independent lateral bony or cartilaginous rods which are attached either to the hæmapophyses (fishes) or to the pleurapophyses, and embrace the part of the body cavity containing the viscera.

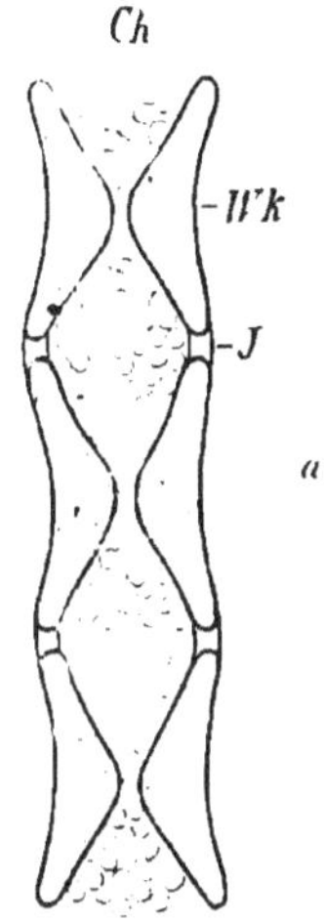

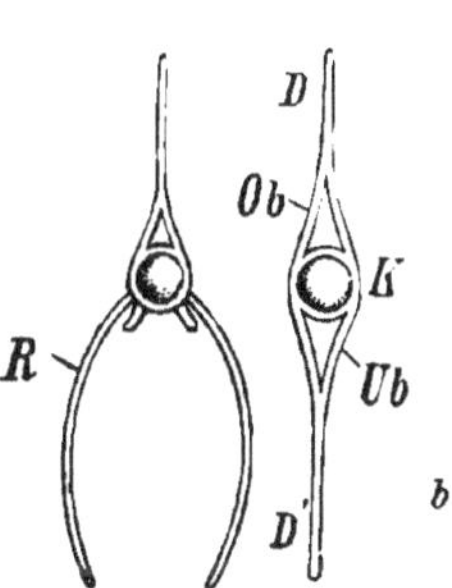

FIG. 570.—*a*, Diagram of the vertebral column of a Teleostean with intervertebral growth of the notochord. *Ch*, Notochord; *Wk*, bony vertebral bodies; *J*, membranous intervertebral portion. *b*, Vertebræ of fish. *K*, Body of vertebra, *Ob*, dorsal arch (*neurapophysis*); *Ub*, ventral arch (*hæmapophysis*); *D*, dorsal spinous process; *D'*, ventral spinous process; *R*, rib.

Regions of the vertebral column.—In the higher Vertebrates the primitive homonomous segmentation of the skeleton gives place to a heteronomous segmentation which leads to the origin of a number of regions. In this point as in others there is a parallel between the segmented Invertebrates and Vertebrates. In the first place an anterior region or **head** can always be distinguished from the posterior uniformly segmented region or **trunk** (fig. 571); and this division corresponds with the enlargement of the anterior part of the central nervous system to form the brain and with the first portion of the alimentary canal. The canal formed by the neural arches is here dilated to form the cranial capsule, on the ventral side of which are placed cartilaginous arches known as the visceral arches, of which the anterior pair constitutes the mandibular apparatus, is armed with teeth and surrounds the entrance to the alimentary canal (fig. 571). The mandibular arch is followed by a number of arches which surround the pharynx; the first of these is the hyoid arch and the rest are the branchial arches.

The part of the body behind the head may be divided into two regions: (1) an anterior region—the **trunk** proper—in which the peritoneal or body cavity lined by the peritoneal membrane is placed; the vertebræ in this region bear ribs; (2) a posterior region or **tail**,

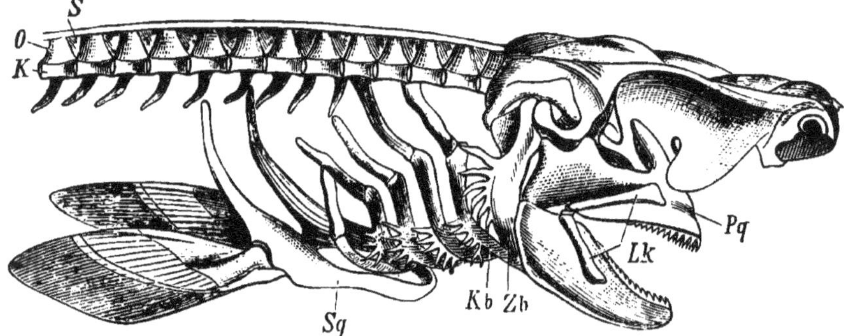

FIG. 571.—Head and anterior region of the vertebral column of *Acanthias* (after Owen). *K*, Body of vertebra; *O*, neural arch; *S*, intercalated piece; *Pq*, Palatoquadrate; *Lk*, labial cartilage; *Zb*, hyoid arch; *Kb*, branchial arch; *Sg*, shoulder-girdle.

in which there is no body cavity, and the hæmapophyses unite with each other to enclose a canal (containing the caudal vessels). This, the most simple form of segmentation of the trunk, is confined to the lower Vertebrates which propel themselves by the flexion and undulatory movements of the vertebral column, and, like the Annelids, live in water, in mud, or in the earth, or even creep after the manner of snakes on the surface of the earth.

In the higher Vertebrates, however, in which, as in the Arthropods, the function of locomotion is discharged by paired appendages, the movements of the chief axis are reduced and in many regions are altogether absent. In the Vertebrata there are only two pairs of limbs, an anterior pair and a posterior pair. In the lower Vertebrata they have the form of fins and play but a subordinate part in locomotion. In such cases, therefore, the vertebral column retains its mobility and the uniformity of its segmentation. It is only in those cases in which the method of locomotion requires a greater expenditure of force on the part of the limbs, and a firmer connection between them and the axial skeleton, and the limbs are more strongly developed, that the vertebral column is divided into successive regions, each of which is characterised by the special form of the vertebræ composing it.

Since the posterior limbs constitute the chief supports of the body, and are the principal seat of the propulsive power, their girdle is usually immoveably fused with a region of the vertebral column, which is distinguished by the firm and rigid connection of its vertebræ (fig. 572). This region, which is situated between the trunk and the tail, is called the **sacral region**, and is formed in the *Amphibia* by a single vertebra, in *Reptilia* by two, and in the higher Vertebrates by a number of vertebræ, the transverse processes of which are specially large and are firmly united to the iliac bones of the pelvic girdle by means of their corresponding ribs (fig. 572, *S*). With the development of the anterior limbs, and the need of a firmer connection between them and the trunk, a more rigid region of the vertebral column makes its appearance in the anterior part of the body. This region is known as the **thoracic** region and its vertebræ as the **thoracic** or **dorsal** vertebræ (fig.

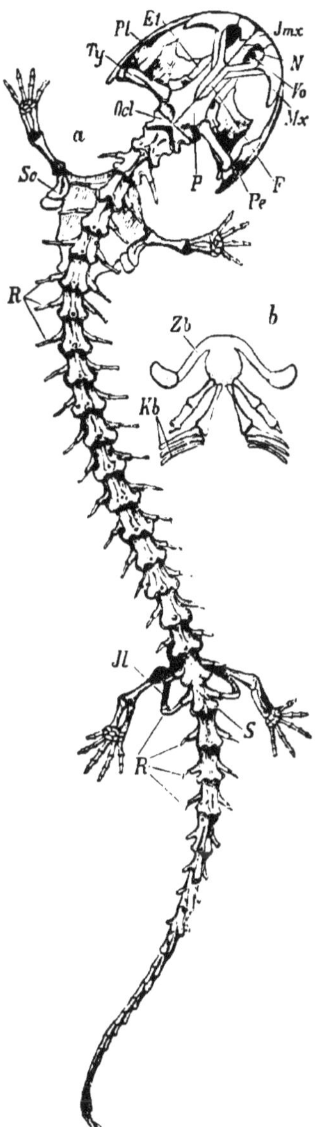

FIG. 572. Skeleton of *Menopoma alleghaniense*. *Ocl*, Exoccipital bone; *P*, parietal bone; *F*, frontal; *Ty*, tympanic; *Pe*, petrosal; *Mx*, maxillary; *Jmx*, inter-maxillary (præmaxillary); *N*, nasal; *Vo*, vomer; *Et*, girdle bone; *Pt*, pterygoid; *Sc*, pectoral girdle; *Il*, pelvic girdle; *S*, sacral vertebra; *R*, ribs; *b*, hyoid arch (*Zb*), and branchial arches (*Kb*) of the same.

573, *D*). Its ribs are distinguished by their special length, and they are connected ventrally with a system of cartilaginous or bony pieces (**sternum**) placed in the middle line of the ventral body-wall. Between the head and thoracic region on the one hand, and between the thoracic and sacral region on the other, there is a region the vertebræ of which are more movable upon one another. The region which connects the head with the thorax—the neck (**cervical region**)—is characterised by the greater freedom of movement possessed by its vertebræ, on which the rudiments of ribs are retained. The region between the thorax and sacrum—the **lumbar region** (fig. 573, *L*)—is distinguished by the great size of its transverse processes, and at the same time by a greater mobility of its vertebræ which are as a rule without ribs.

Accordingly the trunk of the higher Vertebrates is divided into **cervical, thoracic (dorsal), lumbar** and **sacral** regions, which are followed by the **caudal** region (fig. 573, *C*).

The **limbs,** which have perhaps been derived from lateral folds of the skin or possibly from parts of the visceral arches, present very considerable differences in their form and function. They may have the form of **legs** and support the body as in terrestrial animals, or serve for flight as the **wings** of aerial animals, or they may be used in swimming, as the **fins** of aquatic animals. It can be shown, however, that in every case they are composed of the same essential parts, the variation, suppression, and reduction of which determines the differences between them.

Just as legs, wings, and fins are homologous organs, so the anterior and posterior pairs of limbs seem to be repetitions of the same arrangement. In both we can recognise the **girdle** for the connection with the vertebral column, the shaft composed of long tubular bones and the terminal region. In tracing the development of the extremities Gegenbaur takes the skeleton of the fin of *Ceratodus* and of the *Crossopterygians* (**archipterygium**) as his starting-point, from which, by the reduction of certain regions and the modification of others, the limbs of the higher Vertebrates may be derived. The **pectoral girdle,** that is the girdle of the anterior pair of limbs, consists of three paired pieces—the dorsal shoulder-blade (**scapula**) and two ventral pieces placed one behind the other, known as the **præcoracoid** (with the **clavicle**) and the **coracoid.** The **pelvic girdle,** or girdle of the posterior limbs, corresponds to the pectoral girdle, and is likewise composed of three paired elements—a dorsal element attached to the sacrum and known as the **ilium,** and two ventral elements which join their fellows in the middle ventral line, and are known as the **pubis** and **ischium**. The limbs are divided into three segments—the two proximal of which are long and contain long hollow bones articulated together, the third segment being shorter and terminal. These segments are called **brachium, antebrachium** and **manus** in the fore-limb; **femur, crus** and **pes** in the hind-limb.

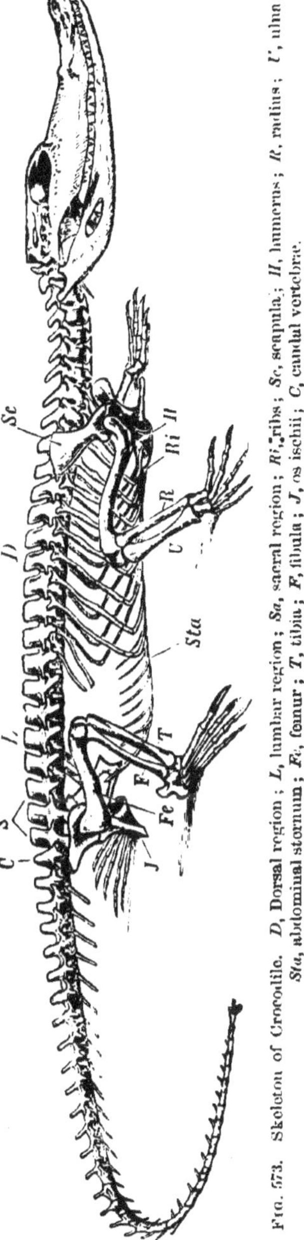

Fig. 573. Skeleton of Crocodile. *D*, Dorsal region; *L*, lumbar region; *Sa*, sacral region; *Ri*, ribs; *Sc*, scapula; *H*, humerus; *R*, radius; *U*, ulna; *Sta*, abdominal sternum; *Fe*, femur; *T*, tibia; *F*, fibula; *J*, os ischii; *C*, caudal vertebræ.

The proximal segments (*i.e.*, the brachium and femur) each contain one bone—the **humerus** (*H*) and **femur** (*Fe*) respectively. The middle segments (*i.e.*, antebrachium and crus) each contain two bones—the **radius** and **ulna** in the former (*R*, *U*), the **tibia** and **fibula** (*T*, *F*) in the latter. The distal or terminal segments (*i.e.*, the manus and pes) each contain a large number of elements placed close together. These elements consist of two proximal rows of bones, known in the hand as the **carpus**, and in the foot as the **tarsus**; of a middle row, known respectively as the **meta-carpus** and **metatarsus**; and of a number of distal bones known as the **phalanges**, and constituting the skeleton of the fingers and toes.

The **skull** varies considerably in form and structure. When the vertebral column is membranous and cartilaginous, the skull likewise consists of a continuous membrano-cartilaginous capsule, which in essential points agrees with the embryonic rudiment of the cranium (**primordial cranium**) of the higher Vertebrates (fig. 571). From this primordial cranium the bony skull* is developed partly by ossifications in the cartilaginous capsule or by ossifications proceeding from the membranous perichondrium; partly by the addition of membrane bones, which gradually supplant the cartilaginous parts.

Segmentation of the skull.—It is only when the cranial capsule is bony that any comparison can be instituted between the arrangement of the hard parts of the skull and that of the parts of a vertebra: this comparison has led to the view that the skull is composed of three or four vertebræ or segments. These are from behind forwards, the **occipital, parietal, frontal** and **ethmoid** segments. Each such segment, according to the vertebral theory of (P. Frank) Gœthe and

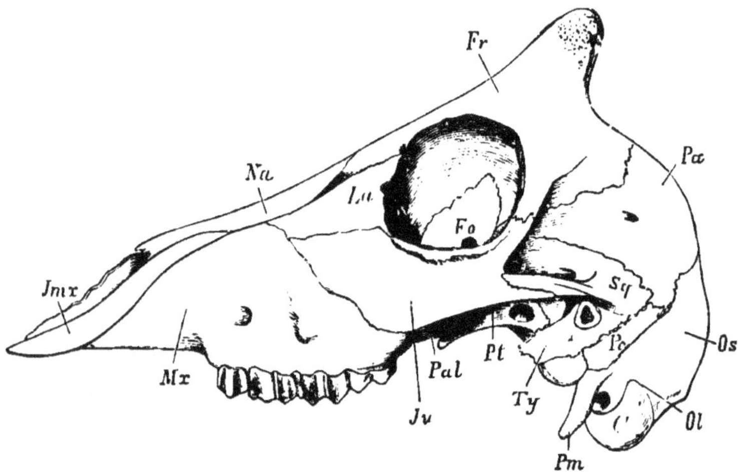

FIG. 574.—Lateral view of a goat's skull; *Ol*, exoccipital bone; *C*, condyle; *Os*, supra-occipital; *Sq*, squamosal; *Ty*, tympanic; *Pe*, petrosal; *Pm*, paramastoid process; *Pa*, parietal; *Fr*, frontal, *La*, lachrymal; *Na*, nasal; *Fo*, optic foramen; *Mx*, maxilla; *Jmx*, inter-maxilla (pre-maxilla); *Ju*, jugal; *Pal*, palatine; *Pt*, pterygoid.

Oken, is supposed to consist of a basal part corresponding to the body of the vertebra, and of a neural arch formed of two lateral pieces and a median dorsal piece (spinous process) (fig. 574). According to this theory the **basi-occipital** bone would correspond to the body of the vertebra, the two **exoccipitals** to the lateral parts of the neural arch, and the **supra-occipital** to the dorsal median parts or spinous process. The bones of the middle or parietal region of the skull consist of a basal bone, the **basisphenoid**, two lateral bones, the **alisphenoids** and two dorsal bones, the **parietals**; the two latter are membrane bones, and complete the arch dorsally. The bones of the

* Compare especially Reichert and Kölliker, Huxley, Parker etc.

anterior or frontal region likewise consist of the basal **præsphenoid**, the two lateral **orbitosphenoids**, and the two dorsal **frontal** bones, which are membrane bones and complete the arch dorsally.

The **ethmoid** may be regarded as representing the body of a fourth or anterior vertebra; it is covered above by the **nasal** bones and below by the **vomer**.

Finally, between these different bones other bones are intercalated, *e.g.*, the **mastoid** and **petrosal** between the occipital and sphenoidal.

Recently essential objections to this vertebral theory have been raised by Huxley and Gegenbaur; and these objections have proved fatal to the theory. According to Gegenbaur, the skull is composed

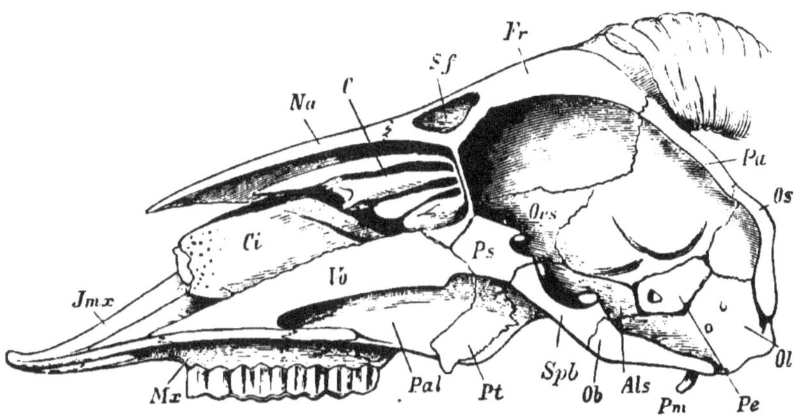

Fig. 575.—Median longitudinal section of a sheep's skull seen from the inside. *Ob*, basi-occipital; *Ol*, exoccipital; *Os*, supraoccipital; *Pe*, petrous bone; *Spb*, basisphenoid; *Ps*, præsphenoid; *Als*, alisphenoid; *Ors*, orbitosphenoid; *Pa*, parietal; *Fr*, frontal; *Sf*, frontal sinus; *Na*, nasal; *C*, turbinal; *Ci*, inferior turbinal; *Pt*, pterygoid; *Pal*, palatine; *Vo*, vomer; *Mx*, maxilla; *Jmx*, inter-maxilla (pre-maxilla).

of a much greater number of segments corresponding to the primary visceral arches, and the resemblances between the cranial bones, especially of the median and anterior regions of the skull, and the parts of a vertebra are entirely secondary.

The rest of the hard parts, which are more or less intimately connected with the skull, consist of a number of arches lying one behind the other, and surrounding the entrance into the visceral cavity.

The anterior of these—the maxillo-palatine apparatus—forms the facial region. In its simplest form it consists of two moveable pieces (**palato-quadrate** and **lower jaw**), which are attached by the hyomandibular (the dorsal element of the second arch) to the auditory region of the skull (fig. 571, *ll*). The upper piece of the first arch, likewise, is sometimes more or less firmly applied along its whole length to the skull, and when ossification takes place it becomes divided on either side into an outer and inner series of pieces, the first including the **jugal**, the **maxilla**, and **præmaxilla**, the latter the **pterygoid** and **palatine** (fig. 575). These series of bones form the upper jaw and the roof of the mouth.

The lower jaw, which is primitively a simple cartilaginous arch (Meckel's cartilage), also becomes replaced on either side by a number of bones (**articulare, angulare, dentary,** etc.), of which the dentary usually bears teeth and is the largest.

The visceral arches which follow the mandibular arch and are also connected with the skull are developed in the wall of the pharynx, to which they bear the same relation that the ribs do to the thorax and body cavity. The anterior arch (**hyoid** arch), the upper portion of which in the lower Vertebrates serves as the suspensorium of the jaw (**hyomandibular**), forms a support for the tongue, and the arch of each side meets a median basal piece (*os linguale*). The latter is followed by a series of median unpaired bones (*copulæ*), which connect the following arches (branchial arches). The branchial arches are most developed in the aquatic Vertebrates, in which they are separated by the pharyngeal slits, and serve to bear the gills. In the air-breathing Vertebrates they become more and more reduced, and finally are only discernible in imperfect number as embryonic structures. The remains of the whole apparatus form the body and cornua of the hyoid bone.

Integument.—The external skin of the Vertebrates is divided into two very distinct layers, the epidermis externally and the cutis internally. The latter is principally composed of a fibrous connective tissue, with which muscular elements come into relation, without however forming a complete dermal-muscular envelope as in the Annelids.

When the dermal muscles have a considerable extension over large surfaces, they serve exclusively to move the skin and its manifold appendages, but are not used for the movements of the trunk, which are produced by a highly-developed muscular system surrounding the skeleton. The cutis is continued into a deeper, more or less loose layer, the subcutaneous connective tissue, but its more superficial part is tolerably compact, and contains not only various pigments, but also blood-vessels and nerves. At its upper surface the cutis is raised into small conical papillæ, which are covered by the epidermis and are of importance not only as special sense organs (tactile organs), but also for the production of various hard structures (scales, teeth). The epidermis is composed of several layers of cells, of which the upper and older layers are cast off, while the lower layers (*stratum Malpighi*) are actively growing and serve as a matrix for the continual renewal of the upper layers, and sometimes contain the cutaneous pigments. Some of the appendages of the skin are epidermal structures, in which case they arise as the result of peculiar and independent growths of the epidermis (hairs and feathers). Some are derived from ossifications of the dermal papillæ which sometimes may even give rise to a hard and complete dermal armour (scales of Fishes and Reptiles, carapace of Armadillos and Tortoises).

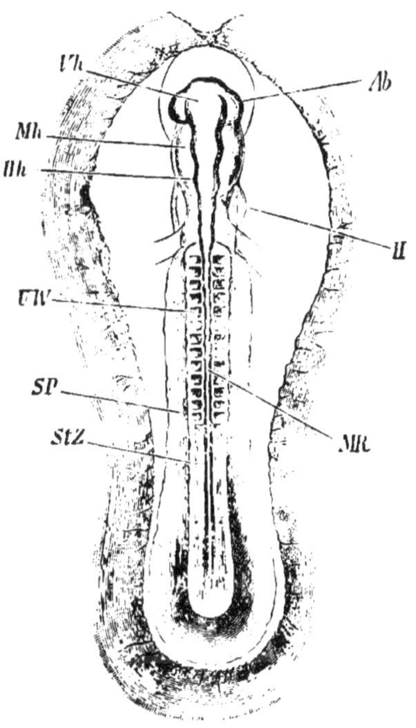

FIG. 576.—Embryo chick at end of second day (after Kölliker). *Vh*, Fore-brain; *Mh*, mid-brain; *Hh*, hind-brain; *Ab*, optic vesicle; *MR*, medullary canal; *UW*, protovertebræ; *StZ*, vertebral plates of the mesoderm; *SP*, lateral mesoblastic plates; *H*, heart.

The **central nervous system** is placed in the dorsal cavity formed by the upper arches of the vertebræ; it consists essentially of a cord—the *spinal cord*—the anterior enlarged and more differentiated part of which is distinguished as the *brain*. The spinal cord contains a narrow central canal, which is continued into the brain, where it widens out and forms the **ventricles** of the brain. The brain and spinal cord are, therefore, parts of the same organ. The brain seems to be the seat of the intellectual faculties and the central organ of the sensory apparatuses; while the spinal cord conducts the impulses to and from the brain, and in particular is the centre of reflex movements, but it also contains the centres of certain automatic actions. The mass of the brain and that of the spinal cord increase as might be expected, as the grade of life is higher. They increase, however, in an unequal ratio, for the brain soon preponderates over the spinal cord. The lower Vertebrates with cold blood have a relatively small brain, the mass of which is still considerably smaller than that of the spinal cord. In the warm-blooded Vertebrates, on the other hand, this proportion is reversed, and the more markedly, the higher the organisation and grade of life of the animal in question.

The **spinal nerves** arise in pairs from the spinal cord: each nerve has two roots—a dorsal sensory root and a ventral motor root. They correspond in number with the vertebræ, between which they pass out, so that the spinal cord repeats in a general manner the segmentation of the vertebral column.

In the brain the arrangement of the *nerves* presents several complications which are further increased by the origin of two sensory nerves—the olfactory and optic. In spite of the differences in form

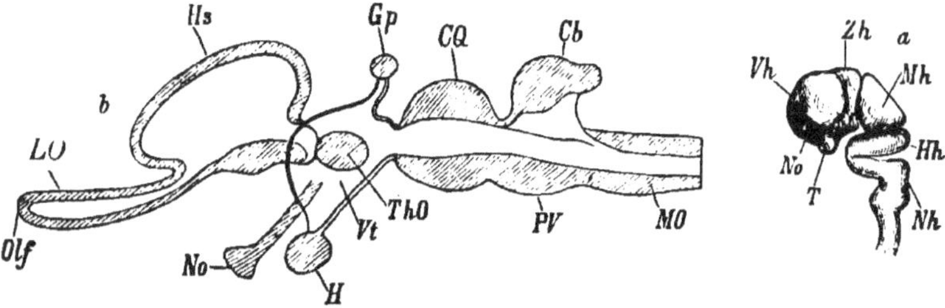

FIG. 577.—*a.* Brain and anterior part of the spinal cord of a human embryo seen from the side (after Kölliker). *Vh*, Fore-brain; *Zh*, thalamencephalon; *Mh*, mid-brain; *Hh*, hind-brain; *Nh*, medulla oblongata; *T*, anterior ventral end of the thalamencephalon; *NO*, optic nerve. *b.* Diagrammatic longitudinal section through a vertebrate brain (after Huxley). *Hs*, Hemispheres; *LO*, olfactory lobes; *Olf*, olfactory nerve; *ThO*, optic thalamus; *Vt*, third ventricle; *No*, optic nerve; *H*, pituitary body (hypophysis); *Gp*, pineal gland; *CQ*, corpora quadrigemina; *Cb*, cerebellum; *MO*, medulla oblongata; *PV*, pons Varolii.

and structure presented by the brain, three principal regions which correspond to the three vesicles found in the embryo can always be distinguished (fig. 576). The anterior vesicle (fore-brain, fig. 576, *Vh*) corresponds to the *cerebral hemispheres* and the *optic thalami* (fig. 577, *Hs*, *ThO*), the middle vesicle (mid-brain, *Mh*) to the *corpora quadrigemina* (fig. 577, *C Q*), and the posterior vesicle (hind-brain, fig. 576, *Hh*) to the *cerebellum* and *medulla oblongata* (fig. 577, *Cb*, *MO*). The anterior vesicle, however, is again divided into two parts—an anterior bilobed part, which constitutes the cerebral hemispheres and contains the lateral ventricles, and a posterior unpaired part which constitutes the so-called thalamencephalon with the thalami optici and the parts surrounding the third ventricle (fig. 577). The third cerebral vesicle is also divided into two parts—anteriorly the cerebellum, and posteriorly the medulla oblongata.

The **sense organs** present the following arrangement. The anterior is the **olfactory** organ, which consists of a pit usually paired, exceptionally unpaired (Cyclostomes); the nerves which pass to these pits arise from the fore-brain and are often swollen at their origin into special lobes (olfactory lobes). In aquatic animals which breathe by gills the nasal cavity consists with rare exceptions (*Myxine*) of a blind sac. In all lung-breathing Vertebrates, on the contrary, it communicates with the cavity of the mouth by the nasal passages, and serves for the entrance and exit of the pulmonary air.

Next come the **eyes** with the optic nerves which arise from the thalamencephalon and mid-brain. They are always paired (for the structure of the eye *vide* p. 73, vol. i.). In *Amphioxus* alone they are represented by an unpaired pigment spot placed on the anterior end of the central nervous system.

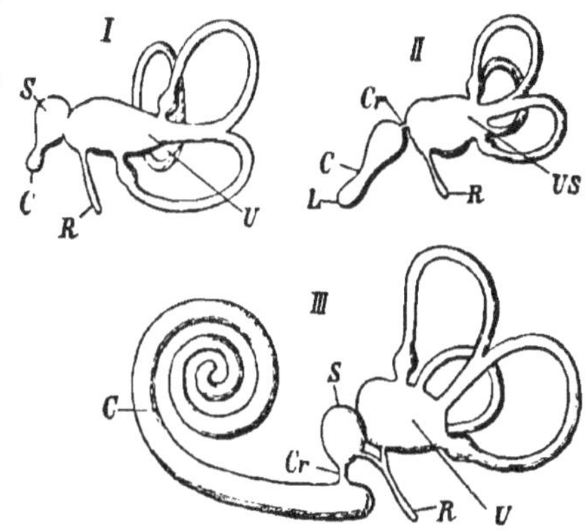

Fig. 578.—Diagram of the auditory labyrinth (after Waldeyer). *I*, of fish; *II*, of bird; *III*, of mammal; *U*, utricle with the three semicircular canals; *S*, saccule; *US*, alveus communis; *C*, cochlea; *L*, lagena; *Cr*, canalis reuniens; *R*, aquaeductus vestibuli.

The **auditory** organ, the nerve of which belongs to the hind-brain (probably derived from the sensory root of a spinal-like cranial nerve), is entirely absent in *Amphioxus*. In its simplest form it is a membranous sac (membranous labyrinth) containing fluid and otoliths. The posterior part of this sac is usually prolonged into three semi-circular canals, while the anterior part, which in many cases is separated as the *saccule*, gives off a prolongation which forms the *cochlea* (fig. 578, *S*, *C*.).

The sense of **taste** is located in the palate and the root of the tongue. The organs of taste consist of peculiarly-modified groups of epithelial cells (taste buds), and are supplied by a spinal-like cerebral nerve (*glossopharyngeal*). The general sensibility, which is distributed over the whole surface of the body, and the tactile sense are also connected with the terminations of the sensory fibres of spinal nerves.

In addition to the cerebro-spinal nervous system there is (except in *Amphioxus* and the *Cyclostomes*) a special visceral nervous system

—the **sympathetic.** This is formed by special branches of the spinal nerves and spinal-like cranial nerves, which are connected with special ganglia and give off nervous plexuses to the viscera (fig. 80).

The organs of nourishment, circulation, and reproduction are placed in the body cavity which extends beneath (ventral to) the skeletal axis. The **digestive canal** is a more or less elongated tube which in the region of the skull is encircled by the visceral arches: it begins with the mouth and ends with the anus, which latter is placed on the ventral surface at various distances from the hinder end of the body (according to the length of the caudal region of the vertebral column). The alimentary canal is invested in the greater part of its course by a fold of the peritoneum which lines the body cavity, and is fastened to the under surface of the vertebral column by the two lamellæ of this fold, which are closely applied to one another and form the mesentery. As a rule the alimentary canal is much longer than the distance between the mouth and anus, and therefore forms more or less numerous coils in the body cavity.

The digestive canal is almost always divided into three regions, the œsophagus and stomach, the small intestine with liver and pancreas, and the large intestine. The œsophagus always begins with a buccal cavity, on the floor of which a muscular fold, the tongue, projects. Although this organ, which is richly supplied with nerves, is in general rightly regarded as an organ of taste, it nevertheless plays a considerable part in the reception of the food, and may even in some cases altogether lose its importance as an organ of taste. The buccal cavity, except in *Amphioxus* and the *Cyclostomes*, is enclosed by the skeletal arch known as the maxillo-palatine apparatus and the lower jaw, of which the latter is always capable of powerful movements, while the parts of the former are either more or less firmly united together and attached to the bones of the skull, or are capable of movement on the latter. The two jaws, unlike those of the Arthropoda, work upon one another in the direction from below upwards. They are usually furnished with teeth. The teeth are derived from ossified papillæ (**dentine**) of the mucous membrane of the mouth (fig. 579), which are covered with an epidermal structure—the **enamel**; they are either directly fused with the bones of the jaw or inserted into special alveoli in the latter. The teeth in the higher Vertebrates are confined to the upper and lower jaws, but in the lower Vertebrates they may appear on all the bones which surround the buccal cavity. Teeth are, however, often altogether absent. In Birds and Tortoises they are replaced by a horny covering of the sharp edges of the jaws (beak), and certain toothless Whales bear horny plates (the so-called whalebone) on their palate.

In almost all cases the alimentary canal is provided in its different regions with independent glands which mix their secretion with its contents. In the cavity of the mouth the saliva secreted by a greater or less number of salivary glands is mingled with the food. In many aquatic animals these salivary glands may be reduced or be wholly absent. Into the first part of the small intestine the bile and the secretion of the pancreas, which is of great importance for the digestion of the food, is poured. The bile is secreted by the liver—a large gland through which the venous blood returning from the viscera passes on its course to the heart (portal circulation). In *Amphioxus* the liver is represented by a simple cæcal diverticulum of the intestine. In *Amphioxus* and some other fishes the pancreas is wanting. The small intestine in which the juices are absorbed is distinguished not only by its great length—it is in fact this portion of the alimentary canal which is arranged in coils—but also by the presence of internal folds and papillæ which considerably increase the extent of absorbing surface. The terminal region (large intestine, rectum) of the digestive canal is principally distinguished by its width and its powerful muscles.

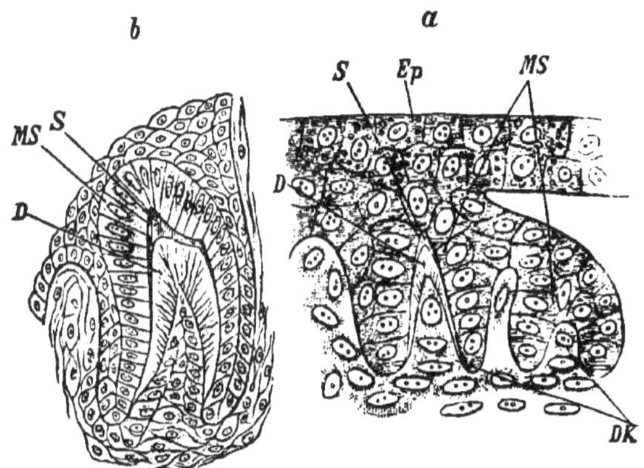

FIG. 579. – The development of the tooth in *Triton* (after O. Hertwig). *a*, The first stages of the development of a tooth; on the right hand is the earliest rudiment. *b*, Later stage of development. *DK*, papilla of the cutis which later becomes the dentine of the tooth *MS*, enamel membrane (epithelial growth which forms the enamel); *D*, dentine; *S*, enamel; *Ep*, epithelium of the mouth.

Special **respiratory organs**, as lungs or gills, are always present. The gills usually consist of double rows of lancet-shaped lamellæ, which are arranged on the sides of the pharynx behind the mandibular arch, and except in the *Cyclostomes* are borne by visceral arches. Between these arches there are always narrower or wider slit-like openings, which lead directly into the pharynx and allow the water which bathes the gills and serves for respiration to pass from the pharynx into the branchial cavity. On the external side the gills are often protected by a cutaneous fold or by an operculum, at the lower or posterior margin of which there is a long slit for the passage outwards of the water from the branchial cavity. The gills may, however, project as uncovered external appendages (external gills of Amphibians and Selachian embryos).

In the lower Vertebrata lungs and gills may coexist in the same animal, and in fishes the lungs are represented by a morphologically equivalent organ—the swimming bladder. Lungs, however, in their more complete development are only found in the higher and for the most part warm-blooded Vertebrates. In their simplest form they appear as two sacs filled with air and opening by a common air passage (*trachea*) into the pharynx. The walls of the pulmonary sacs contain the respiratory capillaries; their surface is usually increased by folds and projections which give them the appearance of a spongy organ or of an organ traversed by tubes. The two lungs often extend far into the body cavity, but in the higher Vertebrates they are confined to the anterior part of the latter which may be more or less completely separated off from the hinder part of the body cavity by a transverse partition—the **diaphragm**—and is then called the thoracic cavity.

Aerial respiration also requires a continual change of the medium serving for respiration; the exchange of the used-up air saturated with carbonic acid gas for the atmospheric air rich in oxygen. This exchange is effected by various mechanical arrangements on which the so-called respiratory movements are dependent. These movements take place in all those Vertebrates which breathe by means of lungs, but are most complete in the Mammalia, in which they consist in alternating rhythmical contractions and dilatations of the thorax.

At the entrance of the trachea and in connection with the organs of respiration is the **vocal organ** (**larynx**), which is usually formed by a modification of the upper portion of the trachea. The larynx contains vocal chords, and opens into the pharynx by a narrow slit (*glottis*) which is usually capable of being closed by an epiglottis.

The **circulatory** organs are in close relation with the respiratory organs. The vascular system is always closed and contains red blood (except in *Amphioxus* and the *Leptocephalida*, where the blood is white). The red colour of the blood, which was formerly held to be the essential character of blood (Aristotle), is due to the presence of an enormous number of red blood corpuscles, which are flat, disc-shaped globules, contain the colouring matter (**hæmoglobin**) and carry the oxygen to the tissues. In addition to the red blood-corpuscles there are small colourless cells in the blood—the amœboid white blood-corpuscles (vol. i., fig. 19).

Except in *Amphioxus,* in which the larger vascular trunks pulsate, a definite part of the vascular system is always developed to form a heart. The heart lies in the anterior part of the body cavity, and is primitively placed exactly in the middle line. It has a conical shape and is enclosed in a pericardium. The position of the principal vessels and their connection with the heart are in the simplest case as follows: A large artery—the dorsal aorta—runs along the vertebral column and gives off numerous lateral branches, corresponding to the segmentation of the vertebral column, to the right and left. Beneath this there is in the caudal region, an unpaired vein—the caudal vein,—in the body cavity on the contrary a pair of veins—the inferior cardinal veins. These veins receive their blood from lateral venous branches which proceed directly from the capillary network of the arterial branches. Another principal vein—the vena cava inferior—separated from the cardinal veins by the hepatic portal system, and connected with two superior cardinal veins, conveys the venous blood back to that portion of the heart which is known as the auricle. From the auricle the blood flows into the the muscular ventricle and is forced thence into an ascending artery (*aorta ascendens* or cardiac aorta). The latter divides into lateral arterial arches which pass towards the dorsal side and unite beneath the vertebral column to form the anterior part of the dorsal aorta (*aorta descendens*) (vol. i., fig. 57).

This system of the aortic arches is, however, complicated in various ways by the insertion of the respiratory organs in the course of the circulation (compare vol. i., p. 63 *et seq*).

In all Vertebrates there is a system of **lymphatic** vessels. These are a special part of the vascular system and contain a clear nutritive fluid (chyle and lymph) which is filled with colourless corpuscles (lymph corpuscles). They conduct the lymph (containing plastic materials for the renewal of the parts of the blood which have been consumed in metabolism) to the blood. The principal trunk of the lymphatic system (the thoracic duct) runs along the vertebral column and in the higher Vertebrates opens into the upper part of the vena cava superior. In the lower Vertebrates there are several communications between the lymphatic and vascular systems. Special gland-like organs—the so-called **vascular glands**, **spleen**—are inserted into the course of the lymphatic vessels.

Urinary organs or **kidneys** are generally present. They have the form of paired glands and lie beneath the vertebral column. The first rudiments of the kidneys appear in the form of organs resem-

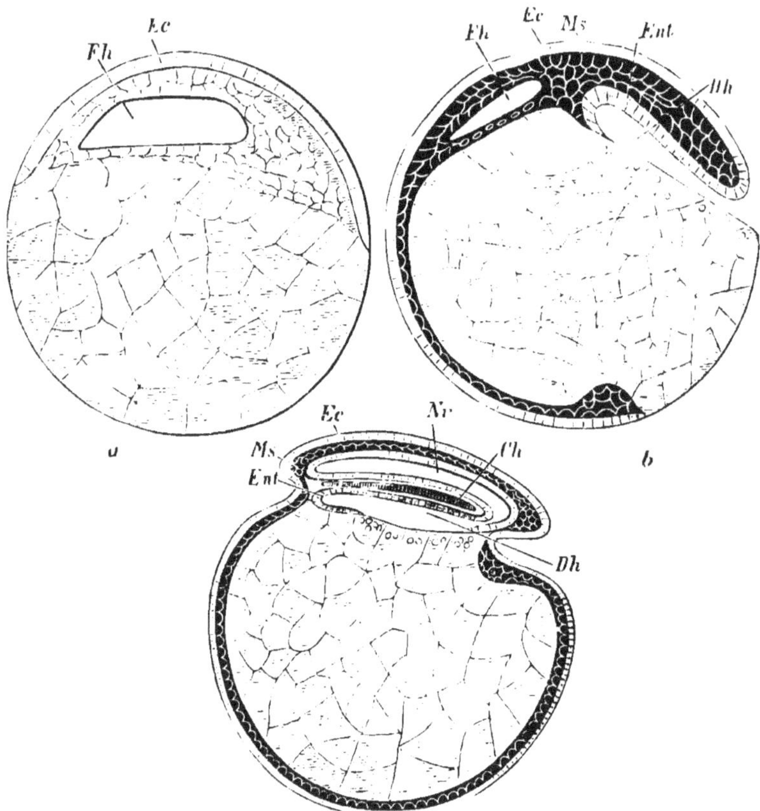

FIG. 580.—Diagrammatic longitudinal section through an ideal vertebrate embryo (after Balfour). *a*, After the completion of segmentation. *b*, Later stage in which the mesenteron is being formed at the hind end of the embryo (gastrula). *c*, Stage in which the neural canal is closed and communicates with the alimentary canal. *Ec*, ectoderm; *Ent*, entoderm; *Ms*, mesoderm; *Fh*, segmentation cavity; *Dh*, alimentary cavity; *Nr*, neural canal; *Ch*, notochord.

bling the segmental organs of Annelids. Peritoneal invaginations (urinary tubules), which communicate with the body cavity by funnel-shaped openings, come into connection with the primitive kidney-duct (archinephric duct) which is the first part of the system to appear (compare vol. i., p. 76, fig. 71). The ducts of the kidneys—the *ureters*

—usually unite to form an unpaired terminal section—the *urethra*, which, in *Teleosteans* only, opens behind the anus; very often it opens into the cloaca, and in Mammals almost always unites with the terminal parts of the genital ducts to form a common urogenital canal. A vesicular reservoir—the urinary bladder—is often inserted into the course of the efferent ducts. In fishes only does the bladder lie behind the intestine.

Reproduction is always sexual, and separate sexes are the rule. A few fishes only (species of *Serranus*) are hermaphrodite. In male Amphibians however traces of ovaries are found.

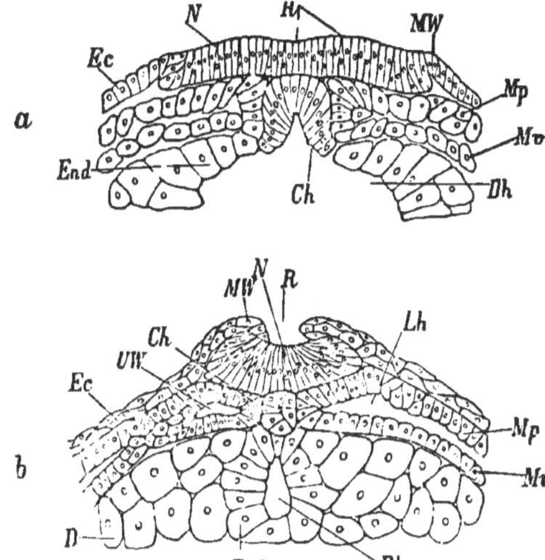

FIG. 581.—Transverse section through a young embryo of *Triton tæniatus* (after O. Hertwig). *a*, First appearance of the medullary folds and formation of the notochord. *b*, Closing of the medullary groove. The notochord is completely separated off from the entoderm. The constriction of the mesoderm into the protovertebræ is beginning (left hand side of the figure). *Ec*, ectoderm; *N*, nervous system; *R*, dorsal groove; *MW*, medullary folds; *Mp*, somatic mesoblast; *Mv*, splanchnic mesoblast; *Ch*, notochord; *End*, intestinal endoderm; *Dh*, lumen of gut; *Lh*, body cavity (pleuroperitoneal cavity); *UW*, protovertebra; *D*, yolk.

Both kinds of sexual glands lie as paired organs in the body cavity, and send off paired ducts which in the lower Vertebrates open into the cloaca and often join to form an unpaired canal. Sometimes indeed the ducts are absent and the genital products fall into the body cavity and pass out thence to the exterior by a genital pore. The division of the generative ducts into different regions, and their connection with accessory glands and external copulatory apparatuses determines the great variations in the structure of the generative organs which are most complicated in the *Mammalia*.

In many Fishes and Amphibia copulation is confined to an external union of the two sexes, and the eggs are fertilized in the water. Most Fishes, many Amphibia and Reptiles, and all Birds lay their eggs. All the *Mammalia* are viviparous and their small ova undergo embryonic development in the female generative ducts.

The **development** of the embryo (fig. 580) begins with a total or partial (discoidal) segmentation. The first rudiment of the embryo is usually a germinal disc or blastoderm lying upon the yolk. From the posterior end of this disc the alimentary cavity is developed. A primitive streak which marks the long axis of the embryo is developed by a thickening of the layers of the blastoderm. Two laterally placed longitudinal folds give rise to an ectodermal groove—the medullary groove or first rudiment of the central nervous system—beneath which is placed the notochord which is developed from the endoderm (fig. 581).

The medullary groove which is dilated anteriorly is closed by the growing together of its edges, and the tube so formed gives rise to the spinal cord and to the brain. Its lumen is for some time in

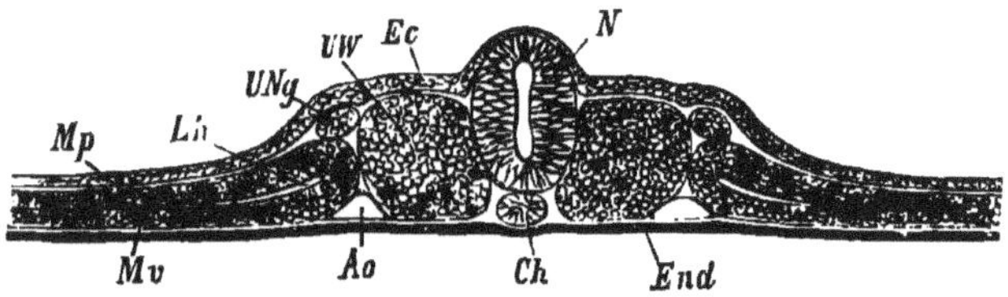

FIG. 582.—Transverse section through a chick embryo of the second day (after Kölliker). *Ec*, ectoderm; *N*, medullary canal; *End*, endoderm; *Ch*, notochord; *UW*, proto-vertebra; *UNg*, Wolffian duct (primitive duct of kidney); *Mp*, somatic mesoblast; *Mv*, splanchnic mesoblast; *Lh*, body cavity; *Ao*, primitive aorta.

communication with the alimentary cavity by the neurenteric canal. At the sides of these structures the mesoderm extends in the form of two bands, the median portions of which (protovertebral plates) become segmented in the course of the further development and give rise to the protovertebræ (figs. 576 and 582). The archinephric duct is separated off at the boundary between the protovertebræ and the unsegmented lateral plates of mesoblast, while the generative glands arise nearer the median line from the peritoneum of the lateral plates of mesoblast. While the dorsal part of the embryo is thus being formed the alimentary canal becomes further developed on the ventral side of the blastoderm, and gradually absorbs the yolk, often leaving an external yolk sac. The young animals only undergo a metamorphosis in the naked Amphibia and several Fishes.

The division of the Vertebrata into the four classes of Fishes, Amphibia, Birds, and Mammals was first established by Linnæus, though it had been already indicated in the system of Aristotle.

The Fishes and Amphibia are cold-blooded animals (*i.e.*, animals with a varying temperature): Aves and Mammals are warm-blooded (*i.e.*, with a constant temperature), and as they attain a much higher grade of life they are distinguished as the higher Vertebrates. Recently the naked *Amphibia* have rightly been separated from the scaly animals or *Reptilia*, and together with the Fishes have been distinguished as lower Vertebrates, in distinction to the Reptiles, Birds, and Mammals, which have been classed as higher Vertebrates. Fishes and Amphibia have, in fact, many characters in common, and seem to be less sharply marked off from one another (*Dipnoi*) than are the *Amphibia* from the *Reptilia*. The two former groups not only resemble one another in the branchial respiration and in the frequent persistence of the notocord, but also in the simpler course of the embryonic development and in the absence of the embryonic organs characteristic of the higher Vertebrates—the **amnion** and the **allantois.** On these grounds, and in consideration of the many relations between Reptiles and Birds, Huxley distinguishes three principal groups of Vertebrata—the **Ichthyopsida** (Pisces and Amphibia), the **Sauropsida** (Reptilia and Aves), and the **Mammalia.** Among the Fishes there are certainly such wide differences of organisation that we are justified in dividing them into several classes. The *Leptocardia* might be separated not only from all the Fishes but also from all other classes of Vertebrates as **Acraniata**; also the *Selachians*, the *Cyclostomes* and the *Dipnoi* might be regarded as separate classes if it were not more convenient to preserve the unity of the class **Pisces.**

CHAPTER V.

Class I.—PISCES.*

Cold-blooded, generally scaly, aquatic animals with unpaired fins and paired pectoral and pelvic fins. They breathe exclusively by means of gills, and have a simple heart consisting of auricle and ventricle. They are without anterior urinary bladder.

The peculiarities which the structure and internal organisation of these animals present result in general from the requirements of their

* Cuvier et Valenciennes, "Histoire Naturelle des Poissons." 22 Vols., Paris, 1828-49.
Joh. Müller, "Vergleichende Anatomie der Myxinoiden." Berlin, 1835-45.
L. Agassiz, "Recherches sur les poissons fossiles." Neufchâtel, 1833-44.
Günther, "Catalogue of the fishes in the British Museum." London.
C. E. v. Baer, "Entwickelungsgeschichte der Fische." Leipzig. 1835.

aquatic habits. Although there are in all classes of Vertebrates forms which move and live in water, yet nowhere is the whole organisation so completely adapted to an aquatic life as in Fishes.

The body is in general spindle-shaped and more or less compressed, but in details presents numerous modifications. There are cylindrical, snake-like fishes (*Lampreys*) as well as fishes with a spherical, balloon-like form (*Gymnodonta*). Others are elongated and band-shaped, and others again are very short, flat and unsymmetrical (*Pleuronectidæ*). Finally a dorsoventral flattening may lead to a flat discoidal form (*Rays*).

Locomotion is effected mainly by lateral flexions of the vertebral column, which are caused by the powerful body muscles. The effect of these movements may be greatly increased by the unpaired dorsal and ventral fins, which are capable of being elevated and depressed. The two pairs of extremities—the pectoral and pelvic fins—appear, on the contrary, to be used more as rudders to direct the course of

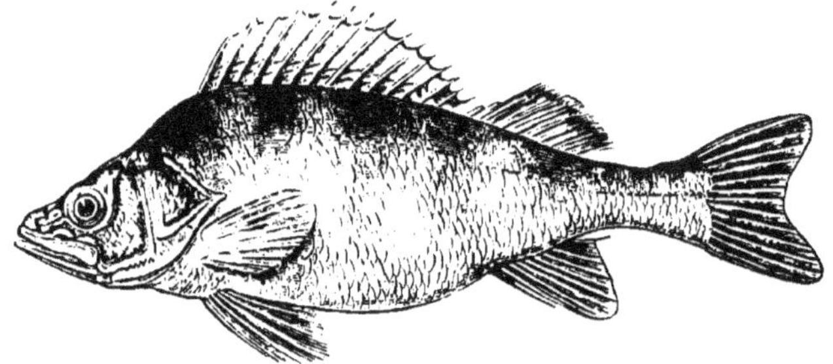

FIG. 583.—*Perca fluviatilis* (règne animal).

the animal. The structure of the vertebral column, which is not divided into many regions, corresponds to the mode of locomotion. The head is directly attached to the trunk, and is usually rigidly connected with it. A moveable cervical region, which would be a hindrance in swimming, is completely absent. The anterior part of the body is rigid, but behind it becomes more flexible and passes gradually into the the tail, the vertebræ of which permit of the most complete movements on one another, and which on that account constitutes the principal organ of locomotion.

Fins.—The system of unpaired fins is developed from a median cutaneous fold of the embryo, extending over the back and tail as far as the anus. Subsequently this fold becomes broken up into parts, the definite unpaired fins. There are usually three such parts, constituting the dorsal fin (*pinna dorsalis*), the caudal fin (*pinna caudalis*), and anal fin (*pinna analis*) (fig. 583). These ridges of skin are supported as a rule by firm rays—the fin-rays; in the Teleosteans either by hard, bony, pointed spines—the so-called spine-rays (*Acanthopteri*)—or by soft jointed rays (*Malacopteri*). The caudal fin is as a rule composed of a part of the dorsal and a part of the ventral fin-fold, but it varies much in its form. When the dorsal and ventral lobes are symmetrical the caudal fin is said to be **homocercal**; when the ventral lobe is the larger, in which case the caudal part of the vertebral column is usually bent dorsalwards, the caudal fin is said to be **heterocercal**. It sometimes happens, however, that while the caudal fin is externally homocercal the axial skeleton is bent dorsalwards so that the fin is internally heterocercal.

The paired **pectoral** and **pelvic** fins correspond to the anterior and posterior limbs of other Vertebrates. The former are attached to the head immediately behind the gills by means of an arched shoulder-girdle, while the two pelvic fins are approached to the middle line and placed further back, usually on the abdomen (ventral fins); sometimes, however, they lie between the pectoral fins (thoracic fins), and more rarely in front of the latter on the throat (jugular fins).

The integument of fishes is seldom completely naked (*Cyclostomi*). As a rule scales—ossifications of dermal papillæ, which are completely covered by epidermis—are embedded in it. The scales are often so small that they are hidden beneath the skin and seem to be completely absent (Eels). As a rule, however, they are present as firm, more or less flexible plates, which are covered with a number of concentric lines and radial striations and lie on one another like slates on a roof. Scales may be distinguished according to the structure of their free edges as **cycloid** scales with smooth edges, and **ctenoid** scales with serrated edges. Scales, which overlap but little and are generally rhomboidal, more rarely cycloidal in shape, and have an outer layer of enamel, are called **ganoid** scales, while the term **placoid** scale is applied to the small bony granules (composed of enamel and dentine) of different shapes, which lend to the surface of the skin the appearance of shagreen (these are the primitive form of teeth). Agassiz divided the Fishes according to the shape of their scales into *Cycloids*, *Ctenoids*, *Ganoids*, and *Placoids*.

In the skin there are peculiar cutaneous canals communicating with the exterior by lateral rows of pores. These are called the **lateral lines** and were considered to be slime-secreting glands till Leydig* discovered that they contain a sense organ.

* Compare Leydig, "Ueber das Organ eines sechsten Sinnes." Dresden, 1868. Fr. E. Schulze. "Ueber die Sinnesorgane der Seitenlinie bei Fischen und Amphibien." *Arch. für mikrosk. Anatomie*, Tom. VI., 1870.

In the *Myxinoids* and *Acipenseridæ* these organs have the form of short sacs; in the Rays, Skates, and Chimæras they are simple tubes, which begin as ampullæ and extend also over the head in several rows. In the *Teleostei* there are branching tubes which pierce the scales of the lateral lines as pores, and are also present on the head in several rows (fig. 583). Nerves run in the walls of these tubes and end in knob-like swellings. The epithelial covering of the latter contains in the centre short piriform cells, which at the free end are prolonged into a fine stiff hair, while at the base they pass into a varicose process—the axis cylinder of a nerve fibre (fig. 584).

The **skeleton** in its simplest form consists only of the notochord (*Amphioxus*). The notochord also persists in the Myxinoids, which possess a cartilagino-membranous cranial capsule. In the *Petromyzontidæ** there appear for the first time, above the notochord, cartilaginous neural arches, and similarly beneath it paired cartilaginous bands. These are the first rudiments of the dorsal and ventral vertebral arches. These vertebral arches are more perfect in sturgeons (*Acipenser*), and in the sea-cats (*Chimæra*), in which the notochord persists, surrounded by a very compact connective-tissue sheath. A differentiation of the axial skeleton into separate vertebræ is first found in the Skates and Rays, where dorsal and ventral arches are united with annular portions of the notochordal sheath which become cartilaginous vertebral bodies. The notochord is constricted by the growth of the latter in the centre of each vertebra, in such a manner that biconcave (*amphicœlous*) vertebral bodies are formed,

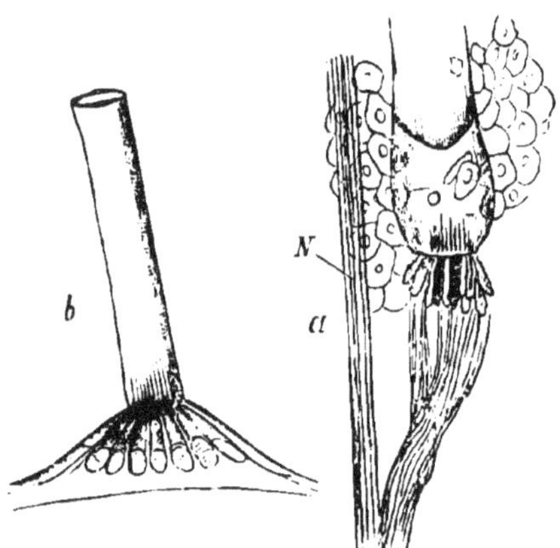

FIG. 584.—*a*, Lateral organ in the tail of a fish (roach); *N*, nerve. *b*, lateral organ in the head of a young fish (bream) (after F. E. Schulze).

* Compare Joh. Müller l.c., Reichert, "Ueber die Visceralbögen im Allgemeinen, etc." *Müller's Archiv*, 1837.

A. Kölliker, "Ueber die Beziehungen der Chorda dorsalis zur Bildung der Wirbel der Selachier und einiger anderer Fische." Würzburg, 1866.

C. Gegenbaur, "Ueber die Entwickelung der Wirbelsaüle des Lepidosteus mit vergleichenden anatomischen Bemerkungen." *Jen. naturwissensch. Zeitschr.*, Tom. III.

the conical cavities of which contain a part of the remains of the notochord. The notochord as a rule persists also in the centre of the vertebral body as a thin cord (connecting the dilated inter-vertebral portions, fig. 570 *a*). In the bony Ganoids and the Teleosteans the biconcave* vertebral bodies are completely ossified and fuse with the corresponding upper and lower bony arches, so as to form a complete vertebra. In some parts of the trunk ribs are attached to the pieces of the ventral arches (haemapophyses) which here diverge from one another; and there are often in addition ossifications of the inter-muscular ligaments.

The structure of the **skull** in Fishes presents a series of grades of development culminating in the complicated skull of the *Teleostei.* The primordial skull of the *Cyclostomes* is the simplest. It consists of a cartilagino-membranous cranial capsule, in the hard basilar part

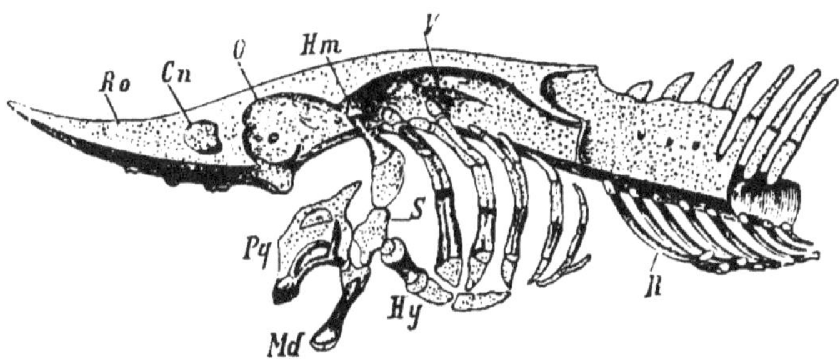

FIG. 585.—Cephalic skeleton of the Sturgeon (after Wiedersheim). *Ro*, rostrum; *Cn*, nasal pit; *O*, orbit; *Hm*, hyomandibular; *S*, symplectic; *Pq*, palatoquadrate; *Md*, lower jaw; *Hy*, hyoid bone; *V*, foramen for the vagus; *R*, ribs.

of which the notochord ends. Two bony capsules—lateral appendages of the bony basilar region—enclose the auditory organ, while two anterior pieces are connected with the complicated apparatus of the facial and palatal cartilages. The primordial skull of the Selachians (fig. 571) shows a further advance in development. It has the form of a simple cartilaginous capsule which is not further divided into separate pieces. The notochord ends in its base. In the sturgeon (fig. 585), there are bony pieces as well as the cartilaginous cranial capsule. These consist of a flat basilar bone—the parasphenoid—and a system of dermal membrane bones. A true bony cranial investment is first developed round the primordial skull of the *Dipnoi.* In the bony skulls of the *Ganoidei* and *Teleostei* there still remain continuous portions of the primordial cartilaginous cranium (Pike

* In the genus *Lepidosteus* alone is there an anterior articular surface on the vertebral bodies; the centra being convex in front and concave behind.

and Salmon). The remains of cartilage are retained longest in the ethmoid region (*Silurus*, *Cyprinus*), while on the roof and base of the skull all remains of cartilage are replaced, partly by membrane bones and partly by the primarily ossifying occipitals (basi- and exoccipital) and petrosals (periotic) as well as by the alisphenoids.

The posterior part of the skull is connected with the vertebral column without any special articulation (except in the Rays and

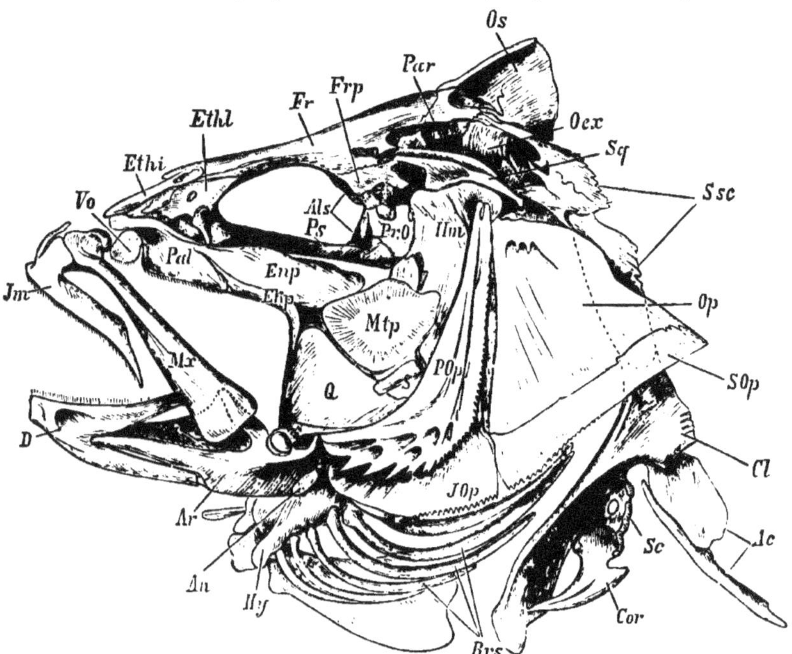

FIG. 586.—Cephalic skeleton of *Perca fluviatilis* (règne animal). *Os*, supraoccipital; *Oex*, epiotic; *Par*, parietal; *Sq*, squamosal (pterotic); *Fr*, frontal; *Frp*, postfrontal (sphenotic); *PrO*, prootic; *Als*, alisphenoid; *Ps*, parasphenoid; *Ethi*, median ethmoid; *Ethl*, lateral ethmoid (præ-frontal); *Hm*, hyomandibular; *S*, symplectic; *Q*, quadrate; *Mtp*, metapterygoid; *Enp*, endopterygoid; *Ekp*, ectopterygoid; *Pal*, palatine; *Vo*, vomer; *Jm*, intermaxillary (premaxillary); *Mx*, maxillary; *D*, dentary; *Ar*, articulare; *An*, angulare; *Op*, operculum; *POp*, præ-operculum; *SOp*, sub-operculum; *JOp*, inter-operculum; *Hy*, hyoid arch; *Brs*, branchiostegal rays; *Cl*, clavicle; *Sc*, scapula *Cor*, coracoid; *Ssc*, supraclavicle; *Ac*, accessory bone.

Chimæra), the os basilare having the conical depression and form of a vertebral body. Between the exoccipitals (which contain the foramina for the exit of the vagus and glosso-pharyngeal nerves) and the supra-occipital, which is distinguished by a strong ridge, an epiotic bone (occipitale externum) is inserted on either side (fig. 586, *Oex*). Close to the epiotic bone is the opisthotic (Huxley), which varies greatly in size and form (being very large in *Gadus* and small in

Esox), and the prootic (*PrO*), which surrounds the anterior semicircular canal and is pierced for the exit of the trigeminal nerve. There is also an external bone, the squamosal (pterotic) (*Sq*), to which the hyomandibular is articulated. The lower surface of the cranial capsule is covered by the long parasphenoid (*Ps*). The lateral walls of the skull are formed by two pairs of wing-like bones—the orbitosphenoids and the alisphenoids (fig. 586). Of these the alisphenoids are applied to the sides of the parasphenoid, and are almost always discernible with their openings for the exit of the optic nerves and the orbital branch of the trigeminal. The two orbitosphenoids are often united on the floor of the skull so as to form a median bone, which, when the cranial cavity is reduced, may be represented by a cartilaginous or membranous septum.

The roof of the skull is formed of bony plates, below which remains of the primordial cartilaginous cranium are only rarely retained. Close in front of the occipital are two parietal bones (*Par*), and in front of these again the great frontal bone (*Fr*), on each side of which is developed a post-frontal (*FrP*), which reaches to the squamosal (pterotic), and takes part in the articulation with the hyomandibular.

In the ethmoid region there is in the prolongation of the base of the cranium an unpaired cartilage or bone,—the median or unpaired ethmoid. This is covered ventrally by the large vomer, which is attached to the parasphenoid. There are also two paired lateral bones—the lateral ethmoids or præfrontals—which are perforated by the olfactory nerves and form the supports of the nasal pits (nasal capsules). There are finally accessory membrane bones—the infra-orbital and supra-temporal—which protect the cranial (sensory) canals.

A true maxillary apparatus appears for the first time in the Selachians and Sturgeons, where a hyomandibular attached to the auditory region serves to support the mandibular and hyoid arches (figs. 571 *H* and 585 *Hm*). The upper part of the mandibular arch (the palatoquadrate) is usually moveably attached to the skull by ligaments. In the *Teleostei* the mandibular suspensorium is divided into several parts, and the branchial operculum is attached to it. The upper part is formed of a hyomandibular, and two bones called by Cuvier the symplectic and tympanic (metapterygoid); the præoperculum forms the middle part, and finally the lower part, which bears the articulation of the lower jaw, is formed by the quadrate or quadrato-jugal. The flat osseous plates applied to the hinder edge of the præoperculum constitute the branchial operculum, and are distinguished as operculum, suboperculum, and interoperculum. A bone extending from the metapterygoid and quadrate to the upper jaw corresponds to the pterygoid, and is, as a rule, formed of an external (ectopterygoid) and an internal piece (endopterygoid). Then come the palatine bone and the apparatus of the upper jaw, with the præmaxilla (intermaxilla), which is placed at the front of the snout and is usually moveable, and the very variable, usually toothless maxilla. The two limbs of the lower jaw are only rarely fused together in the middle line, and are divided at least into a posterior

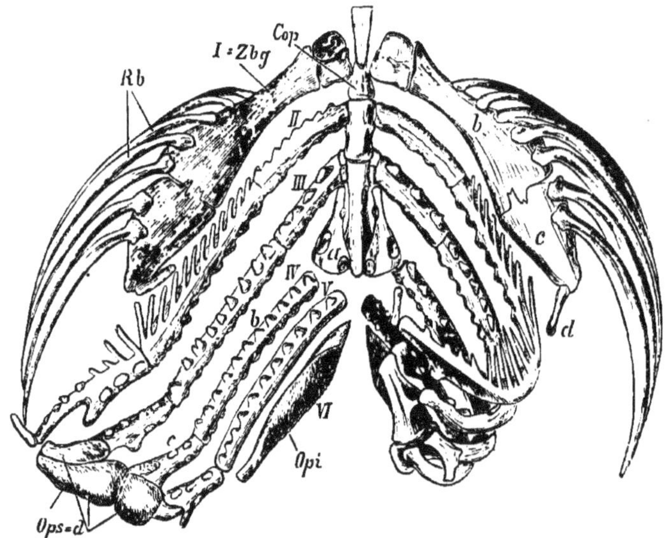

Fig. 587.—Hyoid apparatus and branchial arches of *Perca fluviatilis* (règne animal). *I*, hyoid apparatus; *II—V*, branchial arches; *a*, *b*, *c*, *d*, joints of the branchial arches, the upper joints (*Ops*) are the superior pharyngeal bones (pharyngo-branchials); *VI*, (*Opi*) the inferior pharyngeal bones (reduced 5th branchial); *Cop*, copulæ; *Rb*, branchiostegal rays.

os articulare and an anterior dentary; there may, however, also be an angulare and an operculare.

Behind the mandibular arch there follows a system of equivalent arches surrounding the pharyngeal cavity. Of these the anterior—the hyoid arch—bears on its outer edge a number of cartilaginous rods, which serve to support the opercular membrane and are called the branchiostegal rays (fig. 587, *Rb*), while the remaining arches are the branchial arches and serve for the support of the branchial lamellæ (fig. 587). In the Teleosteans four (seldom three) arches bear gills, while the posterior arch is reduced so that only its ventral part (ceratobranchial) remains and forms the so-called inferior pharyngeal bones (*pharyngealia inferiora*). The upper segments of the branchial arches, which are applied to the base of the skull, are distinguished as the superior pharyngeal bones (pharyngobranchials or *pharyngealia superiora*).

Paired Fins.* The pectoral fins are in the Teleosteans attached to the skull by means of the shoulder girdle. In the cartilaginous fishes the shoulder girdle is a simple cartilaginous arch, which unites with that of the other side in the middle ventral line. In the cartilaginous Ganoids the shoulder-girdle is transitional between this primary form and the secondary form, which is characteristic of the Teleosteans (fig. 586), inasmuch as membrane bones (clavicle) are applied to the primary cartilaginous girdle. Ossifications also arise in the cartilage itself and give rise to bones known as the scapula and coracoid, or the præcoracoid.

The skeleton of the fins, which is articulated to the shoulder-girdle, can be derived from the primitive form of fin known as the archipterygium, which still persists in *Ceratodus* as an axial row of cartilaginous pieces beset with jointed lateral rays (*radii*).

The **nervous system** (fig. 588) presents the lowest and simplest form found in any Vertebrate. In general the brain is small and consists of several swellings lying one behind another. Of these the small anterior, as the *lobi olfactorii*, pass into the olfactory nerves. The larger anterior lobes correspond to the *hemispheres*, the median globular swellings to the lobe of the *third ventricle* with the corpora quadrigemina. From this part of the brain the optic nerves are given off anteriorly, while on its lower surface the infundibulum, to which the pituitary body is attached, arises from the floor of the third ventricle.

The posterior region corresponds to the cerebellum and the medulla oblongata. The cerebellum, which varies considerably in size and form, constitutes a transverse bridge, which covers the anterior part of the fourth ventricle. Lateral swellings—the so-called *lobi posteriores*—are often developed in this region; in the Sturgeons and Squalidæ at the origin of the trigeminal nerve, as the *lobi nervi trigemini*; in Torpedo as the large *lobi electrici*, projecting over the fourth ventricle.

A separate visceral (*sympathetic*) nervous system is absent in the

* Compare C. Gegenbaur, "Untersuchungen zur vergleichenden Anatomie der Wirbelthiere." 2 Heft, Leipzig. 1865.
C. Gegenbaur, "Ueber das Skelet der Gliedmassen." *Jen. naturwiss. Zeitsch.*, Tom. V.

Cyclostomes alone, where it is represented by the vagus and by fibres of the spinal nerves. The spinal cord, the mass of which is con-

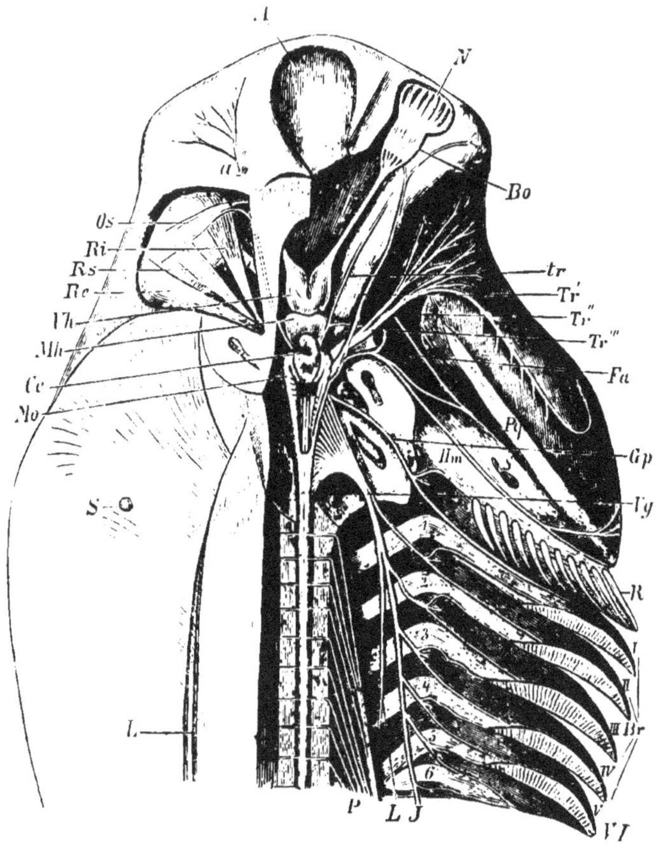

FIG. 588.—Brain and anterior part of the spinal cord and nerves of *Hexanchus griseus* (after Gegenbaur). The nerves are dissected out on the right side; the right eye removed. *A*, Anterior cavity of the skull; *N*, nasal capsule; *Vh*, fore-brain (cerebral hemispheres); *Mh*, mid-brain (optic lobes); *Ce*, cerebellum; *Mo*, medulla oblongata; *Bo*, olfactory bulb; *tr*, trochlear nerve (fourth nerve); *Tr′*, first (ophthalmic) branch of the trigeminal or fifth; *a*, terminal branches of the same in the ethmoid region; *Tr″*, second branch; *Tr‴*, third branch; *Fa*, facial (seventh); *Gp*, glossopharyngeal (ninth); *Vg*, vagus (tenth); *L*, lateral branch of vagus (to lateral line); *J*, intestinal branch; *Os*, superior oblique muscle of eye; *Ri*, internal rectus muscle; *Re*, external rectus; *Rs*, superior rectus; *S*, spiracle; *Pq*, palatoquadrate; *Hm*, hyomandibular; *R*, branchial rays; *I—VI*, branchial arches; *Br*, branchiæ; *P*, spinal nerves.

siderably greater than that of the brain, extends tolerably uniformly throughout the whole length of the neural canal, and usually does not form a so-called *cauda equina*. Rarely its upper part presents paired or unpaired swellings (*Trigla, Orthagoriscus*) at the origin of the spinal nerves.

The **eyes** are seldom hidden beneath the skin and the muscles (*Myxine, Petromyzon, Amblyopsis*). In *Amphioxus* they are represented by a pigment spot lying directly on the central nervous system. In all other fishes they are characterised by possessing a flat cornea and a large, almost spherical crystalline lens, the anterior surface of which projects far out of the pupil (fig. 589). As peculiar structures of the eyes of fishes are further to be mentioned the so-called *choroideal gland*—a vascular body (rete mirabile) usually projecting at the entrance of the optic nerve, as well as a fold of the choroid known as the *processus falciformis*, which traverses the retina, and the *campanula Halleri* which is attached to the lens.

The **auditory organ** * (absent only in *Amphioxus*) consists only of the labyrinth (fig. 578, *l*), and in Teleosteans, Ganoids, and *Chimæra* lies partly in the cranial cavity, surrounded by fatty tissue. It is worthy of notice that in *Cyprinoidæ Characinæ, Siluridæ*, and others, the labyrinth is connected with the swimming bladder by a chain of small bones.

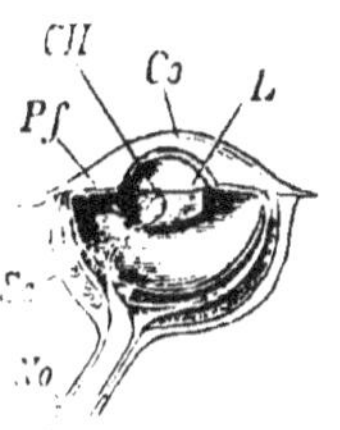

Fig. 589. – Horizontal section through the eye of *Esox lucius*. *Co*, cornea; *L*, lens; *Pf*, processus falciformis; *CH*, campanula Halleri; *No*, optic nerve; *Sc*, ossifications of the sclerotic.

The **olfactory organ** in *Amphioxus* consists of a simple unsymmetrical pit at the anterior end of the nervous centre. In *Cyclostomes* also it consists of a simple tube, with an unpaired median opening. All other fishes possess double, and indeed with the exception of the *Dipnoi* blindly-closed nasal cavities, the internal surface of which is considerably increased by folds of the mucous membrane.

The sense of **taste** seems to be less developed. It is located in the buccal cavity, and especially in the richly innervated part of the soft palate. For the **tactile** sense, lips and their appendages—the frequently appearing barbules—probably serve. Certain isolated rays of the ventral fin may also, on account of their rich nerve supply, be regarded as tactile organs (*Trigla*). The nervous organs of the so-called mucous canals, which we have before mentioned, constitute an organ of a special sense.

* Compare E. H. Weber, "De aure et auditu hominis et animalium." P. I., "De aure animalium aquatilium." Lipsiæ, 1820.
C. Hasse. "Anatomische Studien." Heft 3: "Das Gehörorgan der Fische." Leipzig, 1872.

The **electrical organs*** may be mentioned as a peripheral appendage of the nervous system (*Torpedo, Gymnotus, Malapterurus, Mormyrus*). They are nervous apparatuses which in the arrangement of their parts may be compared to a Voltaic pile. They develop electricity, and give electrical discharges when their opposite poles are

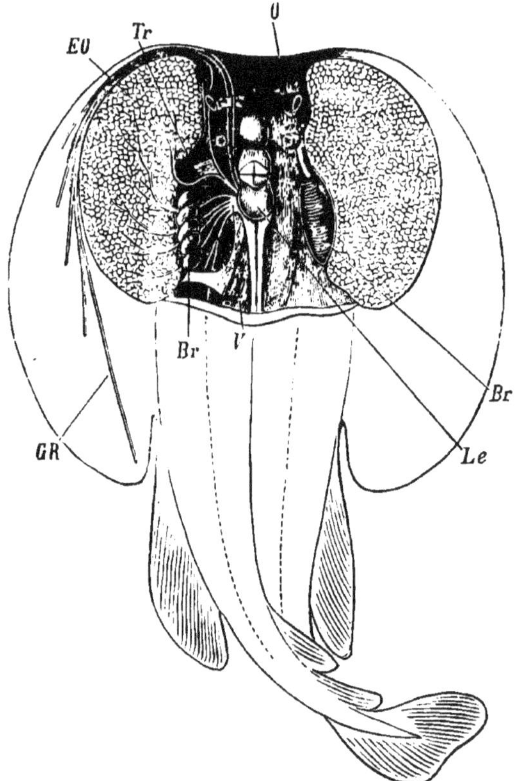

Fig. 590.—Torpedo with electric organ dissected out (*EO*) (after Gegenbaur). On the right side the dorsal surface only of the organ is exposed; on the left side the nerves which go to it are shown. *Le*, electric lobe; *Tr*, trigeminal nerve; *V*, vagus nerve; *O*, eye; *Br*, gills; on the left the individual branchial sacs; on the right the latter are shown covered with a common muscular layer. *Gr*, Gelatinous tubes of the skin (sense canals).

connected. In *Torpedo* these organs are situated (fig. 590) between

* Compare Savi, "Recherches anatomiques sur le système nerveux et sur l'organe électrique de la torpille." Paris, 1844.

Bilharz, "Das elektrische Organ des Zitterwelses." Leipzig, 1857.

Max Schultze, "Zur Kenntniss des elektrischen Organs der Fische." 1, 2, Halle, 1858 and 1859.

Max Schultze, "Zur Kenntniss des den elektrischen Organen verwandten Schwanzorganes von Raja clavata." *Müller's Archiv*, 1858.

Sachs, "Untersuchungen am Zitteraal." Leipzig, 1881.

the branchial pouches and the anterior cartilages of the pectoral fins, and consist of a number of perpendicular columns enclosed by walls of connective tissue. The columns are divided by a great number of membranous transverse partitions into a series of compartments placed one above another. Each of the latter contains a layer of gelatinous tissue, and a finely granular plate containing nerve endings and large nuclei (*electrical plate*). The latter corresponds in a certain degree to the copper and zinc elements of the Voltaic pile, the former to the moist intermediate layers: while the connective tissue framework seems to serve only to carry the nerves and blood-vessels. Each transverse partition contains a rich network of nerves, which is distributed on the electrical plates. The face on which the nerves ramify is the same in all the columns of the same

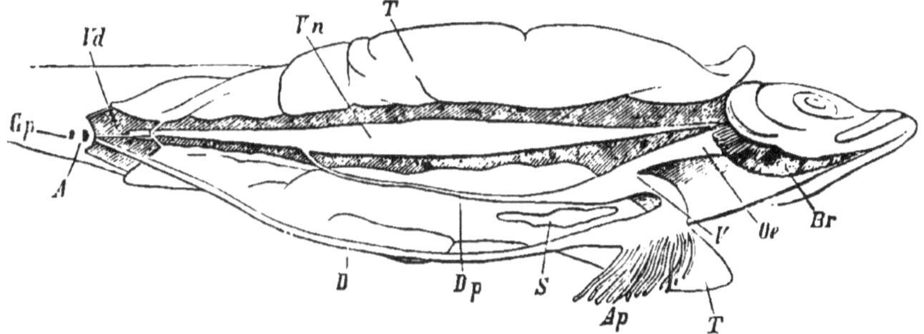

FIG. 591.—Alimentary canal and generative organs of *Clupea Harengus* (after Brandt). *Br*, gills; *Oe*, œsophagus; *V*, stomach; *Ap*, pyloric appendages; *D*, intestine; *A*, anus; *Vn*, swimming bladder; *Dp*, pneumatic duct; *S*, spleen; *T*, testis; *Vd*, vas deferens; *Gp*, genital pore.

organ, and is always electro-negative, the opposite free surface being positive. In *Malapterurus*, the other surface of the plate (the posterior surface) on which the nerves enter is electro-positive, but this apparent exception is explained by the fact that the nerves pass through the plate and are distributed on the anterior surface, which is electro-negative. In the electric Eel (*Gymnotus electricus*) the electric organ lies at the side of the tail and consists of long horizontal columns; in *Malapterurus* it lies along the body beneath the skin. Similar organs in *Mormyrus* are distinguished as pseudelectric organs, since although they have a similar structure, they give rise to no electric phenomena.

The **digestive organs** vary much in structure. The mouth, which is placed at the anterior end of the head, usually has the form of a transverse slit, and can sometimes be extended forward by means of the moveable supporting bones of the upper and lower jaws (*Labroidea*). The buccal cavity is distinguished by its width, and by the great number of teeth it contains, which are developed from the papillæ of the mucous membrane by dentinal ossification. There are often two curved parallel rows of teeth on the upper jaw; an outer row on the premaxilla, and an inner row on the palatine, and there may also be a median unpaired row on the vomer. On the lower jaw there is only one curved row of teeth. There may also be teeth on the hyoid arch and on the upper jaw (maxillæ) and parasphenoid, and, as a rule, on the branchial arches also, especially on the upper and lower pharyngeal bones. The teeth are distinguished according to their shape into pointed conical *prehensile* teeth and *grinding* teeth.

A small, hardly moveable tongue is developed on the floor of the buccal cavity, and the lateral walls of the pharynx are pierced by the gill slits. Following the pharyngeal cavity, there is a usually short, funnel-shaped œsophagus, and a large stomach, which is frequently drawn out into a cæcum of considerable size (fig 591). Cæcal appendages (*pyloric appendages*) are not unfrequently met with at the entrance to the longer mid-gut (small intestine) which is marked off by a valve; they probably serve the purpose of increasing the extent of the secreting surface of the alimentary canal. The intestine is usually several times coiled, and its internal surface is remarkable for the longitudinal folds of the mucous membrane; villi such as are found in the higher Vertebrates are only rarely present; but in the Selachians, Ganoids, and Dipnoi there is a peculiar spirally-coiled longitudinal fold—the so-called spiral valve—which contributes essentially to the enlargement of the absorbent surfaces. A rectum is not always clearly marked off, and when present is always short, and in the *Selachians* it is furnished with a cæcal appendage. The anus is usually situated far back, and is always ventral and in front of the urinary and generative openings. In fishes with jugular fins, and in some Teleosteans without ventral

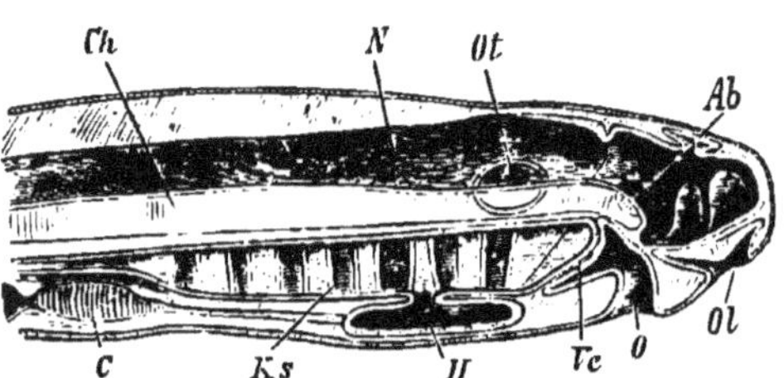

Fig. 592.—Diagrammatic longitudinal section through the head of a larva of *Petromyzon* (after Balfour). *N*, nervous system; *Ch*, notochord; *Ot*, auditory vesicle (represented as visible); *O*, mouth; *Vc*, velum; *H*, thyroid involution; *Ks*, branchial pouches; *C*, heart; *Ab*, optic vesicle; *Ol*, olfactory pit.

fins, it is situated very far forward, and may even be on the throat.

Salivary glands are absent in Fishes, but there is a large liver which is rich in fat and is usually provided with a gall-bladder; there is also usually a pancreas, which is by no means replaced by the pyloric appendages as was formerly believed.

In many fishes the **swimming bladder,** an organ which by its mode of origin corresponds to the lungs, is developed as a diverticulum of the alimentary canal. It is almost always an unpaired sac filled with air and placed on the ventral side of the vertebral column, dorsal to the alimentary canal: it is sometimes closed and sometimes

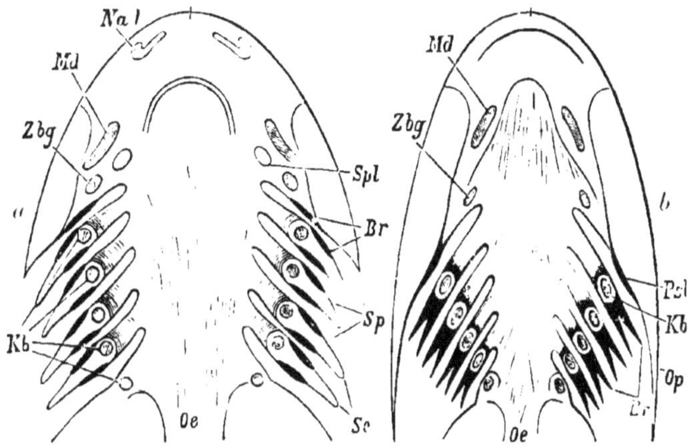

Fig. 593.—Horizontal section through the branchial cavity showing the roof. *a*, of one of the *Squalidæ*, *b*, of a Teleostean, (altered from Gegenbaur). *Nal*, nasal aperture; *Md*, mandible; *Zbg*, hyoid arch; *Kb*, branchial arches; *Oe*, œsophagus; *Spl*, spiracle; *Br*, gills; *Sp*, gill slits; *Se*, septa of branchial pouches; *Psb*, pseudobranch of the branchial operculum (hyoid pseudobranch); *Op*, operculum.

communicates by an air tube—the pneumatic duct—with the interior of the alimentary canal (*Physostomi*) (fig. 591 *Vn*). Its walls are formed of an external elastic membrane which is sometimes invested with muscles, and an internal mucous membrane. Glandular structures are sometimes present in the internal coat, and these may exert an influence on the enclosed air. The internal surface is usually smooth, but sometimes is provided with reticulated projections which lead to the origin of cellular cavities (*Ganoidei*). Physiologically the swimming bladder is a hydrostatic apparatus, the function of which seems to consist essentially in rendering the specific weight of the fish variable, and in facilitating the rapid change in the position of the centre of gravity. The fact that many fishes which swim very well are without the swimming bladder is by no means favourable to the interpretation of its function. When it is present the fish must have the power of compressing it, partly by the muscles in its walls and partly by the muscles of the body, and thus rendering the body specifically heavier so that it sinks. When the compression of the muscles is removed the compressed air will again expand, the specific gravity diminish and the fish will rise. If the pressure is unequal on the anterior and posterior parts then that half of the fish which is rendered specifically heavier will sink. Still more complicated relations, however, seem to exist according to the investigations of Bergmann.*

Respiration is in all cases effected by gills.

In the *Cyclostomes* (fig. 592) which have no visceral arches there are six or seven pairs of branchial pouches. These open into the œsophagus either by internal branchial passages or (*Petromyzon*) by a common canal which receives all the branchial passages. The water is expelled through external branchial passages round which a network of cartilaginous rods is developed.

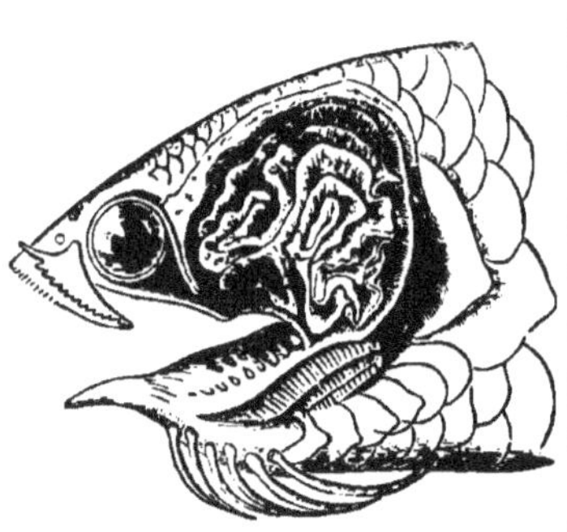

Fig. 591.—Head of *Anabas scandens* (règne animal). The operculum has been removed to shew the spacious upper pharyngeal bones (pharyngo-branchials).

In the *Plagiostomes* (fig. 593 *a*) there are saccular spaces the walls of which are supported by cartilaginous rods. These branchial sacs communicate with the exterior by lateral openings and contain the branchial leaflets which are attached to their walls: they are separated from one another by partition walls which are placed between the two rows of leaflets of each arch, and they are supported by an external framework of cartilaginous rods. In the Selachians there are, as a rule, five pairs of branchial sacs, of which the last has a row of leaflets on its anterior wall only, *i.e.*, on the posterior side of the fourth true branchial arch; while the first pouch has, in addition to the anterior gill of the first branchial arch, a gill on the hyoid arch corresponding to the accessory gill of *Chimæra* and the *Ganoidei*. The mandibular arch, however, sometimes bears a

* Compare Bergmann and Leuckart, "Anat. Phys. Uebersicht des Thierreichs." Stuttgart, 1852.

remnant of a gill—the *pseudobranch* of the spiracle—the vessels of which belong to the arterial circulation and form a rete mirabile.

In the Teleosteans (fig. 593 *b*) and the Ganoids the lancet-shaped lamellæ are arranged in double rows on the four visceral arches which function as branchial arches, and they form four comb-shaped gills on either side. These gills lie in a spacious branchial cavity covered by the branchial operculum and the branchial membrane. There is, however, an accessory gill on the inner side of the branchial operculum; this in many Ganoids and Chimæra functions as a gill, but in the Teleosteans has lost its respiratory function, and is then known as the pseudobranch of the operculum or of the hyoid arch.

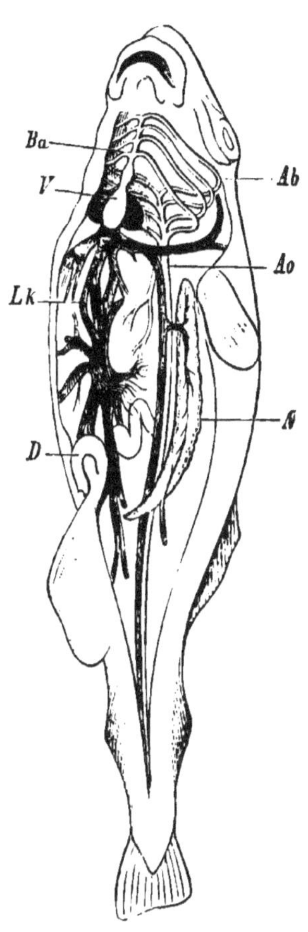

Fig. 595.—Diagram of the circulation of a Teleostean. *V*, ventricle; *Ba*, bulbus arteriosus with the arterial arches which carry the blood to the gills; *Ab*, arterial arches; *Ao*, aorta descendens into which the epibranchial arteries passing out from the gills unite; *N*, kidneys; *D*, intestine; *Lk*, portal circulation.

External gills projecting from the slits of the branchial pouches are found only in the embryos of the *Plagiostomes*. Rudiments of external gills are found in *Rhinocryptis annectens*.

Finally the secondary cavities, which are sometimes found annexed to the branchial cavity and increase the respiratory surfaces by the development of a capillary network, must be regarded as accessory organs of respiration. They consist either of labyrinthine cavities in the superior pharyngeal bones (fig. 594) or of saccular appendages of the branchial cavity (*Saccobranchus*, *Amphipnous*). True lungs derived from the swimming bladder, with internal cellular spaces, a short air-tube and glottis-like opening into the pharynx, are only found in the *Dipnoi* (according to Hyrtl the swimming bladder of *Gymnarchus* is also a lung).

Vascular system.—The blood is generally red; it is white only in *Amphioxus* and the *Leptocephalidæ*; it circulates in a closed vascular system, in which, except in *Amphioxus*, a muscular pulsating region or heart is present. The heart (fig. 595) is placed far forward on the throat, ventral to the branchial framework, and is enclosed in a pericardium, the cavity of which communicates with the body cavity in some *Plagiostomes*, *Chimæra*, *Acipenser*, etc. It is a simple venous branchial heart, and is composed of a thin-walled large auricle and a very powerful muscular ventricle. The auricle receives the venous blood returning from the body, and the ventricle forces it through an ascending aorta to the respiratory organs. The aorta begins with a bulbous swelling (*bulbus arteriosus*), which in the *Ganoids*, *Plagiostomes*, and *Dipnoi* is replaced by an independently pulsating part of the heart with rows of semi-lunar valves (*conus arteriosus*). While the fishes with a simple non-muscular *bulbus arteriosus* have but two semi-lunar valves at its origin, the above mentioned orders usually have two to four, or rarely five rows of three, four or more valves each in the conus arteriosus. The aorta at once divides into a number of paired vascular arches corresponding to the embryonic aortic arches. These are the branchial arteries; they pass into the branchial arches and give off branches to form the capillary networks of the gills. From the capillary networks small vessels pass out which unite to form a larger branchial vein in each branchial arch (epibranchial artery). The arrangement of these veins corresponds to that of the branchial arteries; they unite to form the large *aorta descendens* or dorsal aorta. Before they unite the cephalic arteries pass off from the epibranchial arteries of the anterior arch. The arrangement of the principal venous trunks in fishes is most nearly related to the embryonic condition. Corresponding to the four cardinal veins of the embryo, two anterior and two posterior vertebral * veins (jugular and cardinal veins) bring back the blood from the anterior and posterior part of the body respectively. These veins unite on each side to form two transverse veins—the *ductus Cuvieri*—which enter the sinus venosus of the heart. The course of the returning venous blood is complicated by the insertion of a double portal circulation. The caudal vein passes directly into the posterior cardinal veins only in Cyclostomes and Selachians: in all other fishes there is a renal-portal circulation, in that the caudal vein breaks up into capillaries in the kidneys, from which the blood passes into the posterior cardinal veins. For the hepatic portal circulation on the other hand the venous blood of the

* Often called the anterior and posterior cardinal veins.

intestine is used; this blood after passing through the capillaries of the liver is returned to the heart by one or more veins which correspond to the inferior vena cava and open into the sinus venosus between the two ductus Cuvieri. Such capillary systems must be a considerable hindrance to the circulation of the blood and explain the development of the so-called accessory hearts on the caudal vein of the eel and on the portal vein of *Myxine*.

The **urinary organs** of Fishes (fig. 596) consist of paired kidneys extending along the backbone from the head to the end of the body cavity, and giving off two ureters which unite into a common duct on which a bladder is usually developed. The urinary bladder and its duct always lie behind the intestinal canal. In most Teleosteans the efferent duct of the bladder opens by a common orifice with the sexual opening, or on a special papilla behind the sexual opening. In the *Plagiostomes* and *Dipnoi* on the other hand a cloaca is developed; in the former the ureters and the generative ducts open into the dilated terminal part of the intestine—*i.e.*, the cloaca—behind the rectum; while in the latter the ureters open into the cloaca separately on each side.

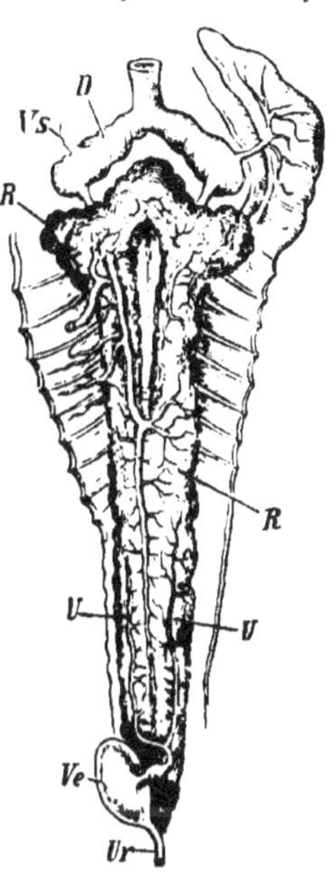

FIG. 596.—Kidneys of *Salmo fario* (after Hyrtl). *R*, kidneys; *U*, ureter; *Ve*, bladder-like dilation; *Ur*, efferent duct of bladder; *D*, ductus Cuvieri; *Vs*, subclavian vein.

Generative organs.—Excepting in certain forms, such as *Serranus* and *Chrysophrys*, which are hermaphrodite (also some carps), Fishes are of separate sexes; the two sexes often present more (*Macropodus*) or less (*Tinca*, *Cobitis*) considerable sexual differences. The male and female reproductive organs (fig. 591) often resemble one another so closely in form and position that it is necessary to investigate their contents in order to distinguish the sex, especially as external sexual differences are frequently absent.

The ovaries are paired (in the *Myxinoids* the *Squalidæ*, and certain Teleosteans, as *Perca*, *Blennius*, *Cobitis*, they are unpaired) elongated sacs, which lie ventral to the kidneys at the sides of the intestine and the liver. The ova originate on the internal transversely folded walls of the ovaries in closed follicles in which they receive a thick egg-capsule (with pores and micropyle), and escape thence into the cavity of the sac which becomes greatly swollen at the breeding time. The testes on the other hand are, except in the *Cyclostomes*, paired, and they are composed of transverse canals or vesicular cavities.

In the simplest case the testes and ovaries have no special ducts, but the genital products are dehisced from the wall of the gland into the body cavity, whence they pass out to the exterior through a genital pore situated behind the anus (in Cyclostomes, Eels, and female Salmon). As a rule, however, generative ducts are present; they may either be direct prolongations of the genital glands as in the Teleosteans, or as in the Ganoids, female Plagiostomes and Dipnoi independent canals which begin with a free funnel-shaped opening into the body cavity (Müllerian ducts). In the Teleosteans the two oviducts as well as the vasa deferentia unite to form an unpaired duct which opens to the exterior on the urogenital papilla between the openings of the anus and the urinary duct; in the Ganoids, on the other hand, as well as in the Plagiostomes and the Dipnoi a common cloaca is formed. Accessory external copulatory organs are only found in the male Plagiostomes, in the form of long grooved cartilaginous appendages of the ventral fins.

Most fishes are oviparous; only a few Teleosteans, as *Anableps*, *Zoarces*, the *Cyprinodonta*, etc., and a great number of the Sharks, bear living offspring, which for the most part undergo their embryonic development in a dilated part of the oviduct which serves as a uterus. Reproduction usually takes place only once in the year, most frequently in spring, more rarely in the summer, and exceptionally, as in many of the *Salmonidæ*, in winter. Many fishes, especially the males, undergo changes of colour and develop growths of skin at the spawning time. The two sexes often assemble in great shoals and seek out shallow places near the banks of rivers or near the sea coast (Herrings) for spawning. Some make more extended migrations and pass in great shoals over great distances along the sea coast (Tunny-Fish). Others leave the sea and pass up the mouths of rivers, and overcoming great obstacles (Salmon leaps) make their way up into the smaller streams in which they deposit their spawn in sheltered places where the food is plentiful (Salmon, Sturgeon, etc.). The Eels on the other hand migrate from the rivers into the sea, and in the following spring the young Eels enter the freshwater by millions and pass up the stream. The spawn is as a rule fertilized in the water, and thus artificial fertilization and pisciculture is rendered possible. In the viviparous fish, and in the Rays, *Chimæra*, and

Dog-fishes, which lay large eggs enclosed in a horny shell, a true copulation and an internal fertilization of the egg takes place. It is worthy of note that in a few exceptional cases the male undertakes the charge of the brood (*Hippocampus*, *Cottus*, *Gasterosteus*).

The **embryonic development** of the fishes is principally distinguished from that of the higher Vertebrates by the fact that neither *amnion* nor *allantois* are developed. Both the small eggs of the Teleosteans, which are provided with a micropyle, and the large eggs of the Plagiostomes, which are surrounded by a hard horny case, contain a large quantity of food yolk, and undergo a partial segmentation. The eggs of Amphioxus and of the Cyclostomes, however, undergo a total segmentation. As a rule the young fishes leave the egg-membranes tolerably early, with more or less distinct remains of the yolk-sac, which is by this time completely taken up into the interior of the body, but projects externally, like a hernia. Although the body-form of the just-hatched fish differs essentially from that of the adult animal, yet no true metamorphosis takes place save in a few exceptional cases.

Most fishes live in the sea, and the number of their species and genera increases as we approach the equator. But they are not all exclusively confined to fresh or salt water. Many, as the Plagiostomes, live almost entirely in the sea; others, as the *Cyprinoidei* and *Esocidæ*, are confined to fresh water, but there are also fish which periodically change their habitat, especially at spawning time. Some fish live in subterranean waters and are blind like the inhabitants of caves (*Amblyopsis spelæus*). Few fish are able to live any length of time out of water; as a rule the wider the gill-slits, the quicker does the fish die on dry land. Fishes with narrow gill-slits (Eels) possess an uncommon tenacity of life out of water. According to Hancock, a species of *Doras* migrates in great shoals over the surface of the ground from one piece of water to another. Except the *Dipnoi*, certain East Indian fresh-water fish, whose upper pharyngeal bones are hollowed out into the form of a labryinth and form a multicellular reservoir for water, are capable of living the longest time out of water (*Anabas scandens*). There are even fishes which can fly (*Exocætus*, *Dactylopterus*).

Fishes are of great importance to our knowledge of the development of animal life on the earth owing to the frequent appearance of their fossil remains in all geological periods. In the palæozoic formations very singular fish-forms, as the *Cephalaspidæ* (*Cepalaspis*, *Coccosteus*, *Pterichthys*), constitute the oldest representatives of the

Vertebrata. From the palæozoic formations to the chalk we find almost exclusively cartilaginous fishes and Ganoids, amongst which the forms with persistent notochord and cartilaginous skull predominate. Ganoids, with a fully-developed bony skeleton, round scales and an externally homocercal caudal fin, appear for the first time in the Jura, where we also find the first Teleosteans. From the chalk onwards, in the more recent formations, the Teleosteans increase in number and variety of forms the nearer we approach to the fauna of the present time.

Order 1.—Leptocardii * (Acrania).

Lanceolate Fishes without paired fins. The notochord is persistent; there is no skull-capsule. The blood is colourless, and there are pulsating vascular trunks.

The body of *Amphioxus* (which was taken by Pallas for a slug) is about two inches long. It is shaped like a lancet, and is provided with dorsal and anal fin-like folds, which, however, are without rays, and are continued into the lancet-shaped caudal fin. In the place of the vertebral column the strong notochord persists; on the dorsal side of this is the spinal cord, the slightly swollen anterior extremity of which represents the rudiment of the brain. There is no capsule corresponding to the skull. There is a rudimentary eye, consisting of an unpaired pigment spot, situated at the anterior end of the central nervous system in the nervous tissue; also a small olfactory pit placed on the left side. There is no auditory organ.

The mouth, which is without jaws, is a long slit supported by a jointed horse-shoe-shaped cartilage, bearing ciliated cirri. It leads into a long and spacious sac (pharynx), which is pierced by a number of lateral slits, and serves the function of respiration. At the entrance of the pharynx there are two folds, and on either side three finger-shaped ciliated projections. The walls on each side are supported by

* Joh. Müller, "Ueber den Bau und die Lebenserscheinungen des Branchiostoma lubricum (Amphioxus lanceolatus)." *Abhandl. der Berliner Akad.*, 1842.

Kowalevski, "Entwickelungsgeschichte von Amphioxus lanceolatus." St. Petersburg. 1867.

Kowalevski, "Weitere Studien, etc." *Arch. für mikr. Anatomie*, Tom. XIII.

W. Rolph. "Untersuchungen über den Bau des Amphioxus lanceolatus." *Morph. Jahr.*, Tom. II., 1876.

P. Langerhans, "Zur Anatomie des Amphioxus lanceolatus." *Arch. für mikrosk. Anatomie*, Tom. XII.

B. Hatschek. "Studien über die Entwickelung des Amphioxus." *Arbeiten aus dem Zool. Institute in Wien*, Tom. IV., 1881.

obliquely directed rods, and form over the rods leaf-shaped, inwardly projecting branchial folds. Between the latter there are slit-like openings for the outflow of the water, which passes into a superficial cavity—the *atrial cavity*—produced secondarily by the growing over of a fold of the integument and opening to the exterior by a pore—the *atrial pore*—on the ventral side. The intestine begins at the posterior end of this branchio-pharyngeal sac, and passes in a straight course as far as the tail, where it opens by a somewhat laterally-placed anus. The intestinal tube is divided into two regions, of which the anterior receives a cæcal hepatic sac, which extends forwards on the left side of the pharynx.

The **vascular system** is without an independent heart, but in its place the principal vessels pulsate. The arrangement of the vessels permits of comparison with the vascular apparatus of the Invertebrata (Annelids), and at the same time it represents, in the simplest form, the arrangement typical of Vertebrates. A longitudinal trunk running beneath the respiratory sac gives off numerous vessels, which are contractile at their origin, to the gills. The anterior pair of these branchial arteries forms a contractile vascular arch placed behind the mouth, the two parts of which unite beneath the notochord to form the aorta, which receives the next following branchial arteries. The venous blood returning from the organs is collected in a vessel placed above the hepatic cæcum; this vessel becomes the subpharyngeal longitudinal trunk. The blood returned from the intestinal canal is collected in a vessel—the hepatic vein—which breaks up into fine branches on the hepatic cæcum. A second contractile vessel (vena cava) receives the blood from these branches, and conducts it back into the subpharyngeal longitudinal trunk. The blood corpuscles are colourless.

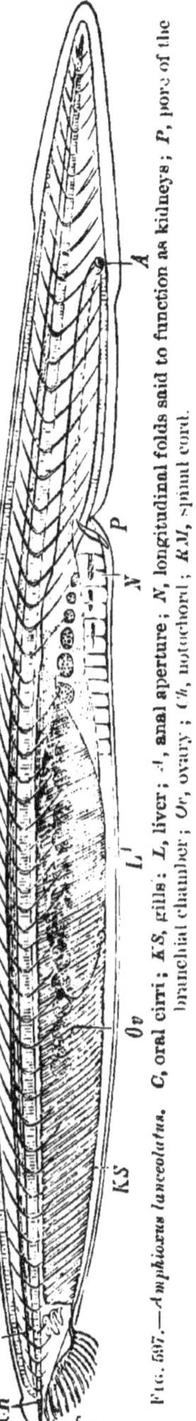

FIG. 597.—*Amphioxus lanceolatus.* **C, oral cirri; *KS*, gills; *L*, liver; *A*, anal aperture; *N*, longitudinal folds said to function as kidneys; *P*,** pore of the branchial chamber; *Ov*, ovary; *Ch*, notochord; *RM*, spinal cord.

Generative Organs.—The animals are diœcious. The ovaries and testes resemble each other externally, and consist of a series of paired bodies. They are arranged segmentally (in prolongations of the body cavity), one pair being found in each segment over the

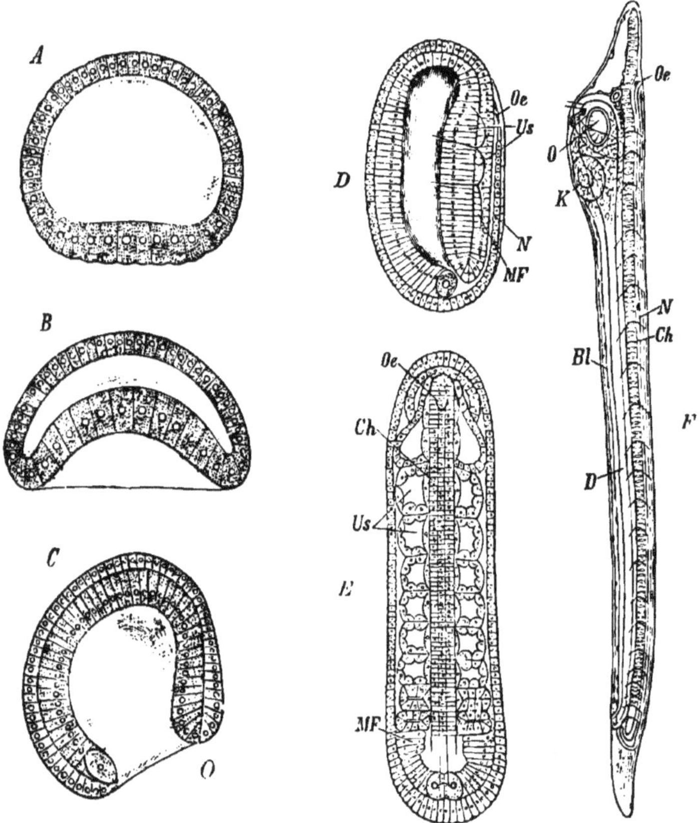

FIG. 598.—Development of *Amphioxus* (after B. Hatschek). *A*, Blastosphere. *B*, commencing invagination of the entoderm (gastrula). *C*, Later gastrula, the cilia of the ectoderm cells are not represented. *D*, Stage with two somites (primitive segments), seen in optical longitudinal section. *US*, Primitive segments or somites; *MF*, mesoderm folds, *N*, medullary canal; *Oe*, external opening of the latter. *E*, Stage with nine somites seen from the dorsal surface to shew the asymmetry of the somites, the notochord (*Ch*) is shown in section. *F*, Larva with mouth (*O*) and first gill slit (*K*) seen from the left side; *D*, intestine; *Bl*, ventral blood-vessel.

greater part of the length of the branchial sac (fig. 597, *Ov*). The generative products are dehisced into the atrial cavity, and pass thence through the pharynx and mouth to the exterior.

For a short distance in front of the atrial pore the epithelium of the ventral wall of the atrial cavity is thrown into a number of peculiar longitudinal folds which have been interpreted as **kidneys.**

Development.—The eggs undergo a total segmentation. The cells resulting from segmentation form a blastosphere, which by invagination is transformed into a ciliated gastrula larva (fig. 598, *A, B, C*). The mesoderm is developed from lateral folds of the entoderm, and at once segments into somites; and at the same time the medullary canal, which communicates with the alimentary canal behind and opens freely to the exterior in front (fig. 598, *D*), is formed from the ectoderm. Soon after the notochord arises from the endoderm. The changes, which take place in the larval life, are introduced by a considerable elongation of the body. In the further development the larva is remarkable for a striking asymmetry (of somites, mouth, anterior gill-slit, anus, olfactory organ). The branchial apparatus, which is at first free, is afterwards covered by a reduplication of the skin (formation of the atrial or peribranchial cavity).

The only genus of the *Leptocardii* is *Amphioxus* Yarrel (*Branchiostoma* Costa) including a single species distributed on the sandy coasts of the North Sea, of the Mediterranean, and of South America. *A. lanceolatus* Yarrel. Lancelet. The forms described as *A. Belcheri* Gray, from the Indian Ocean, and *A. elongatus* Sundev. probably belong to the same species.

Order 2.—CYCLOSTOMI* (MARSIPOBRANCHII).

Vermiform Fishes without pectoral or pelvic fins; with cartilaginous skeleton and persistent notochord. There are six or seven pairs of pouch-like gills. The olfactory fossa is unpaired, and the circular or semicircular suctorial mouth is without jaws.

FIG. 599.—*Myxine glutinosa* (règne animal).

The *Cyclostomi* have a cylindrical vermiform shape (fig. 599), and

* Joh. Müller. "Vergleichende Anatomie der Myxinoiden." Berlin. 1835-45.
Aug. Müller. "Ueber die Entwickelung der Neunaugen." *Müller's Archiv.*, 1856.
Max Schultze, "Die Entwickelungsgeschichte von P. Planeri." Haarlem, 1856.
P. Langerhans. "Untersuchungen über Petromyzon Planeri." Freiburg. 1873.
W. Müller. "Ueber das Urogenitalsystem des Amphioxus und der Cyclostomen." *Jen. naturwiss. Zeitschr.*, Tom. IX., 1875.
A. Schneider. "Beiträge zur vergleichenden Anatomie und Entwickelungsgeschichte der Wirbelthiere." Berlin, 1879.
Calberla, "Zur Entwickelung des Medullarrohrs und der Chorda dorsalis der Teleostier und der Petromyzonten." *Morphol. Jahrb.*, Tom. III., 1877.

their skin is without scales. They have no paired fins but the system of vertical fins is developed over the whole length of the dorsal surface and of the tail, and is usually supported by cartilaginous rays. The skeleton is confined to a cartilaginous rudiment of the vertebral column and skull. The notochord persists as the axial skeleton: its sheath presents traces of segmentation in the presence of rudimentary cartilaginous neural arches (fig. 600, *b*), and in the caudal region (*Petromyzon*) of the lower vertebral arches also.

At the anterior end of the notochord there is a cartilagino-membranous cranial capsule enclosing the brain. It has a bony basal region and lateral cartilaginous vesicles in which the auditory organs are enclosed (fig. 600). In place of the visceral skeleton there are cartilaginous pieces surrounding the palate and pharynx, various labial cartilages and a complicated frame work of cartilaginous rods, which form the so-called branchial basket round the branchial sacs, and are in part attached to the vertebral column.

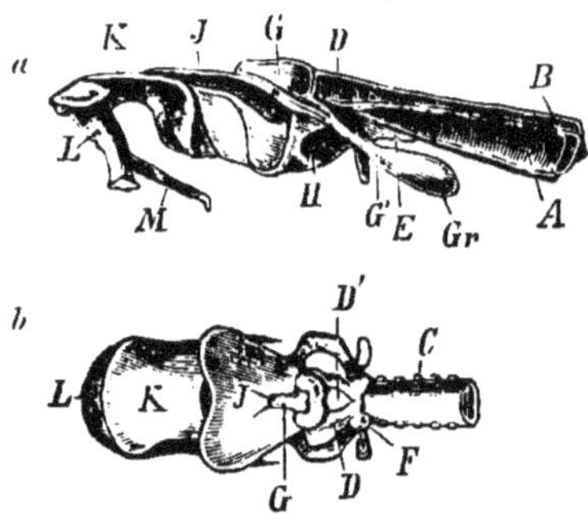

FIG. 600.—Skull and beginning of the vertebral column of *Petromyzon marinus* (after Joh. Müller). *a*, In longitudinal vertical section. *b*, Seen from above. *A*, notochord; *B*, neural canal; *C*, rudimentary vertebral arches; *D*, cartilaginous part, and *D'*, membranous part of the cranial roof; *E*, base of skull; *F*, auditory capsule; *G*, nasal capsule; *G'*, naso-palatine duct; *Gr*, blind end of *G'*; *H*, process of the bony palate; *J*, posterior plate covering the mouth; *K*, anterior plate covering the mouth; *L*, labial ring; *M*, styliform appendage of *L*.

The *Cyclostomi* possess a **brain** of the piscine type with three principal sense nerves and a reduced number of spinal-like nerves. Two **eyes** are always present, but they may be hidden under the skin or even covered by muscles (*Myxine*, larva of *Petromyzon*). The **olfactory organ** is an unpaired sac opening in the median line between the eyes. In the Myxinoids the olfactory capsule has in addition a posterior opening which pierces the palate and can be closed by a valvular apparatus. This communication between the nasal and pharyngeal cavities serves for the introduction of water into the branchial sac; for the mouth when performing its function as a suctorial organ is closed so far as the passage of water is concerned. The **auditory organ** is reduced to a simple membranous labyrinth which consists of the vestibulum and one or two semicircular canals.

Alimentary canal.—The mouth, which is surrounded by fleshy lips and often by filamentous processes, is circular in shape, though the lips can be applied together so as to form a median longitudinal slit. It leads into a funnel-shaped buccal cavity, which is without jaws and is armed on the soft palate as well as on the floor with horny teeth (fig. 601). At the bottom of the funnel is the tongue, which, moving up and down like a piston, enables the animal to attach itself by its mouth as by a sucker. The pharynx, which follows the mouth, communicates with the branchial sacs either directly or by a special passage (*Petromyzon*). The intestinal canal passes straight to the rectum and is divided into stomach and intestine by a narrow region, the walls of which project so as to form a sort of valve. The liver is always well developed, but there is no swimming bladder.

The **gills** (fig. 592) lie at the sides of the œsophagus in six or seven pairs of branchial sacs. These open on either side by external branchial passages into the same number of separate respiratory apertures. In Myxine on the other hand there is on each side, almost on the ventral surface, only one opening, into which all the external branchial passages of the same side open.

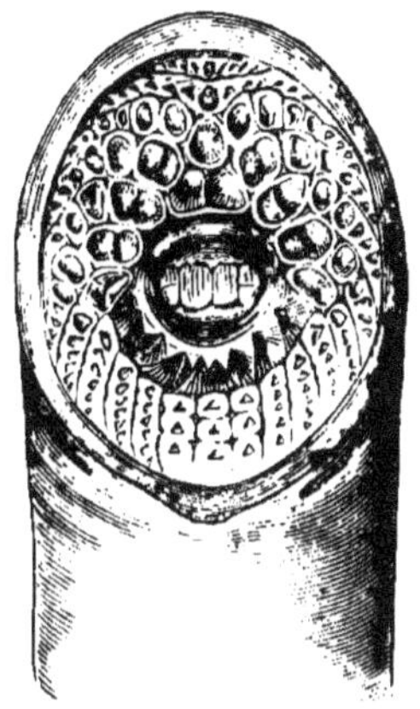

FIG. 601.—Head of *Petromyzon marinus*, seen from below showing the horny teeth of the buccal cavity (after Heckel and Kner).

On the other side the sacs communicate with the œsophagus, but, except in *Ammocœtes*, never directly by simple openings but by internal branchial passages or, as in *Petromyzon*, by a common passage lying beneath the œsophagus into which passage all the other branchial passages open. The water flows in from the exterior through the external branchial openings or in Myxine through the nasal passage, and is driven by the contraction of the constrictor muscles of the branchial sacs either out by the same way (*Petromyzon*), or into the œsophagus, and from this to the exterior through a special unpaired canal on the left side.

The **heart** lies beneath and behind the branchial skeleton. Some of the vascular trunks pulsate, *e.g.*, the portal vein in *Myxine*. The aortic bulb has no muscular layer, and contains, as in the Teleosteans, only two valves.

The **urinary** and **genital** organs are of simple structure. In *Myxine* the kidneys retain the primitive segmental arrangement, there being a urinary tubule and Malpighian body in every segment. In *Myxine* the urinary ducts open with the genital pore, in *Petromyzon* into the intestine. In front of the kidneys, in the region of the heart, there is another part of the kidney which in the adult animal is no longer functional. This is the head-kidney or pronephros (Nebenniere of Joh. Müller). It consists of a number of glandular ducts, which begin with funnel-shaped openings into the body cavity (pericardial cavity), and in the young animal open into the urinary duct.

The genital glands are unpaired in both sexes. In *Myxine* they lie on the right side; in *Petromyzon* in the middle line. They never possess ducts, but the eggs and spermatozoa are at the breeding time dehisced into the body cavity, whence they pass out through a pair of genital pores placed behind the anus.

The *Petromyzontidæ* undergo a kind of metamorphosis, which was

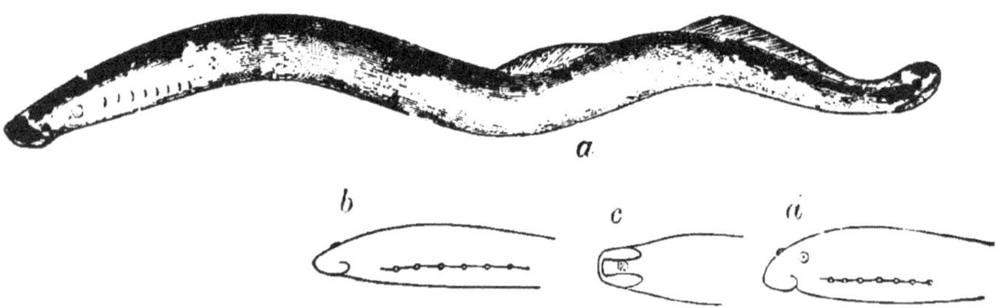

FIG. 602.—*a*, *Petromyzon fluviatilis* (after Heckel and Kner). *b*, *c*, *d*, stages in the transformation of *Ammocœtes branchialis* into *Petromyzon Planeri* (after v. Siebold). *b*, Head of an eyeless larva seen from the side; *c*, the same seen from underneath; *d*, later stage with small eyes, seen from the side.

discovered two hundred years ago by Baldner, a fisherman of Strasburg, but has only recently been rediscovered by Aug. Müller. The young larvæ (fig. 602, *b*, *c*, *d*) are blind and without teeth. They possess a small mouth, surrounded by a horseshoe-shaped upper lip, and were for a long time placed in a special genus—*Ammocœtes.*

The Cyclostomes live partly in the sea; they ascend rivers at spawning time, sometimes carried by the Salmon and Shad (*Alausa vulgaris*), and deposit their eggs in holes in the river-bed. Others are river-fish. They attach themselves to stones and to dead and living fish, which latter they may in this way kill. They also eat worms and small aquatic animals. The genus *Myxine* is exclusively parasitic on other fish and even makes its way into their body cavity, thus affording an example of an endoparasitic Vertebrate.

Fam. **Myxinoidæ** (Hags). Head obliquely truncated; suctorial mouth without lips, and surrounded by labial processes; eyes hidden beneath the skin. There is an opening from the nasal cavity into the mouth through the posterior part of the palate. The branchial pouches open to the exterior either by a common ventral aperture on each side (*Myxine*, *Gastrobranchus*) or by seven apertures on each side or asymmetrically by six apertures on one side and seven on the other (*Bdellostoma*). Marine. *Myxine glutinosa* L. (fig. 599). *Bdellostoma heptatrema* Joh. Müll., found at the Cape.

Fam. **Petromyzontidæ**. Lampreys—Nine-eyes. With seven external gill slits on each side of the neck, and a common internal branchial passage which opens anteriorly into the pharynx. The nasal cavity ends blindly. The round mouth, without labial processes, with fleshy lips, which can be approached so as to leave a slit-like opening. *Petromyzon marinus* L. Lamprey, two feet in length, ascends rivers with the Shad, in the spring, to spawn. *P. fluviatilis* L., River Nine-eye (fig. 602 *a*). *P. Planeri* Bloch, small river Nine-eye, with *Ammocœtes branchialis* as larva. It attains a length of 5 to 6 inches.

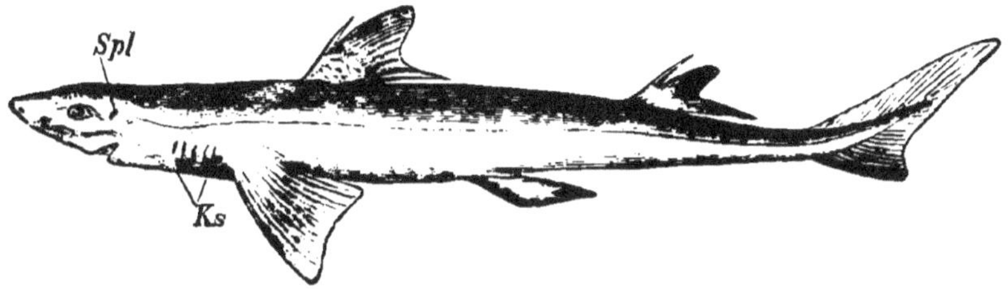

FIG. 603.—*Acanthias vulgaris*. *Spl*, spiracle; *Ks*, gill slits.

Order 3.—SELACHII* (CHONDROPTERYGII).

Cartilaginous Fishes with large pectoral and pelvic fins, with transverse ventrally-placed mouth, usually with five (rarely six or seven) pairs of branchial pouches and branchial slits. They have a muscular conus arteriosus which contains several rows of valves. The intestine has a spiral valve.

The Selachians differ strikingly in their outward appearance from all other Fishes (fig. 603), and present even among themselves great variations. The form and position of the mouth, which is a broad transverse slit placed on the under surface of the snout, is an important distinguishing character. The skin usually contains a number of bony granules (ossified dermal papillæ, *placoid* scales), and obtains

* Compare Joh. Müller and J. Henle, "Systematische Beschreibung der Plagiostomen," mit 60 Steindrucktafeln. Berlin. 1841.
Fr. Leydig, "Beiträge zur mikroskopischen Anatomie und Entwickelungsgeschichte der Rochen und Haie." Leipzig. 1852.
C. Gegenbaur, "Untersuchungen zur vergleichenden Anatomie der Wirbelthiere." Leipzig. 1872.
F. M. Balfour, "A monograph on the development of Elasmobranch Fishes." London. 1878.
C. Hasse, "Das natürliche System der Elasmobranchier." Jena, 1879.

thereby a rough shagreen-like surface. Sometimes, especially on the tail (*Raiidæ*), there are larger bony plates, arranged in rows and provided with pointed spinous processes, which serve for protection (*ichthyodorulites*). All the Selachians have large pectoral and pelvic fins. The former are attached by a cartilaginous shoulder-girdle to the posterior part of the skull, or to the anterior region of the vertebral column; they are either sharply marked off and have an almost vertical position on the anterior part of the body (*Chimæra* and *Squalidæ*), or they have the form of very large, horizontally-placed lateral expansions of the body (*Rays*). In the latter case they reach by means of the so-called cranial fin cartilages to the anterior end of the snout, and lean by posterior suspensors on the pelvic framework of the ventral fins: the latter are always placed near the anus, and in the male bear peculiar grooved cartilaginous appendages, which are the accessory copulatory organs (*claspers*). The unpaired fins also may be well developed, and, as their number and position varies in the different forms, they may be of systematic importance. A sharp bony spine is sometimes present in front of the dorsal fins, or completely isolated on the dorsal surface of the tail (*Trygon*), and this as well as the spinous and hooked processes of the dermal bony plates serve as a weapon of defence. The caudal fin is always markedly heterocercal externally.

The skull is an undivided, cartilaginous capsule, the base of which sometimes is articulated to the vertebral column (*Chimæra* and *Raiidæ*), while sometimes it is excavated like the body of a vertebra (fig. 571). On the facial region the cartilaginous mandibular arch persists, and is attached to the auditory region of its skull by the *hyomandibular*. The palatoquadrate bar is moveably connected to the cranial capsule (except in *Chimæra*). The palatoquadrate and the lower jaw are always cartilaginous, and as a rule are abundantly furnished with teeth. The vertebral column with its remains of the notochord is also principally cartilaginous, but separate biconcave vertebræ are developed, the form of which offers numerous variations.

In all cases there are dorsal and ventral arches, which sometimes remain separate and sometimes fuse with the vertebral bodies. Ribs only appear as cartilaginous rudiments.

In the structure of the **gills** (fig. 593), the Selachians differ essentially from the Teleosteans in possessing five branchial pouches on either side; the branchial lamellæ are attached in their whole length to the partition walls, which are supported by the lateral cartilaginous rays of the branchial arches. The branchial pouches are placed relatively far back, and each of them has a separate external opening. These openings are in the *Squalidæ* on the sides, in the *Raiidæ* on the ventral surface of the body. In the *Chimæridæ* the branchial pouches open on either side into a common gill-slit, over which a cutaneous fold, arising from the suspensorium of the jaw and serving as a branchial operculum, is spread.

The **dentition** presents many variations. Sometimes (*Hexanchus*, *Acanthias*) the whole of the buccal cavity as far as the entrance to the œsophagus is covered with small teeth of the mucous membrane (*placoid scales* *); sometimes there are larger teeth, which also always belong to the mucous membrane, and are arranged in rows on the rounded edge of the jaw in such a manner that the younger posterior rows of teeth have their points turned inwards, while the teeth of the anterior rows, which are older and more or less worn, have their points turned upwards and outwards.

In the *Squalides*, dagger-shaped or saw-shaped serrated teeth preponderate, while conical or flat pavement-like molar teeth are characteristic of the greater number of *Raiides*.

Spiracles are frequently present on the upper surface of the head behind the eyes: they are used for the expulsion of the water from the pharyngeal cavity. The **digestive canal** is dilated to a spacious stomach, but is relatively short; the small intestine is furnished with a spirally coiled fold of the mucous membrane—the so-called spiral valve—which considerably increases the extent of the absorbing surface. A swimming bladder is always absent, though the rudiment of it is often discernible.

The **heart** * has a muscular *conus arteriosus*; it contains two to five rows of valves, and represents a part of the ventricle which has become independent.

In the structure of the **brain** and of the **sense organs**, the Selachians hold the highest place amongst the fishes (fig. 588). The hemispheres are of relatively considerable size, present longitudinal and transverse impressions, and traces of convolutions on their surface. The cerebellum, also, may be so well developed that the fourth ventricle is almost entirely covered by it. The two optic nerves always form a chiasma and some of their fibres cross. The eyes in the *Squalides* are not only protected by free lids, but often also by a moveable nictitating membrane.

* O. Hertwig. *Jen. naturwiss. Zeitschr.* Tom. VIII., 1874.

* C. Gegenbaur, "Zur vergleichenden Anatomie des Herzens." *Jen. naturwiss. Zeitschr.* Tom. II.

The **urinary organs** of the *Plagiostomi* are paired kidneys, which sometimes retain the ciliated funnels (*nephrostomata*).

The sexes can be easily distinguished by the form of the pelvic fins. A true copulation always takes place. The female genital organs consist of a large, single or double ovary and paired glandular oviducts, which are separate from the ovaries and begin with a common funnel-shaped ostium, and in their further course each of them possesses a uterus-like dilatation. The two oviducts open by a common aperture (in the *Chimæridæ* only by separate orifices) into the cloaca. The ova have a large amount of food-yolk, and are enclosed by a mass of albumen, and sometimes by a thin membranous folded chorion, sometimes by a tough, parchment-like, flat shell, which is prolonged into four horns, or into twisted strings, which serve to attach it to marine plants. In the latter case the eggs are laid (the true Rays and Dogfish): in the former (electric Rays and viviparous *Squalides*), on the other hand, they develop in the uterus.

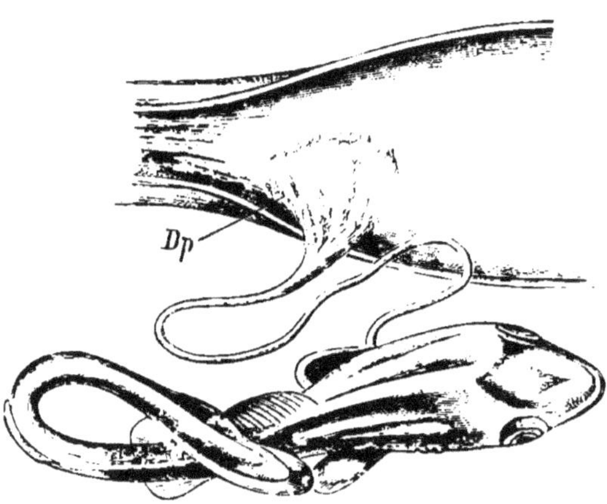

FIG. 604.—Embryo of *Mustelus lævis*, connected with the uterus by the umbilical placenta (*Dp*) (after Joh. Müller).

In this case the eggs are closely applied to the walls of the uterus during the development, the folds of the chorion interlocking with the ridges of the uterine walls. Thus the addition of nutriment is rendered possible. Sometimes the connection between the mother and the embryo is more intimate, and is effected by means of a true umbilical placenta, which was known to Aristotle in *Mustelus lævis* (fig. 604). As Joh. Müller * has shown, the long-stalked yolk-sac of the embryos of *Mustelis lævis* and species of *Carcharias* develops a great number of villi, which are covered by the delicate egg membrane, and like the cotyledons of Ruminants fit into corresponding depressions in the uterine mucous membrane. In other respects,

* Compare Joh. Müller, "Ueber den glatten Hai des Aristoteles." *Abhandl. der Berliner Akad.*, 1840.

also, the embryos of the Plagiostomes exhibit notable peculiarities, especially in the possession of external branchial filaments (fig. 605), which are lost long before birth.

Almost all the Plagiostomes are marine; only a few of them are found in the larger rivers of America and India. They are all carnivorous, and feed on large fishes, or crustacea and mollusca. Some few (*Torpedo*) possess an electric organ.

With the exception of *Pleuracanthus*, remains of spines and teeth only are preserved in the Palaeozoic formations. From the secondary period onwards the remains are more complete and numerous.

Sub-order 1. **Holocephali.**

Selachians with maxillo-palatine apparatus firmly fused to the skull, with single external gill slit on each side and small opercular membrane.

The thick strangely formed head is provided with large eyes which

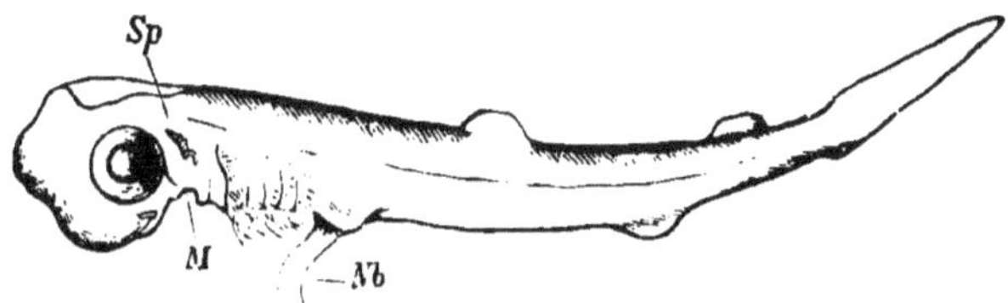

Fig. 605.—Embryo of *Acanthias* with external gills. *Sp*, spiracle; *M*, mouth; *Nb*, stalk of yolk-sac.

are without lids. The mouth is small and lies on the under surface of the snout. The maxillo-palatine (palato-quadrate) bar is firmly fused with the skull, while the lower jaw articulates with a styliform process of the skull (hyomandibular). The mandible has but few teeth (four above, two below). The naked skin is traversed by the large passages of the lateral sense organs. There are no spiracles. The vertebral bodies are replaced by thin calcareous annular incrustations in the sheath of the notochord. They lay eggs with horny shells.

Fam. **Chimæridæ** (Sea-cats). *Chimæra monstrosa* L. (fig. 606), Northern Seas and Mediterranean: *Callorhynchus antarcticus* Lac., Cape and Pacific.

Sub-order 2. **Plagiostomi.**

Selachians with wide transverse mouth, which is placed far back, separate vertebral bodies, and a more or less reduced notochord. There are five (exceptionally six or seven) external gill slits on each side.

The nasal apertures are placed on the under surface of the snout, a little in front of the transversely arched mouth. The skin is rarely naked; it is usually shagreen-like in consequence of the osseous bodies which are embedded in it, or it may also be covered with osseous plates and scutes. The palato-quadrate bar is moveable and is separate from the cartilaginous cranial capsule.

Tribe 1. **Squalides** (Sharks). Spindle-shaped Plagiostomes, with lateral gill slits; eyelids with free edges; incomplete shoulder girdle, without cranial fin cartilages.

The body is spindle-shaped, carries the pectoral fins more or less vertically, and ends with a powerful tail, which is bent dorsalwards at the end. There are, however, forms which, with regard to their body shape, are allied to the Rays, and constitute forms of transition to the latter group, *e.g.*, the genus *Squatina*. The teeth are usually pointed and dagger-shaped, and placed in numerous rows. The

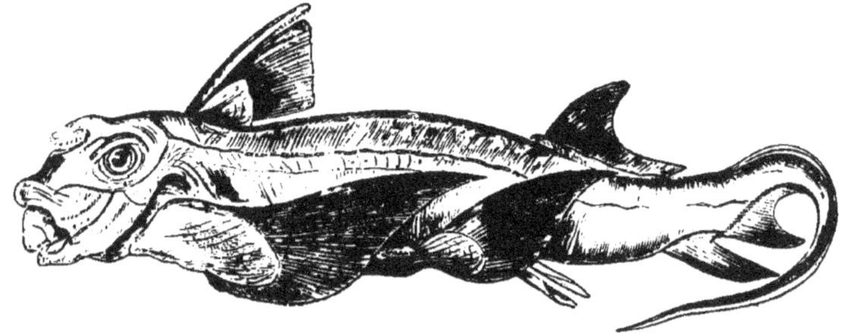

Fig. 606.—*Chimæra monstrosa* (règne animal).

families are distinguished principally by the number and position of the fins, by the presence or absence of spiracles and of a nictitating membrane, and also by the form and structure of the teeth.

Fam. **Scyllidæ** (Dog-fishes). *Scyllium canicula* L., the coasts of Europe.

Fam. **Cestraciontidæ.** *Cestracion Philippi* Blainv.

Fam. **Lamnidæ** (Porbeagles). *Lamna glauca* Müll., Henle; *Selache maxima* Gunn., reaches a length of thirty-two feet.

Fam. **Carchariidæ.** *Carcharias glaucus* Rond, the Blue Shark, with umbilical placenta. *C. lamia* Risso. These two last are found in the Mediterranean and the Ocean. *Zygæna malleus* Risso, the Hammer-headed shark.

Fam. **Galeidæ** (Topes). *Galeus canis* Rond., European seas; *Mustelus vulgaris* and *lævis* Rond., with umbilical placenta, both are found in the Mediterranean.

Fam. **Notidanidæ.** *Notidanus* (*Hexanchus*) *griseus* Gm. and *N.* (*Heptanchus*) *cinereus* Gm., Mediterranean and Ocean.

Fam. **Spinacidæ** (Spiny Dog-fishes). *Acanthias vulgaris* Risso (fig. 603), found from the northern seas to the South Sea.

Fam. **Squatinidæ** (Angel- or Monk-fishes). *Squatina vulgaris* Risso (*Squalus squatina* L.) European seas.

Tribe 2. **Rajides** (Skates and Rays). Plagiostomes, with flat bodies; with five gill slits opening on the ventral surface internal to the pectoral fins; with complete pectoral girdle and cranial fin cartilages, without anal fins.

In consequence of the size and horizontal expansion of the thoracic fins the flat body presents the form of a large disc, prolonged behind into the long thin tail, which is frequently armed with spines, rarely with one or two serrated stings. The mandibles are short and stout, and are furnished with teeth which may be either small and conical, and arranged near one another in rows, or broad and plate-like. The Rays live for the most part at the bottom of the sea, and feed principally on Crustaceans and Molluscs. The Torpedos have an electrical apparatus between the fin cartilages and the branchial pouches. By means of this organ (fig. 590) they can stun even larger fishes. Many Rays reach the considerable size of ten to twelve feet.

Fam. **Squatinorajidæ.** *Pristis antiquorum* Lath. Sawfish. Ocean and Mediterranean: *Rhinobatus granulatus* Cuv.

Fam. **Torpedidæ.** Electric Rays. *Torpedo marmorata* Risso, Mediterranean and Ocean: *Narcine brasiliensis* v. Ott.

Fam. **Rajidæ.** Skates and Rays. *Raja clavata* L.; *R. miraletus* L.

Fam. **Trygonidæ.** Sting Rays. *Trygon pastinaca* L. (*Pastinaca marina* Bell). Atlantic Ocean.

Fam. **Myliobatidæ.** Eagle Rays or Sea Devils. *Myliobatis aquilla* L., Mediterranean.

Order 4.—Ganoidei.*

Cartilaginous and bony Fishes, with enamelled scales, or with osseous dermal plates and fulcra, with muscular conus arteriosus containing rows of valves; with comb-shaped gills and spiral valve in the intestine.

In former periods of the world's history this order was richly and variously represented (*Sauroidæ, Lepidoidæ, Pycnodonta*), while at the present day it contains only a few forms (*Lepidosteus, Polypterus, Calamoichthys, Amia, Acipenser, Scaphirhynchus, Spatularia*). It is difficult to establish the limit towards the Teleosteans, since there is

* Joh. Müller, "Ueber den Bau und die Grenzen der Ganoiden." *Abhandl. der Berliner Akad.*, 1846.

J. Hyrtl, "Ueber den Zusammenhang der Geschlechts-und Harnwerkzeuge bei den Ganoiden. *Denkschr. der k. Akad. der Wissensch.*, Tom. VIII., Wien, 1854.

Lütken, "Ueber die Begrenzung und Eintheilung der Ganoiden." Uebersetzt von Willemoes-Suhm. Palæontographica. 1872.

no single differential character common to all the Ganoids (even the spiral valve of the intestine is rudimentary in *Amia* and *Lepidosteus*).

The scales from which the name of the order is derived are for the most part of a rhomboidal form, and are always covered with a smooth layer of enamel. They are connected together by articular processes, and encircle the body in obliquely directed rings (fig. 607).

As regards the structure of the skeleton, the Ganoids are partly cartilaginous and partly bony fishes. Both among the fossil Ganoids and those living at the present time (Sturgeon) there are forms which, by the persistence of the notochord and the formation of bony arches, are allied to the *Chimaeridae*. The cartilaginous cranial capsule is always covered with external membrane bones, and the mandibular suspensorium, the jaws, the branchial arches, and the operculum possess a bony consistency. In the so-called bony Ganoids, the primordial cranium is more or less completely replaced by a bony skull, and the vertebral column gradually becomes bony, inasmuch as the vertebræ acquire, through various intermediate steps,

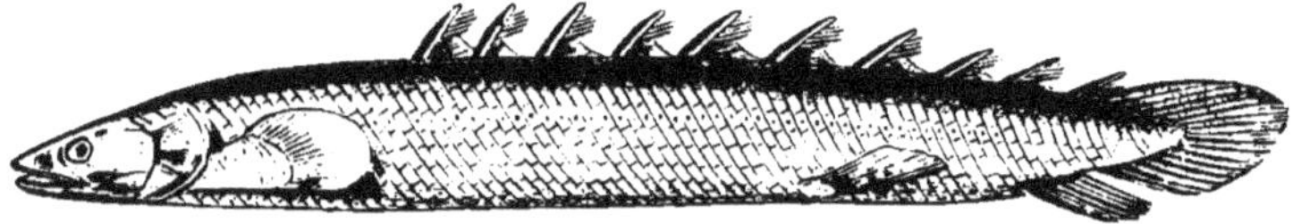

FIG. 607.—*Polypterus bichir*.

the biconcave form of the Teleostean vertebræ, and in *Lepidosteus* reach a phase of development in which, by the presence of an anterior articulating head, they resemble the opisthocœlous vertebræ of Amphibia. Bony ribs, also, are fairly frequently present.

The caudal fin is usually heterocercal, and the end of the vertebral column is sometimes continued into its superior lobe: there are, however, forms which are transitional in this respect, and lead to the homocereal (*diphycercal*) form. The spine-like splints known as *fulcra*, which are arranged in a single or double row on the upper edge and the first ray of the fins, particularly the caudal fin, are peculiar to Ganoids. ("Every fish with fulcra on the anterior edge of one or more fins is a Ganoid."—Joh. Müller.)

Anatomically the Ganoids present many points of resemblance to the Selachians. The anterior region of the ventricle is separated off as a rhythmically contractile *conus arteriosus*, and contains several longitudinal rows of valves, which extend as far as the anterior limit of the muscular investment, and prevent the blood flowing back from the artery into the conus during the diastole. The comb-shaped gills, on the other hand, lie, as in the Teleosteans, freely in a branchial cavity beneath a branchial operculum, to which a large gill containing venous blood is often attached. This respiratory accessory gill (opercular gill) is wanting in *Amia* and *Spatularia*, and must be distinguished from the pseudobranch of the spiracle, which may be present together with it.

All the Ganoids possess a swimming bladder with a *ductus pneumaticus* and two peritoneal canals (*abdominal pores*), which open at the sides of the anus (as in *Chimæra* and *Plagiostomi*). The optic nerves do not simply cross over one another, but form a chiasma with partial exchange of the fibres. The generative organs present many noteworthy peculiarities. There are two ovaries and the ripe eggs escape into the abdominal cavity. Thence they pass into an oviduct [Müllerian duct] which begins with a funnel-shaped opening into the body cavity and opens behind into the urinary duct or into the corresponding cornu of the urinary bladder (*Spatularia, Lepidosteus*), or unites with the oviduct of the opposite side and opens behind the anus by a single genital pore into which the short urethra also opens. (Hyrtl.) In the two first cases a urogenital canal leads from the bladder to a urogenital pore placed behind the anus. In the male it is remarkable that the same abdominal funnels [Müllerian ducts] also function as seminal ducts. [It has been shown by Balfour and Parker (Structure and Development of Lepidosteus, Phil. Trans., 1882), that in *Lepidosteus* at any rate the testis is connected with the Wolffian body by a testicular network.]

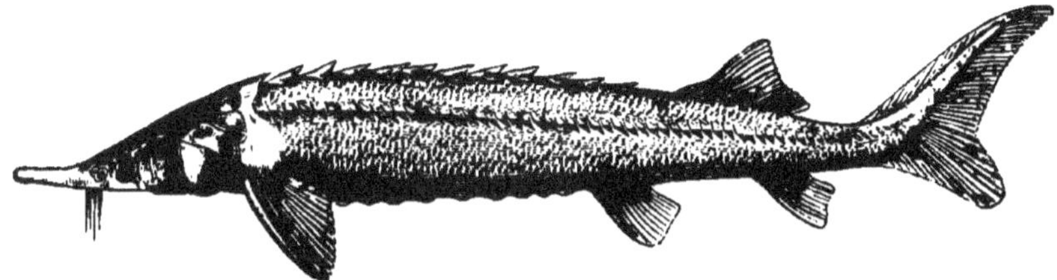

FIG. 608.—*Acipenser ruthenus* (after Heckel and Kner).

Tribe 1. **Chondrostei.** Cartilaginous Ganoids with persistent notochord. Branchiostegal rays scanty or absent. Caudal fin heterocercal, with fulcra. Cranium cartilaginous, covered by dermal bones. The teeth are small or altogether absent. The skin is naked or has osseous plates instead of scales.

Fam. **Acipenseridæ** (Sturgeons). *Acipenser sturio* L., Sturgeon; *A. ruthenus* L., Sterlet (fig. 608); *A. huso* L. (Hausen). *Scaphirhynchus cataphractus* Gray, Mississippi.

Fam. **Spatularidæ** (Löffelstöre). *Spatularia* [*Polyodon*] *folium* Lac., Mississippi: *Sp. gladius* Martens, Yantsekiang.

Tribe 2. **Crossopterygii.** Ganoids with two broad jugular plates instead of the branchiostegal rays, and usually with a pointed (diphycercal) caudal fin. The shafts of the pectoral as well as of the pelvic fins, which are placed far back, are invested with scales, which also cover the rays. The scales are sometimes thin and cycloid, sometimes strong and rhomboid. The Crossopterygii lead to the *Dipnoi* and *Amphibia*.

Fam. **Polypteridæ.** With rhomboid scales, and dorsal fins divided up into a number of small fins. *Polypterus bichir* Geoffr. (fig. 607), with from eight to sixteen small fins, rivers of Tropical Africa: *Calamoichthys calabaricus* Smith, Old Calabar.

Tribe 3. **Euganoides** (Bony Ganoids). Ganoids with rhomboidal scales, and usually with fulcra on the anterior border of the fins. They have numerous branchiostegal rays. The pelvic fins are placed between the pectoral and anal fins.

Fam. **Lepidosteidæ** (Gar-Pikes, Bony Pikes). Form of body elongated, pike-like. The dorsal fins are placed far back: the caudal fin heterocercal and sharply cut off. Fresh waters of Cuba, Central and North America. *Lepidosteus platystomus* Raf.; *L. osseus* L.; *L. spatula* Lac.

Tribe 4. **Amiades.** Bony Ganoids, with large round enamelled scales, bony branchiostegal rays, and heterocercal caudal fin. There are no fulcra.

Fam. **Amiadæ.** *Amia calva* Bonap., rivers of Carolina: most nearly allied to the bony fishes (*Clupeidæ* and *Salmonidæ*).

Order 5.—TELEOSTEI (BONY FISHES).

Fishes with bony skeleton, with free gills (usually four on each side) and an external branchial operculum. There is a bulbus arteriosus with two valves at its base. The optic nerves do not form a chiasma.

The Teleosteans comprise by far the greatest number of all fishes, and are distinguished from the cartilaginous fishes and Ganoids by a number of anatomical characters. They possess a simple bulbus arteriosus with only two valves, which are placed opposite to each other at the origin of the bulbus. The bulbus arteriosus is not a separate part of the heart with independent pulsation, but the thickened commencement of the cardiac aorta. Spiracles and a spiral valve of the intestine are never found. The optic nerves simply cross one another, or the fibres of the one pass between the fibres of the other without forming a chiasma. The gills are usually comb-shaped, and, as in the Ganoids, lie freely in a branchial cavity under a branchial operculum, to which is added a a branchiostegal membrane, supported by branchiostegal rays. The skeleton is characterised by the well separated, usually bony vertebræ, and by the bony skull, beneath which remains of the primitive cartilaginous cranium often persist. The skin is only rarely naked or apparently without scales. In such cases the scales are very small and do not project from the surface; more frequently bony plates and scutes are present in it, especially behind the head. As a rule the skin is covered by cycloid or ctenoid scales which overlap one another.

The urinary and genital organs open behind the anus either separately or by a common aperture on a urogenital papilla. [The kidney is dilated in front to form a head-kidney, which, however, is in the adult, sometimes if not always, largely composed of a tissue resembling lymphatic tissue (Balfour). The generative ducts are continuous with the investments of the generative glands in *both* sexes, and in the male there is no connection between the testis and the kidney.]

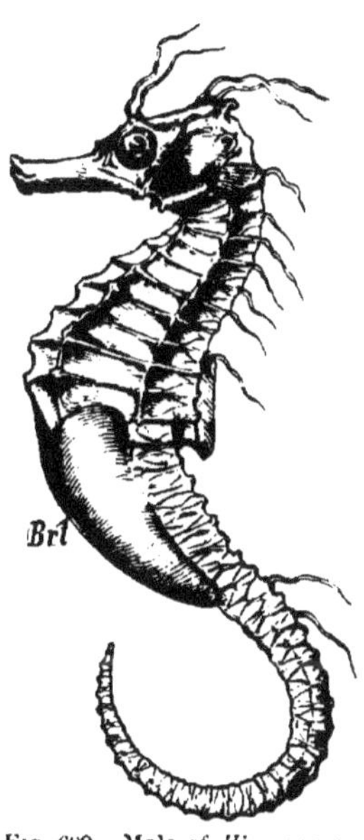

FIG. 609.—Male of *Hippocampus* with the brood-pouch (*Brt*).

Only a few Teleosteans are viviparous; they almost all lay small eggs in enormous numbers in protected places.

Sub-order 1. **Lophobranchii.** Teleosteans with armoured skin, elongated tubular snout which is without teeth. The gills are in the form of tufts and the gill slits are very narrow.

Fam. **Pegasidæ.** The body is flattened: pectoral fins large, spread out like wings; pelvic fins small. *Pegasus volans* L., East Indies.

Fam. **Syngnathidæ.** The body is cylindrical or laterally compressed. The gill openings narrow, and pectoral fins small; males with brood-pouches (fig. 609). *Syngnathus acus* L., Pipe-fish; *Hippocampus antiquorum* Leach., Sea-horse, Mediterranean.

Sub-order 2. **Plectognathi.** Globular or laterally compressed Teleosteans, with immovably fused maxilla and præmaxilla, and narrow mouth. The dermal armour is strong and often bears spines. There are usually no pelvic fins. The gills are comb-shaped.

Tribe 1. **Sclerodermi.** Jaws with separated teeth.

Fam. **Ostracionidæ** (Trunk-fishes). Body coffer-like, triangular or quadrangular, often prolonged into horn-like processes: with firm dermal armour consisting of polyhedral bony plates, on which only the fins and tail are movable. *Ostracion triqueter* L. (fig. 610). West Indies; *O. quadricornis* L., West Africa.

Fam. **Balistidæ** (File-fishes). The body is laterally compressed, and the skin is covered with rough granules, or with hard rhomboid scales, and is often beautifully coloured. *Balistes maculatus* L., Atlantic and Indian Oceans.

Tribe 2. **Gymnodontes.** The jaws modified into a beak, with cutting undivided or double dental plate. Dorsal spines absent.

Fam. **Molidæ.** *Orthagoriscus mola* Bl. Sunfish.

Fam. **Tetrodontidæ** [Globefishes, Sea-Hedgehogs.] *Diodon hystrix* L., Atlantic and Indian Oceans; *Tetrodon cutaneus* Gthr., St. Helena.

Sub-order 3. **Physostomi.** With soft fins (malacopterygians), with

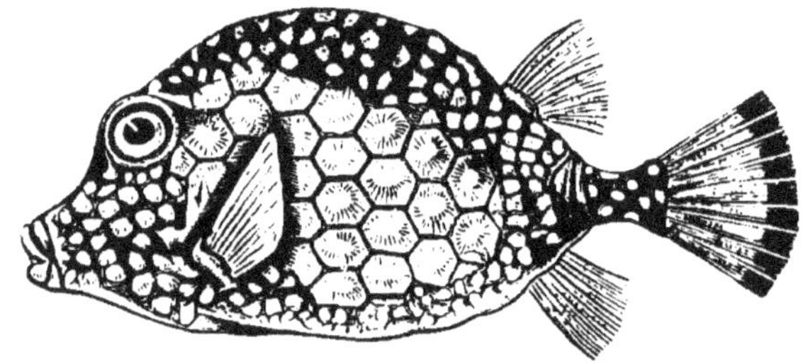

FIG. 610. —*Ostracion triqueter* (règne animal).

comb-shaped gills and separated jaw bones. Pelvic fins abdominal or absent. Swimming bladder always with a ductus pneumaticus.

Fam. **Murænidæ** (Eels). *Muræna helena* L.; *Anguilla anguilla* L. (*vulgaris*), Europe. At the breeding season in autumn they migrate from the rivers into the sea, and there first attain sexual maturity. The reproductive processes are not perfectly known, though male and female have been distinguished from one another, and the presence of both kinds of sexual organs has been shown. In the spring the young eels migrate from the sea into the rivers. *Conger vulgaris* Cuv., coasts of Europe.

Fam. **Gymnotidæ.** *Gymnotus electricus* L. (Electric eel). Lives in the swamps and rivers of South America, attains a length of six feet, and can, by means of its electric discharge, knock down even large animals, *e.g.*, horses. Celebrated by the experiments of A. v. Humboldt.

Fam. **Clupeidæ** (Herrings). With tolerably compressed body, which with the exception of the head is covered by large, thin, easily-detached scales. *Clupea harengus* L., the Herring of the northern seas. It appears every year at certain times, in enormous shoals, on the Scottish and Norwegian coasts. The principal takes occur in September and October. *C.* (*Harengula*) *sprattus* L., the Sprat, North Sea and Baltic; *Engraulis encrasicholus* Rond., Anchovy;

Alausa vulgaris Cuv., Val., the Shad; migrates in May at the spawning season from the sea into the rivers, *e.g.*, up the Rhine to Basel, and in the Main to Würzburg. Attains a length of three feet. *A. pilchardus* Bloch. Sardine. Mediterranean.

Fam. **Esocidæ** (Pikes). The head is broad and depressed; the dorsal fins are placed far back. Pseudobranch glandular, hidden. Voracious carnivorous fish, with wide throat and powerful dental armature. *Esox lucius* L., Pike; *Umbra Krameri* Joh. Müll.

Fam. **Salmonidæ.** With adipose fin, simple swimming bladder, and numerous pyloric appendages. The ovaries are sacs from which the eggs fall into the abdominal cavity. At spawning time, which is usually in the winter months, the two sexes often exhibit striking differences. They are large predatory fishes, and belong principally to the rivers, mountain streams, and lakes of the northern regions. They like clear cold waters with stony bottom; but they have, also, representatives in the sea, which ascend the rivers and their tributaries to spawn. *Coregonus Wartmanni* Bloch. Blaufelchen; in the Alpine lakes. *Thymallus vulgaris* Nilss. (*vexillifer*), Grayling: *Salmo salvelinus* L., Saibling: *S. hucho* L., Huchen, in the region of the Danube, a large predatory fish. *S. salar*, Salmon: *S. lacustris* L. (Seeforelle, Schwebforelle), in the lakes of the Alps of Central Europe. *S. trutta* L., Salmon or Sea trout; *S. fario* L., Trout.

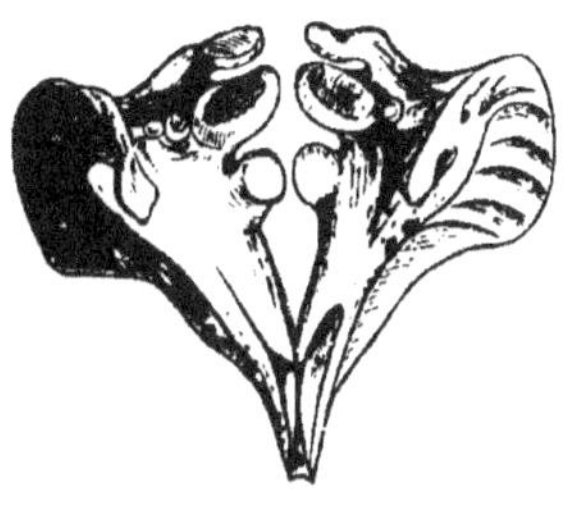

FIG. 611.—Lower pharyngeal bones with the teeth of a carp (after Heckel and Kner).

Fam. **Cyprinidæ** (Carps). Fresh-water fish, with narrow mouth, often provided with barbules. The jaws are weak and without teeth, but the lower pharyngeal bones are abundantly furnished with teeth (fig. 611). *Cyprinus carpio* L., the Carp; *Carassius vulgaris* Nilss., Crucian and Prussian Carps (Karausche); *Tinca vulgaris* Cuv., Tench; *Barbus fluviatilis* Ag., the Barbel;

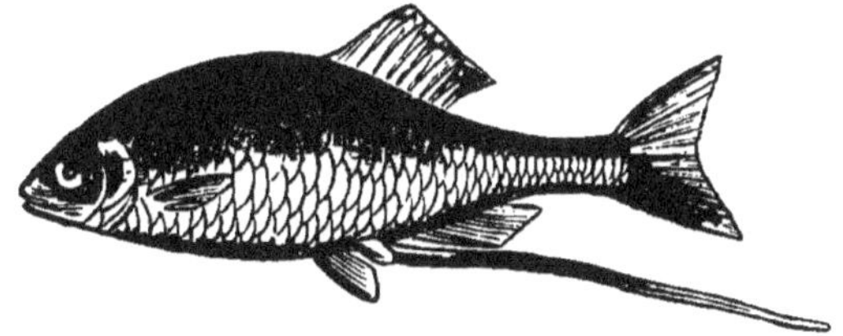

FIG. 612.—*Rhodeus amarus.* Female (after v. Siebold).

Gobio fluviatilis Flem., the Gudgeon; *Rhodeus amarus* Bloch. (Bitterling). The female has an ovipositor with which she deposits the ova in the gills of the fresh-water mussel (fig. 612). *Alburnus lucidus* Heck. Kner, the Bleak; *Leuciscus rutilus* L., the Roach; *L. cephalus* L., the Chub; *Chondrostoma nasus* L., (Näsling); *Abramis brama* Flem., Bream; *Phoxinus lævis* L. Ag., Minnow.

Fam. **Acanthopsidæ.** The swimming bladder is contained in a bony capsule. *Cobitis fossilis* L.; *C. barbatula* L., Loach; *C. tænia* L., Spined Loach or Groundling.

Fam. **Cyprinodontidæ** (Toothed Carps). Viviparous. *Cyprinodon* (*Lebias* Cuv.) *calaritanus* Cuv., South Europe: *Anableps tetrophthalmus* Bl., Guiana.

Fam. **Siluridæ.** Fresh-water fish, usually with broad depressed head, strong dental armature, and skin naked or covered with an armour of bony plates. *Silurus glanis* L. (Wels, Waller). The largest river fish of Europe. *Hypostomus* Lac. (Panzerwels): *Malapterurus electricus* L. (Zitterwells), Nile.

FIG. 613.—*Exocoetus Rondeletii* (after Cuvier and Valenciennes).

Sub-order 4. **Anacanthini.** Malacopterygians (soft fins), which with regard to their internal anatomy are allied to the Acanthopteri by the absence of a ductus pneumaticus; usually with jugular pelvic fins.

Fam. *Ophidiidæ*. *Ophidium barbatum* L., Mediterranean: *Ammodytes tobianus* L., Sand-eel, North Sea.

Fam. **Gadidæ.** *Gadus morrhua* (the Cod). In Germany dried cod is called Stockfisch, salted cod Laberdan. Cod-liver oil is prepared from its liver. Its

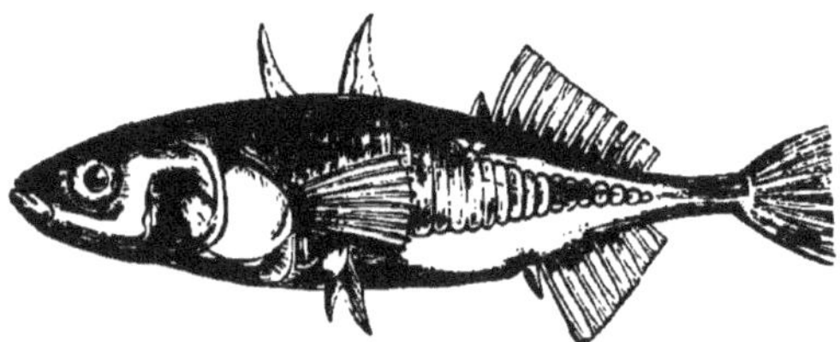

FIG. 614.—*Gasterosteus aculeatus* (after Heckel and Kner).

young (Dorsch) were for a long time considered as a separate species (*G. callarias*). *G. æglefinus* L. Haddock, with a black spot behind the pectoral fin. *G. merlangus* L., Whiting, coasts of North Europe: *Merluccius vulgaris* Flem., Hake, Mediterranean and Northern Seas: *Lota vulgaris* Cuv., Burbot, Eel-pout, Cony fish. Predatory fresh-water fish.

Fam. **Pleuronectidæ.** Flat fishes. The body is compressed, disc-shaped, and strikingly asymmetrical. The side which is directed upwards towards the light is pigmented (with change of colours); the other is free from pigment. Both eyes are placed on the pigmented side, towards which the head is turned and the arrangement of its bones shifted to correspond with this asymmetry. *Hippoglossus vulgaris*, Flem. the Holibut, coasts of North Europe: *Rhombus maximus* L., the Turbot; *Rh. lævis* Rond., the Brill, European coasts; *Pleuronectes platessa* L., the Plaice; *Pl. limanda* L., the Dab; *Pl. flesus* L., the Flounder, ascends rivers: *Solea vulgaris* Quens., the Sole.

Fam. **Scomberesocidæ**. Marine Malacopterygians, with cycloid scales. The lower pharyngeal bones are fused (*Pharyngognathi*). *Belone acus* Rond., Gar-pike; *Scomberesox saurus* Walb.: *Exocœtus evolans* L., the Flying fish. The pectoral fins are strengthened so as to form flying organs. *E. exiliens* L., European Seas: *E. Rondeletii* Cuv. Val., Mediterranean (fig. 613).

Sub-order 5. **Acanthopteri.** Spiny-rayed fishes with comb-shaped gills; lower pharyngeal bones usually separate; thoracic, rarely jugular or abdominal pelvic fins. Swimming bladder closed, without ductus pneumaticus.

Tribe 1. **Pharyngognathi.** The lower pharyngeal bones are fused.

Fam. **Pomacentridæ**. *Amphiprion bifasciatus* Bl., New Guinea; *Pomacentrus fasciatus* Bloch., East Indies.

Fam. **Labridæ**. The Wrasses (Lippfische). Brightly-coloured fish, with fleshy protrusible lips. *Labrus maculatus* Bl., coasts of Europe; *Crenilabrus pavo* Brünn: *Julis pavo* Hassq., Mediterranean; *Scarus creteusis* Aldr., Parrot-fish, Mediterranean.

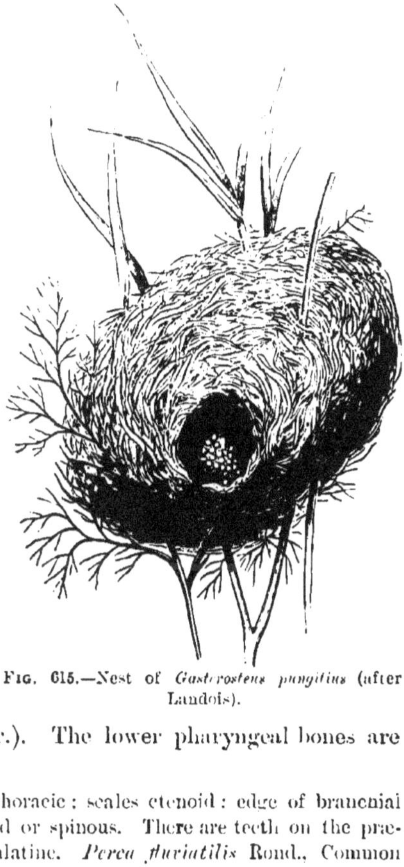

FIG. 615.—Nest of *Gasterosteus pungitius* (after Landois).

Tribe 2. **Acanthopteri** (s. str.). The lower pharyngeal bones are not fused.

Fam. **Percidæ**. Perches. Fins thoracic: scales ctenoid: edge of branchial operculum or præoperculum serrated or spinous. There are teeth on the præmaxilla, lower jaw, vomer and palatine. *Perca fluviatilis* Rond., Common Perch (fig. 583). A voracious fish, especially pursues small Cyprinoids. *Labrax lupus* Cuv., the Bass, Mediterranean: *Acerina cernua* L., the Pope, river fish; *Lucioperca sandra* Cuv., river fish of South Europe; *Serranus scriba* L., hermaphrodite, Mediterranean; *Gasterosteus aculeatus* L., the Stickleback (fig. 614), remarkable for forming a nest and protecting the eggs and young: *G. pungitius* L., ten-spined Stickleback (the Tinker) (fig. 615); *G. spinachia* L., fifteen-spined Stickleback.

Fam. **Mullidæ**. Mullets. *Mullus barbatus* L., red Mullet.

Fam. **Sparidæ**. Sea-breams. *Sargus Rondeletii* Cuv. Val.; *Pagellus erythrinus* L.; *Chrysophrys aurata* L., Mediterranean.

Fam. **Triglidæ**. *Cottus gobio* L., River Bullhead or Miller's Thumb. A small fish found in clear brooks and streams. It hides beneath stones, and defends itself by expanding its branchial operculum. The male undertakes the care of the brood. *C. scorpius* L., Sea-scorpion; *Trigla gunardus* L., Grey

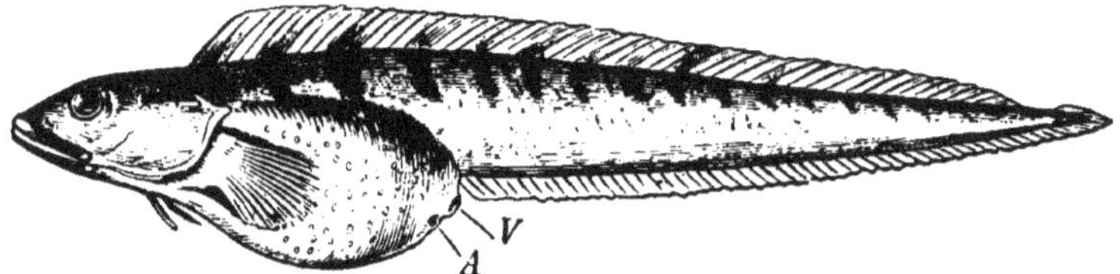

FIG. 616.—*Zoarces viviparus*. *A*, anus; *V*, urogenital opening.

Gurnard; *Dactylopterus volitans* L., Flying Gurnard: *Uranoscopus scaber* L. (Sternseher), Mediterranean; *Scorpæna porcus* L.: *Trachinus draco* L.

Fam. **Sciænidæ** (Umberfische). *Umbrina cirrhosa* L., Mediterranean; *Corvina nigra* Salv., Mediterranean; *Sciæna aquilla* Risso., Mediterranean.

Fam. **Scomberidæ**. Mackerels. Body elongated, more or less compressed, sometimes very high. The skin is often silvery, and sometimes naked, sometimes covered with small scales. There are keeled bony plates in places,

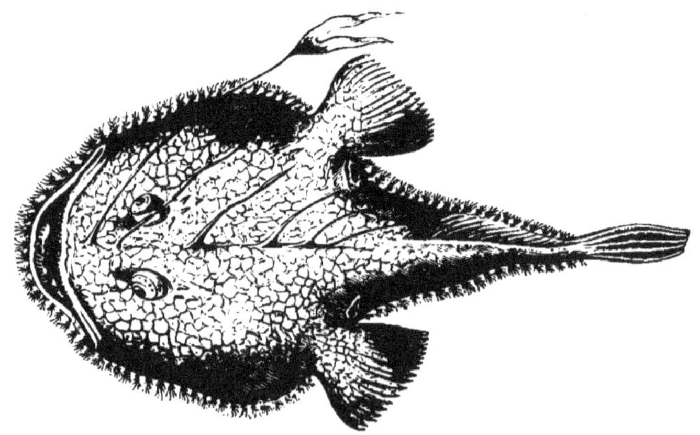

FIG. 617.—*Lophius piscatorius* (after Cuvier and Valenciennes).

especially near the lateral line. The caudal fin usually has a semilunar shape. They constitute, on account of their tasteful flesh, an important object of the fishing industry—the Mackerel in the North Sea and the Channel, the Tunny Fish in the Mediterranean. *Scomber scombrus* L., Mackerel; *Zeus faber* L., the Dory; *Thynnus vulgaris* Cuv., Val., Tunny Fish; *Pelamys sarda* Bl., Mediterranean; *Caranx trachurus* L., Horse-Mackerel, coasts of Europe; *Xiphias gladius* L., Sword-fish; *Echeneis naucrates* L., Sucking-fish.

Fam. **Gobiidæ.** Gobies. *Gobius niger* Rond.; *G. fluviatilis* Pall., Rivers of Italy and of South-west Russia.

Fam. **Blenniidæ.** Blennies. *Anarrhichas lupus* L., Wolf-fish: *Blennius ocellaris* L., Butterfly-fish, Mediterranean; *Zoarces viviparus* Cuv. (fig. 616), viviparous.

Fam. **Tænioidæ.** Silvery marine-fish, with compressed, ribbon-like, elongated-like body. *Trachypterus falx* Cuv., Val. = *Tr. tænia* Bl., Schn., Nice; *Cepola rubescens* L., Band-fish, coasts of Europe.

Fam. **Labyrinthici.** The upper pharyngeal bones are hollowed out so as to have the form of coiled (meandering) lamellæ (fig. 594), in the spaces between which the water required to keep the gills moist is retained. *Anabas scandens* Dald., Climbing Perch, East Indies.

Fam. **Pediculati.** Of stout clumsy shape. The skin is naked, or covered with rough prominences. The pelvic fins, which are small and placed on the throat (jugular), have their so-called carpal pieces elongated, so that they form movable arm-like supports for the body, and are in fact used for hopping and creeping. *Lophius piscatorius* L., Angler, Frog-fish, etc. (βάτραχος of the Greeks), coasts of Europe (fig. 617); *Chironectes pictus* Cuv.

Order 6.—Dipnoi.*

Scaly Fishes with branchial and pulmonary respiration, with persistent notochord, muscular conus arteriosus and spiral valve in the intestine.

The *Dipnoi* (fig. 618) form a group so strikingly transitional between Fishes and Amphibians that their first discoverer regarded

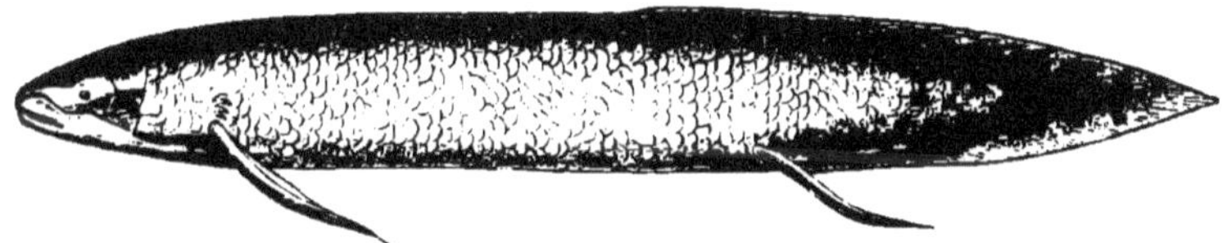

Fig. 618.—*Protopterus annectens.*

them as fish-like Reptiles, and in more recent times they have been regarded as scaly Amphibians. In their external form they decidedly resemble Fishes. The head is broad and flat, and has small, laterally placed eyes and a fairly widely-split snout, at the extremity of which are placed the two nasal openings. Directly behind the head are two thoracic fins, which, like the similarly-formed pelvic fins, possess

* J. Hyrtl, "Lepidosiren paradoxa. Eine Monographie." Prag, 1845.

G. Krefft. "Beschreibung eines gigantischen Amphibiums aus dem Wide-Bay-District in Queensland."

A. Günther. "Ceratodus und seine Stelle im System." *Arch. für Naturgesch.*, Tom. XXXVII., 1871.

A. Günther. "Description of Ceratodus, a genus of Ganoid Fishes." *Phil. Transact.*, 1871.

a membranous border supported by rays, or (*Ceratodus*), like the fins of the Crossopterygians, consist of a central shaft covered by scaly skin, and of a border provided with rays. The pelvic fins are placed far back. In front of the anterior pair of fins there is a gill slit on either side, above which in the African genus *Protopterus* (*Rhinocryptis*) three external gill tufts are retained till late in life. In the Brazilian genus *Lepidosiren* external gills are absent.

The Dipnoi show themselves to be fishes by the possession of gills as well as by the external form. There are either four gills (*Ceratodus*) as in fishes, or their number is reduced. The structure of the skeleton points decidedly to the Ganoids, to which the *Dipnoi* are in other respects closely related. In *Lepidosiren* the notochord persists as a continuous cartilaginous cord, from the fibrous sheath of which dorsal and ventral bony arches with ribs project. In front the notochord is prolonged into the base of the skull, which remains at the stage of the primitive cartilaginous cranium. It is, however, covered by some osseous pieces. The facial bones of the head are much more developed, especially the jaws, whose teeth consist, as in *Chimæra*, of perpendicularly-placed cutting plates or (*Ceratodus*) recall those of *Cestracion*. The intestine contains a spiral valve, which terminates at some distance from the cloaca. The cloaca contains the sexual opening, and the openings of the ureters in its side walls: it opens to the exterior—sometimes to the right side, and sometimes to the left, and on its posterior side there is in *Lepidosiren* an independent urinary bladder.

On the other hand, the respiration by means of lungs and the structure of the heart indicate a relationship to the naked Amphibia. The cartilaginous nasal capsules, as in all lung-breathing animals, open behind into the mouth by apertures, which perforate the roof of the mouth, and are placed far forward, directly behind the extremity of the snout. The swimming-bladder is represented by two sacs (in Ceratodus only one) placed outside the body cavity, ventral to the kidney, and opening into the ventral wall of the pharynx by means of a short common duct. These sacs function as lungs, inasmuch as they obtain venous blood from a branch of the posterior aortic arch and return arterial blood to the heart by pulmonary veins. To this agreement with the Amphibia may be added the similar arrangement of the heart and the principal trunks of the vascular system, the incompletely divided right and left auricles and the double circulation. There is a muscular conus arteriosus which either has an arrangement of valves like that in the Ganoids

(*Ceratodus*) or contains, as in the frogs, two lateral spiral longitudinal folds, which fuse at their anterior end and effect the division of the lumen into two (for the branchial arteries and the pulmonary vessels).

Sub-order 1. **Monopneumona.** The body is covered with large cycloid scales (fig. 619 *a*). Vomer with two oblique incisor-like dental lamellæ. Palate armed with a pair of large and long dental plates (molars), which have a flat undulated surface and five or six sharp prongs on the outer side. Lower jaw with two similar dental plates (fig. 619, *c*). Fins as in the *Crossopterygii*, with scaly shaft and rayed border on each side (fig. 619, *b*). The valves in the conus arteriosus rather resemble those of the Ganoids.

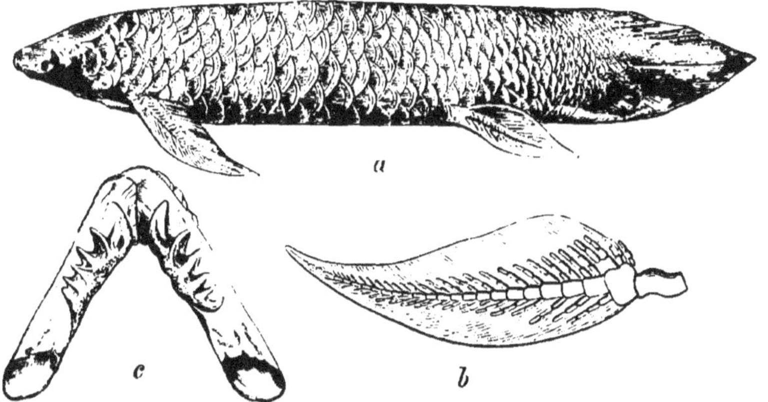

FIG. 619.—*a*, *Ceratodus miolepis*. *b*, its pectoral fin (after Günther). *c*, lower jaw with dental plates of *Ceratodus Forsteri* (after Krefft).

Branchial apparatus formed of five cartilaginous arches and four gills. Pseudobranchs (hyoidean) are present. The lung is composed of two symmetrical cellular halves. The two ureters open on the dorsal side of the cloaca by a common opening. There is a pair of wide peritoneal slits (abdominal pores) behind the anus. The *Monopneumona* feed on leaves, which they tear off with their incisor teeth and masticate with their molars. They make use of the lungs in respiration principally when the muddy water is saturated with gases from organic matter. They have existed since the Triassic period.

Fam. **Ceratodidæ**, with the single genus *Ceratodus* Ag. *C. Forsteri* Krefft, (and *miolepis* Günth.), the Barramunda, Queensland; reaches a length of six feet. Its flesh is salmon-like and much esteemed as food.

Sub-order 2. **Dipneumona.** Fins narrow, with jointed cartilaginous shaft and rays only on one side. Gills more reduced. Valvular arrangement of conus arteriosus like that in the Batrachians. Lungs paired.

Fam. **Lepidosirenidæ.** *Protopterus annectens* (fig. 618) Owen, tropical Africa; *Lepidosiren paradoxus* Fitzg., Brazil.

CHAPTER VI.

Class II.—AMPHIBIA.*

Cold-blooded animals usually with a naked skin, with pulmonary and branchial respiration, and incompletely double circulation. The embryos have neither amnion nor allantois.

The external form of the body is adapted both for an aquatic and a terrestrial life. It presents, however, considerable variations lead-

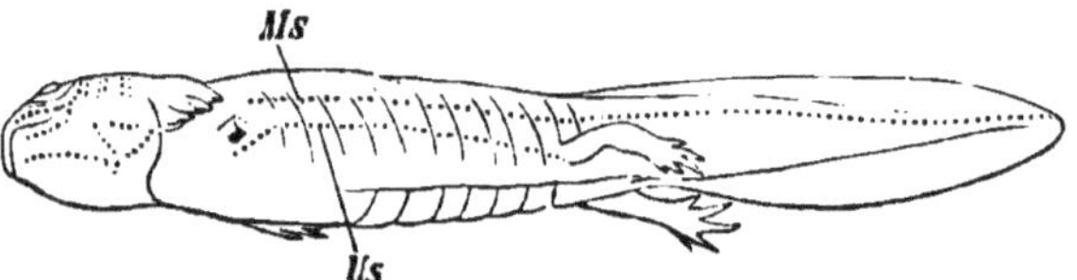

FIG. 620.—Larva of *Salamandra maculata* (after Malbranc). *Ms*, median, *Us*, lower lateral line.

ing to the creeping, climbing, and jumping land animals. An elongated, cylindrical, or more compressed form is the most frequent, and the body often ends with a large compressed swimming tail. Limbs may be absent, as in the cylindrical *Cæciliidæ*, which live underground in damp earth. In other cases there are only short anterior limbs (*Siren*), or anterior and posterior stumps, which have a reduced number of toes and are unable to raise the serpentining body from the ground. Even when the extremities have a considerable size and end with four or five digits, they act rather as pushing organs in the movement of the elongated and flexible body. The Batrachians, which have short and stout bodies and are without a tail in the adult state, alone possess powerful limbs adapted for running and jumping, and even for climbing.

The **skin,*** which is of great importance not only as a secretory

* Wagner, "Natürliches System der Amphibien." Munich, 1830.
Duméril et Bibron. "Erpétologie générale, etc." Paris, 1834-1854.

* Fr. E. Schulze, "Epithel-und Drüsenzellen. I. Die Oberhaut der Fische und Amphibien." *Arch. für mikrosk. Anatomie*, Tom. III.

but also as a respiratory organ, is as a rule naked and slimy. The *Cæciliidæ* alone possess thickened cutaneous rings, in which scales are imbedded. The sense organs of the lateral line (fig. 620) also are present in the aquatic forms, especially in the larval condition. Glands and pigments are very generally present in the integument. The former often secrete strongly smelling and caustic juices, which act as poisons on other organisms (*parotid* glands, as well as glands on the sides and posterior extremities). The various colourings of the skin are principally due to branched pigment cells of the cutis. The change of colour in the Frogs—a phenomenon which has been known for some time—is caused by changes in the form of these cells.

Skeleton.—Although a notochord may persist (*Cæciliidæ, Proteus*), yet bony, at first biconcave vertebræ,* are always developed, and are separated by intervertebral cartilages. In the *Salamandrina* the cartilage in the intervertebral regions grows considerably and gradually supplants the notochord, the remains of which become cartilaginous. As the result of further differentiation of the intervertebral cartilages, the rudiments of an articular head and an articular cup are developed, which, however, are only completely separated in the Batrachians provided with procœlous

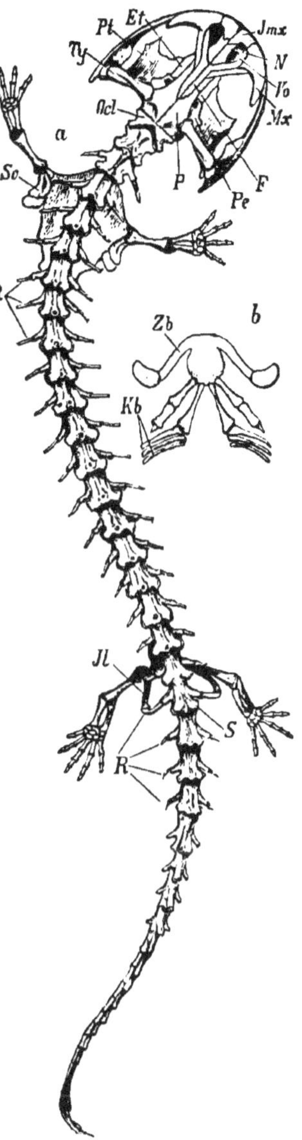

FIG. 621.—Skeleton of *Menopoma alleghaniense*. *Ocl*, Exoccipital; *P*, parietal; *F*, frontal; *Ty*, tympanic; *Pe*, petrous (prootic); *Mx*, maxilla; *Jmx*, præmaxilla; *N*, nasal; *Vo*, vomer; *Et*, girdle bone (sphen-ethmoid); *Pt*, pterygoid; *Sc*, pectoral arch; *Jl*, pelvic arch; *S*, sacral vertebra; *R*, ribs. *b*, hyoid apparatus (remains of hyoid (*Zb*) and branchial arches (*Kb*).

* Compare especially C. Gegenbaur, "Untersuchungen zur vergleichenden Anatomie der Wirbelsaüle bei Amphibien und Reptilien," Leipzig, 1862.

vertebræ.* The number of vertebræ is usually considerable, in accordance with the elongated form of body; but in the *Batrachia* the vertebral column consists of only ten vertebræ with very long transverse processes, which usually at the same time represent the ribs; while, with the exception of the first vertebra which is modified to form the atlas, almost all the vertebræ of the trunk possess small cartilaginous rudiments of ribs. The sacral region is formed by a single vertebra (fig. 621).

Skull.—The primordial cartilaginous cranium persists, but usually loses its roof and floor, and is partly replaced by bony pieces, some of which are ossifications of the cartilaginous capsule (*exoccipitals, auditory capsules, sphen-ethmoid, quadrate*), while others are

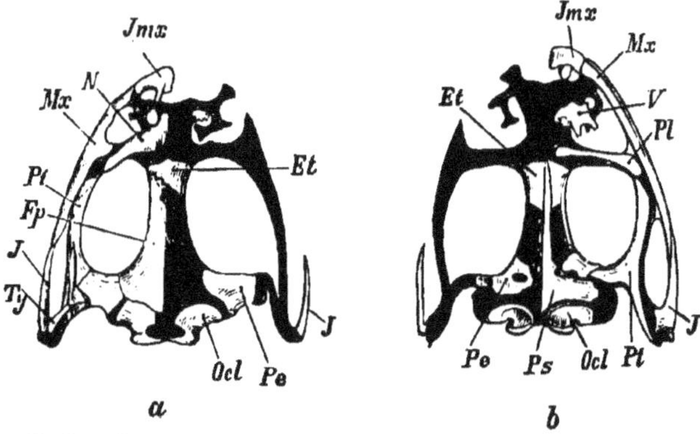

FIG. 622.—Skull of *Rana esculenta* (after Ecker). *a*, from the dorsal, *b*, from the ventral side; [Membrane bones of one side removed in each case]. *Ocl*, exoccipital; *Pe*, petrosal (prootic); *Et*, girdle-bone or sphen-ethmoid; *Ty*, tympanic; *Fp*, fronto-parietal; *J*, quadrato-jugal (jugal); *Mx*, maxillary; *Jmx*, præmaxillary; *N*, nasal; *Ps*, parasphenoid; *Pt*, pterygoid; *Pl*, palatine; *V*, vomer.

investing bones (*parietals, frontals, nasals, vomer, parasphenoid*) (fig. 622). As in *Lepidosiren* the basi- and supra-occipital remain as small cartilaginous tracts. There is also a parasphenoid on the base of the skull (fig. 622, *Ps*). The large exoccipitals (*Ocl*) (fused with the *opisthotic*) articulate by means of two condyles with the first vertebra, as in the Mammalia. The projecting auditory region is pierced by the *fenestra ovalis*, and the bone in its anterior part corresponds to the prootic (*Pe*). The lateral walls of the skull remain cartilaginous, but in the ethmoid region there is a ring-shaped bone—the girdle bone, or sphen-ethmoid.

* [For a fuller account of the development of the Amphibian vertebral column, *vide* Balfour, "Comparative Embryology," vol. ii., p. 456 *et seq.*]

As in *Lepidosiren* the mandibular arch is firmly connected with the skull. The *mandibular suspensorium* and the *palato-quadrate* are in direct connection with the cartilaginous cranium, and form on either side a wide outstanding infra-orbital arch, the anterior end of which either remains free or fuses with the ethmoid cartilage. The ossification appearing at the end of the suspensorium gives rise to the *quadrate*, while a membrane bone, almost hammer-shaped and overlying the suspensorial cartilage, is called the *squamosal* or perhaps more correctly *tympanic* (*Ty*). Two membrane bones extend forward along the lower side of the palato-quadrate bar—the *pterygoid* (*Pt*) behind and the *palatine* (*Pl*) in front. The palatine is transversely placed behind the *vomer*. The outer arch of the upper jaw, formed by the *præmaxillary* and *maxillary* bones (*Jmx*, *Mx*) may by means of a third posterior bone—the *quadrato-jugal* (*J*)—be continued back to the quadrate, but in many *Perennibranchiata* it is incomplete, the maxillaries being absent. The skeleton of the visceral arches is more or less considerably reduced in correspondence with the retrogression of branchial respiration. In the perennibranchiate Amphibia (Amphibia with gills throughout life) the visceral arches are more numerous, and present an arrangement similar to that found only transitorily in the larvæ of the other forms. In the *Salamandrina*, in addition to the hyoid arch, the remains of two branchial arches persist; while in adult Batrachians only a single pair of arches is retained on the hyoid bone. This branchial rudiment is attached to the posterior edge of the body of the hyoid bone, and serves as a suspensorium for the larynx.

In the pectoral girdle three parts may be distinguished—the *scapula*, the *præcoracoid*, and the *coracoid*, to which a dorsal cartilaginous *supra-scapula* is added. While in the tailed *Amphibia* (*Urodela*), the arch is interrupted below, in the *Batrachia* the two halves are joined to each other in the middle ventral line, as well as to a posterior plate which has the value of a *sternum*, and an anterior plate known as the *episternum*. The pelvic girdle is characterised by the narrow form of the iliac bones, which are attached to the strong transverse processes of a single vertebra, and at their posterior end are fused with the ischiac and pubic bones.

The **nervous system** is higher in several respects than that of the fishes. The brain (vol. i., fig. 80) is certainly in all cases small, but the hemispheres are large and the differentiation of the thalamencephalon and mesencephalon is further advanced. The optic lobes reach a considerable size, and the medulla oblongata encloses a wide fourth ventricle. The cranial nerves have the same relations as in the Fishes, since not only are the *facial* nerves and the nerves supplying the muscles of the eye often connected with the *trigeminal*, but the *glossopharyngeal* and the *spinal accessory* are represented by branches of the *vagus*. The *hypoglossal* is, as in the Fishes, the first spinal nerve.

With regard to the sense organs the two eyes may be rudimentary and concealed beneath the skin (*Proteus, Cæciliidæ*). In the *Perennibranchiata* eye-lids are completely absent, while the *Salamandrina* have an upper and lower eye-lid, and the *Batrachians*, except *Pipa*, have, besides the upper eye-lid, a large very movable nictitating membrane, with which a rudimentary lower eye-lid co-exists only in *Bufo*. In the Batrachians there is a retractor muscle by means of which the large bulb of the eye can be drawn back. The structure of the *auditory organ* * of the Amphibia resembles that of Fishes. It is usually confined to the labyrinth with three semi-circular canals; in the Batrachians alone there is a tympanic cavity, which communicates with the pharynx by means of a wide Eustachian tube, and is closed externally by a tympanic membrane, which is sometimes freely exposed on the surface and sometimes covered by the skin. The tympanic membrane is connected with the fenestra ovalis by a small cartilaginous rod (remains of the hyomandibular) with a cartilaginous plate (*columella* with *operculum*). When there is no tympanic cavity these structures are covered by muscles and skin. The cochlea, which was first discovered by Deiters in the frog, is probably present in all *Amphibia*. The *olfactory* organs are always paired nasal cavities, which are provided with folds of the mucous membrane and open internally either anteriorly within the lips, or, in the Batrachians and Salamandrines, further back between the maxillaries and palatines. The external skin, which is richly supplied with nerves, is to be regarded as the seat of the *tactile sense*. The possession of the *sense of taste* is indicated by the presence of taste papillæ on the tongue of the Batrachians. The *Amphibia* certainly swallow their food unmasticated, and the tongue also subserves other functions; for instance, in the *Batrachia* it is used as a prehensile organ.

Alimentary canal.—The mouth is a wide slit. The vomers, palatines, and jaws are usually armed with sharp backwardly curved teeth, which are used not for mastication, but for holding the prey. Teeth are seldom absent, as in *Pipa* and some Toads; but in the Frogs they are always present on the upper jaw and palate.

* Compare especially the works of Deiters, Hasse, and Retzius.

The **respiratory** and **circulatory** organs resemble, in essential points, those of the Dipnoi, and stamp the *Amphibia* as connecting links between the aquatic animals which breathe by gills and the higher Vertebrates with pulmonary respiration. In all cases there are two lung sacs, either simple or provided with cellular spaces: but in addition to these there are, either in the larva or in the adult animal (*Perennibranchiata*, fig. 58), three (or four) pairs of gills, which sometimes lie in a cavity covered by a reduplication of the skin and provided with an external opening, and sometimes project freely on the neck as branched or tufted cutaneous appendages. The respiratory movements are effected, in the absence of a thorax capable of distension and contraction, by the muscles of the hyoid bone and by the abdominal muscles. The unpaired air-tube (trachea), which is supported by cartilaginous rods, is usually exceedingly short and wide, like a larynx, and in the *Anura* alone is developed to form a vocal organ, which produces loud croaking sounds and is in the male sex frequently reinforced by a resonating apparatus, consisting of one or two sacs communicating with the buccal cavity.

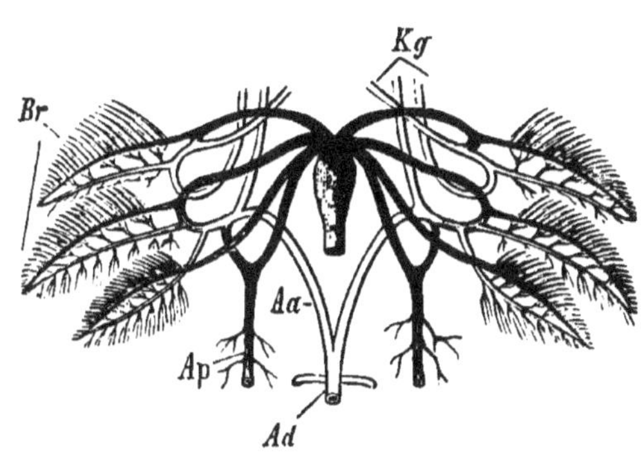

FIG. 623.—Aortic arches of an old frog larva (from Bergmann and Leuckart). *Aa*, the aortic arches uniting into the descending aorta (*Ad*); *Ap*, pulmonary artery; *Kg*, cephalic arteries; *Br*, gills.

As long as the respiration is carried on entirely by means of gills the structure of the heart and the arrangement of the principal arterial trunks are the same as in Fishes. Later, when the pulmonary respiration begins, the circulation becomes double and the auricle becomes divided by a septum into a right and left chamber, of which the right receives the veins from the body, the left those from the lungs. The ventricle, on the contrary, still remains single, and therefore contains mixed blood. It leads by a muscular rhythmically pulsating conus arteriosus into the ascending aorta with the reduced vascular arches.

In the first period of larval life there are four pairs of vascular arches, which surround the pharynx without dividing into capillaries and unite beneath the vertebral column to form the two roots of the descending aorta. With the appearance of gills the three anterior pairs of arches give off vascular loops, which form the system of the branchial capillaries, while the dorsal parts of the arches unite with one another in various ways to form the roots of the descending aorta (fig. 623).

The fourth vascular arch, which, moreover, is frequently a branch of the third (Batrachians), or arises in a common ostium with the latter on the bulbus (Salamander), has no relation to the branchial respiration, and leads directly into the root of the aorta. It is this posterior vascular arch which sends a branch, one on each side, to the developing lungs (fig. 624, *Ap*), and so constitutes the first rudiments of the pulmonary arteries, which soon increase in size and importance. In the Perennibranchiates these arrangements persist in essentials through life, but in Batrachians and Salamanders the disappearance of the gills is followed by further reductions, which lead to the arrangement of vessels found in the higher Vertebrates. With the atrophy of the branchial capillaries the connection between the bulbus arteriosus and the descending aorta is again represented by simple arches, which are in part reduced to narrow canals or even to solid cords of tissue (ductus Botalli) (fig. 624 and fig. 59). The anterior arch sends off branches to the tongue, and also the carotids, at the origin of which there is a swelling—the so-called *carotid gland* (fig. 624). The two middle arches form the roots of the descending aorta and branches may be also given off from them to the head. The posterior arches, which at their origin are often fused with the preced-

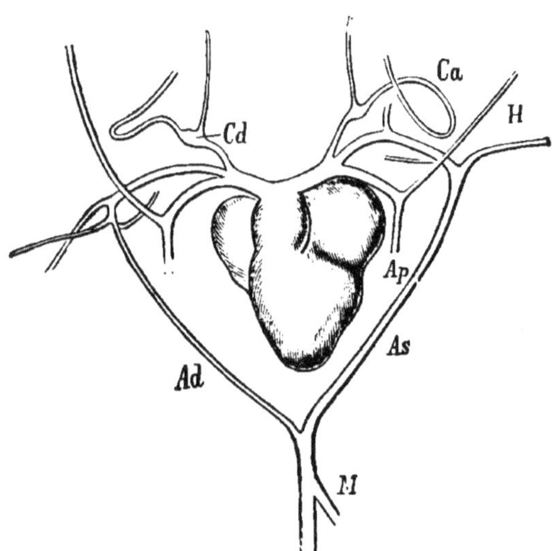

FIG. 624.—Heart and principal arteries of a toad. *Ad*, Right aortic arch; *As*, left aortic arch; *Ca*, carotid; *Cd*, carotid gland; *Ap*, pulmonary artery; *H*, cutaneous artery; *M*, mesenteric artery.

ing, give rise to the pulmonary arteries, a narrow ductus Botalli, the lumen of which is sometimes obliterated, being usually retained. Vessels for the head and occipital region are often given off from the roots of the aorta. In the Batrachians, which, in consequence of the union of the two posterior branchial arches, possess only three vascular arches, the aortic root is the prolongation of the middle arch on each side, and gives off the vessels of the scapular region and the anterior extremity, and often, also, on one side an artery to the viscera (mesenteric artery). The posterior arch sends off the pulmonary arteries and a strong trunk to the skin of the back, but does not retain its connection with the roots of the aorta. As in Fishes, there is a renal-portal system, as well as an hepatic-portal system.

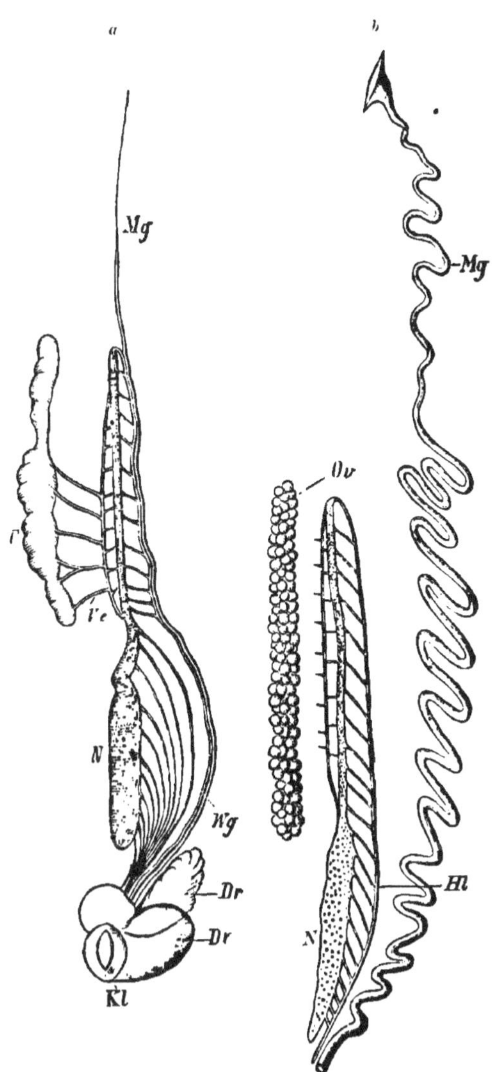

FIG. 625.—*a*, Urinary and genital apparatus of the left side from a male Salamander, partly diagrammatic. *T*, testes; *Ve*, vasa efferentia; *N*, kidney with collecting urinary tubules; *Mg*, Müllerian duct; *Wg*, Wolffian duct or vas deferens; *Kl*, cloaca; *Dr*, prostate glands. *b*, Urinary and genital apparatus of the left side of a female Salamander without the cloacal part. *Ov*, Ovary; *N*, kidney; *Hl*, the urinary duct corresponding to the Wolffian duct; *Mg*, oviduct or Müllerian duct.

The **lymphatic vessels** of the Amphibia accompany the blood-vessels as plexuses, or as wide lymphatic sinuses. In certain places the lymph receptacles are rhythmically contractile, and have the value of lymph hearts. In the Salamanders and Frogs there are two lymph hearts beneath the dorsal integument in the scapular region, and two close behind the ileum. Of the vascular glands the most noteworthy are the *thymus*, which is always paired, and the *spleen*, which is never absent.

The **urinary organs** (fig. 625) are paired kidneys, the numerous collecting tubules of which enter the ducts of the primitive kidney; these open on wart-like protuberances on the dorsal wall of the cloaca. The urinary bladder is an unpaired diverticulum of the ventral wall of the cloaca: it is usually bifid at its free end.

In all cases there is a close relation between the urinary organs and the efferent ducts of the **generative organs** (fig. 625). As in the higher Vertebrates the primitive kidney (Wolffian body or mesonephros) in part becomes the epididymis and the efferent apparatus of the testis, so also in the *Amphibia* a part at least of the primitive kidney, which in these animals persists as a urinary organ, functions as epididymis. The vasa efferentia sink into the kidneys and become connected with the urinary tubules, and thus conduct their contents, usually by means of a common duct, into the terminal portion of the duct of the primitive kidney, which functions as a urogenital duct. In the *Salamanders* there are, in addition, glands called prostate glands on the wall of the cloaca. In the female sex the Müllerian duct, which is rudimentary in the male, assumes the function of *oviduct*. This duct begins with a free funnel-shaped dilated opening into the body cavity, takes a sinuous course, and opens, often after forming a uterus-like dilation, with the urinary duct laterally into the cloaca, in the wall of which, in the *Salamandrina* according to v. Siebold's discovery, saccular glands functioning as seminal receptacles are placed. A complete hermaphroditism seems never to occur, although in the male Toad, especially in *Bufo variabilis*, rudiments of the ovaries have been found near the testes.

Males and females are often distinguished by their size and colour, and also by other peculiarities (vocal sacs), which are especially prominent at the breeding season in spring and summer. In spite of the absence of external organs of copulation, sexual intercourse takes place, but it usually consists merely of an external approximation of the two sexes (Batrachians), and has for its consequence a fertilisation of the eggs outside the body of the mother. The male Salamanders alone have copulatory organs in the form of the swollen lips of the cloaca, which during copulation clasp the cloacal aperture of the female, and thus render an internal fertilization possible. In this case the eggs can undergo their development within the body of the female, and the young be born at a more or less advanced stage of development.

It is only in exceptional cases that the parents have an instinct which leads them to watch over the further fate of their brood, as for example *Alytes* (fig. 626) and the South American Surinam Toad (*Pipa dorsigera*). The male of *Alytes* winds the string of eggs round its hind legs and burrows into the damp earth, and only gets rid of his load when the embryonic development is completed. The male of *Pipa* places the eggs when laid on the back of the female, which then develops a cell-like pouch round each egg. The larvæ are hatched and undergo their metamorphosis in these pouches. In other genera, as *Notodelphys*, the females possess a spacious brood-sac beneath the dorsal integument. Except in these cases, the eggs are either attached singly to water plants (*Tritonidæ*) or laid in strings or irregular clumps.

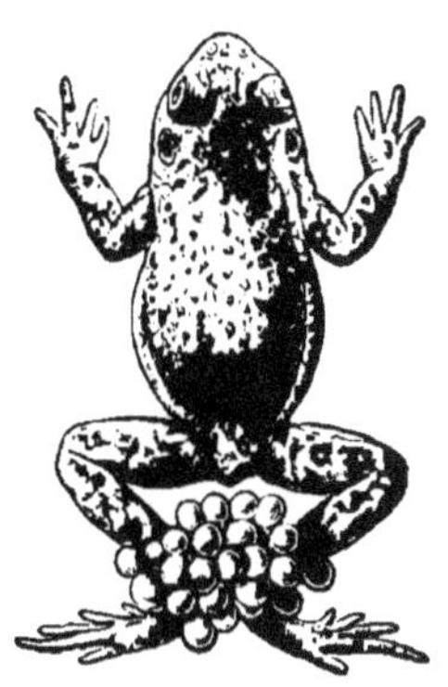

FIG. 626.—*Alytes obstetricans.* Male with the string of eggs.

Development.—The eggs, which are relatively small,* undergo an unequal segmentation (vol. i., fig. 104) after fertilization. The Amphibia agree with the Fishes in not developing an *amnion* or *allantois* — the embryonic membranes of such importance in the higher Vertebrates. In the Amphibians, however, the urinary bladder which arises from the ventral wall of the cloaca is morphologically equivalent to the allantois. The embryos are also without any external yolk-sac constricted off from the body, the yolk being enclosed at an early period by the ventral plates. As respiratory organs gills are

* C. E. v. Baer, "Ueber die Entwickelungsgeschichte der Thiere," II., Königsberg, 1837.

Reichert, "Das Entwickelungsleben im Thierreich," Berlin, 1840.

C. Vogt, "Untersuchungen über die Entwickelungsgeschichte der Geburtshelferkröte," Solothurn, 1842.

Rusconi, "Histoire naturelle, développement et métamorphose de la Salamandre terrestre." Paris, 1854.

A. Götte, "Entwickelungsgeschichte der Unke," Leipzig, 1874.

O. Hertwig, "Die Entwickelung des mittleren Keimblattes der Wirbelthiere," *Jen. naturwiss. Zeitschr.*, Tom. XV., 1881.

developed on the visceral arches; they usually only reach their full development in larval life. The young are always hatched at an early stage, and undergo a metamorphosis. The larva when hatched recalls the piscine type by the laterally compressed swimming tail, and by the possession of external gills (fig. 627); it is still without the two pairs of limbs, which only sprout out as the growth of the body progresses. During these processes the lung sacs which have grown out on the pharynx begin to function, sometimes (*Batrachia*) after the external gills have been replaced by internal branchial leaflets covered by the skin, and a branchial slit has been formed on the side of the neck to allow of the exit of the water (fig. 111). Finally the branchial respiration is completely lost in consequence of the atrophy of the gills and their vessels, the tail becomes shorter and shorter and finally, in the *Batrachia* at least, completely vanishes. In the other groups the later or earlier phases of the developmental series are maintained throughout the whole life: thus in the

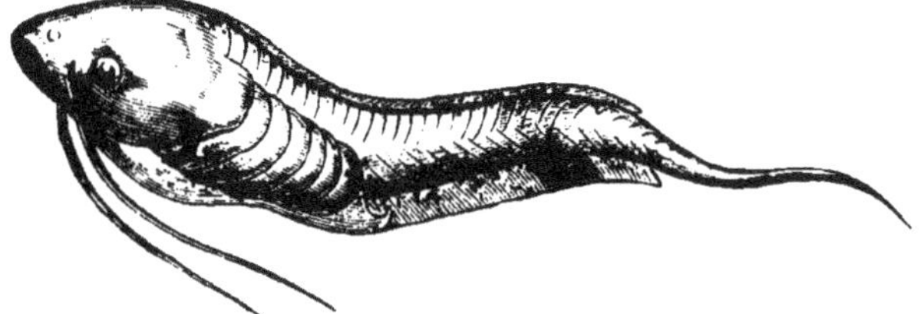

Fig. 627.—Larva of *Dactylethra* (after Parker).

Salamandrinae the tail, and in the *Perennibranchiata* the gills also, or at least the external gill slits (*Derotrema*) persist, and the extremities remain rudimentary, or even the anterior pair alone are developed. Accordingly the series of forms indicated by the classification of these animals offers a strikingly close parallel to the successive phases of the developmental history of the individual forms.

The Amphibia frequently live in water only during larval life; as terrestrial animals in the adult state they choose damp shady places near water, since the cutaneous respiration necessitates in all a moist atmosphere. The food almost always consists of insects and worms, but in larval life principally of vegetable matters. The need of food is, however, relatively small, in correspondence with the low energy of the vital processes, with the sluggishness of their movements and psychical manifestations. The *Amphibia* can live for months without food, and, as for example the *Batrachia*, hibernate buried in the mud.

Fossil remains of this group first appear in the Tertiary period, with the exception of the extinct family of the *Labyrinthodonta* (*Mastodonsaurus*) which belongs to the Trias.

Order 1.—Apoda* (Gymnophiona).

Vermiform Amphibia covered with small scales, without limbs, with biconcave vertebræ.

The external skin of the *Gymnophiona*, which were for a long time classed with the Snakes, contains small scales which are arranged in transverse rings (fig. 628). The internal organisation and the transitory branchial respiration, however, places them amongst the Amphibia, of which group they are in many respects, the most lowly organised. This is especially the case with the skeleton, which is

FIG. 628.—*Siphonops mexicana* (règne animal).

distinguished by the biconcave form of the vertebræ and the persistent notochord. The bony skull, which has two condyles, is firmly united to the facial bones, of which the maxillæ and palatines bear small backwardly-curved teeth. Pectoral and pelvic girdles and limbs are entirely wanting. The small slit-like mouth lies on the lower side of the conical head. The two nares are placed in front on the snout, and near them a blind pit on each side is visible in several genera. These so-called false nares (like the cephalic pits of snakes) lead into canals, which are regarded by Leydig† as sense organs.

The eyes are always small in correspondence with the subterranean mode of life, and are only visible through the skin as small specks. There is neither tympanic membrane nor tympanic cavity.

The *Gymnophiona* live in South America and the East Indies, and feed principally on Worms and Insect-larvæ. Joh. Müller was the first to show that *Cæcilia glutinosa* possesses, in the larval period, a gill slit on each side, which leads to the internal gills. According to Gervais, *Cæcilia compressicauda* is born without a trace of branchial apertures, and Peters has recently confirmed this assertion. Peters,

* Joh. Müller, "Beiträge zur Anatomie und Naturgeschichte der Amphibien," Treviranus: *Zeitschr. für Phys.*, Tom. IV., 1832.
R. Wiedersheim, "Die Anatomie der Gymnophionen." Jena, 1879.

† Fr. Leydig, "Ueber die Schleichlurche (Cæcilia). Ein Beiträg zur anatomischen Kenntniss der Amphibien." *Zeitschr. für wiss. Zool.*, Tom. XVIII.

however, observed on the neck of the recently-born young, which are deposited in water, large vesicles which he regarded as gills.

Fam. **Cœciliidæ.** *Cæcilia lumbricoidea* Daud., South America: *Siphonops mexicana* Dum. Bibr. (fig. 628): *S. annulata* Wagl., Brazil; *Epicrium* Wagl., Ceylon.

The extinct **Labyrinthodonta** of the Triassic, the Permian, and the Carboniferous formations must be regarded as a special order of *Amphibia*. They unite in a remarkable manner the characters of the Ganoids and those of the urodele Amphibians. They possessed an external dermal skeleton, consisting of three broad bony thoracic plates and small scutes on the abdomen, amphicœlous vertebræ and peculiar folded teeth (hence the name of the group) in the Crocodile-like jaws. It has also been shown that they possessed branchial arches in the young state (*Archegosaurus*). The footmarks of gigantic animals (*Chirotherium*), which have been discovered in the Bunter-sandstein in England and Germany (Hildburghausen), and which some have ascribed to *Chelonia* and others to *Marsupials*, are probably due to the *Labyrinthodonta*. Owen has distinguished the oldest forms with armoured skull as *Ganocephala*. *Archegosaurus Dechenii* Goldf., *Labyrinthodon Rütimeyeri* Wied.

Order 2.—Caudata * (Urodela).

Elongated Amphibia with naked skin, usually with four short limbs and persistent tail, with or without external gills.

The body, which is naked, ends with a long, usually laterally compressed, swimming tail, and possesses as a rule two pairs of short extremities far removed from one another. These limbs effect the relatively clumsy movements of the animals on land, but in swimming are used in a much more effective manner as oars. The posterior limbs are completely absent only in exceptional cases (*Siren*) while the anterior limbs remain as short stumps.

Some Urodeles (*Perennibranchiata*) possess throughout life three pairs of branched external gills, in addition to the lungs. Others indeed cast off the gills in the course of their development, but retain throughout life an external gill slit on each side of the neck

* Daudin, "Histoire naturelle gén. et partic. des Reptiles," Paris, 1802 to 1804.

Aug. Duméril, "Observations sur la reproduction dans la ménagerie des Reptiles du Musée d'hist. nat. des Axolots, etc., sur leur développement et sur leurs métamorphoses." *Nouv. Arch. du Musée d'hist. nat. de Paris*, II., 1860.

Alex. Strauch, "Revision der Salamandridengattungen," Petersburg, 1870.

(*Derotrema*). Many, however, even completely lose the latter, and show themselves by their whole organisation to be the highest members of the order (*Salamandrina*). In the two first cases the vertebræ are biconcave, like those of the Fishes, and enclose well-preserved remains of the notochord. The fully-developed *Salamandrina*, on the contrary, have vertebræ with an articular head in front and a concavity behind (*i.e.*, are opisthocœlous).

The eyes, which are small and sometimes rudimentary, are placed beneath the transparent skin, and except in the *Salamandrina* are without distinct lids. In all cases the auditory organ is without a tympanic membrane and tympanic cavity. The nasal apertures are placed at the end of the projecting snout, and lead into slightly developed nasal cavities, which communicate with the buccal cavity by openings placed far forward in the roof of the mouth immediately behind the maxillæ. The buccal cavity is armed with small sharp hooked teeth, which on the lower jaw are arranged in single rows, but on the upper jaw and often on the palatine bone are in double rows. Almost the whole lower surface of the tongue is attached to the floor of the buccal cavity.

The life history of the Axolotl, which was taken by Baird, Cuvier, and others for the larva of a Salamandrine, is very remarkable. According to the observations which were made by Duméril in the Jardin des Plantes at Paris, the young reared from the eggs of the Axolotl under suitable conditions lose the gill tufts and develop into a form which agrees with the Salamandrine genus *Amblystoma*, while the specimens which were originally introduced from Mexico preserve the Perennibranchiate form in the sexually adult condition. Species of *Triton* also have occasionally been found with perfectly developed gill tufts in the sexually adult state.

Sub-order 1. **Ichthyoidea.***

Urodela with three pairs of external gills or without them, but with persisting branchial aperture; with fish-like biconcave vertebræ and well-preserved notochord.

The *Ichthyoidea* represent the lowest grade among the *Urodela* with regard to their respiration, the structure of their skeleton, and their whole organisation; and to a certain extent represent persistent developmental stages of the *Salamandrina*. The eyes are small, and

* Configliachi and Rusconi, "Del Proteo anguino di Laurenti." Paris, 1819.
Hyrtl, "Cryptobranchus japonicus." Wien, 1868.

are covered by the transparent integument. The palatal teeth (*Siren*) are arranged in rows like the teeth of some Fishes, or form a curved arch at the anterior end of the palatine bones. The extremities also are weak and reduced; the anterior end with three or four digits, and the posterior with two to five jointed digits. The digits may, however, be rudimentary and be without distinct joints.

Amongst the Tertiary remains of this group the gigantic *Andrias Scheuchzeri*, which became famous as *Homo diluvii testis*, is worthy of remark.

Tribe 1. **Perennibranchiata.** With persistent gills, usually without maxillary bones. The vomer and palatine bone with rows of teeth.

Fam. **Sirenidæ** (Armmolche). With elongated eel-like body and rudimentary anterior limbs, without posterior limbs. *Siren lacertina* L., South Carolina.

Fam. **Proteidæ** (Olme). Body elongated and cylindrical; anterior limbs short, with three digits; hind limbs placed far back, with two digits. Only

FIG. 629.—*Menobranchus lateralis* (règne animal).

two gill slits on each side. *Proteus anguineus* Laur. Flesh-coloured and living in the subterranean waters of Carniola and Dalmatia.

Fam. **Menobranchidæ**. Body elongated, with tolerably broad head and four-toed limbs. There are four gill slits on each side. *Menobranchus lateralis* Say, Mississippi (fig. 629). Probably holds the same relation to the genus *Batrachoseps* Bonap. that *Siredon* does to *Amblystoma* (Cope). *Siredon pisciformis* Shaw, and *maculatus* Baird., Axolotl. The eggs are laid in the water either singly or in masses. The larvæ, when hatched, are from fourteen to sixteen mm. long, have three pairs of gills, and are still without limbs. Under suitable conditions they lose in the course of further development (according to Duméril, whose observations have been several times confirmed) the gill tufts, the dorsal and caudal crests, and assume the form of *Amblystoma* (second sexual form).

Tribe 2. **Derotrema.** Without gill tufts, usually with a branchial aperture on each side of the neck, with maxillary bones, and teeth which are usually arranged in one row.

Fam. **Amphiumidæ** (Aalmolche). Body elongated eel-shaped, with short extremities far apart from one another. *Amphiuma* L., *A. tridactylum* Cuv. (*A. means* L., with but two digits), Florida.

Fam. **Menopomidæ**. Of Salamander-like appearance, with four anterior and five posterior digits. *Menopoma alleghaniense* Harl., Pennsylvania and Virginia; *Cryptobranchus japonicus* v. d. Hoev., more than three feet long. Japan.

Sub-order 2. **Salamandrina.***

Without gills or gill aperture, with valve-like eye-lids and opisthocœlous vertebræ.

The body, which is shaped more or less like that of a Lizard, is without external gills and gill-slits in the adult state, and always possesses anterior and posterior extremities, of which the first usually have four and the latter five digits. Well-developed eyelids are always present. The palatal teeth form two rows, which unite in the middle line at the posterior margin of the palatine bone. The skin is moist and slimy, and has a more or less uneven warty appearance, owing to the presence of a number of glands which secrete a pungent and irritating milky fluid. These glands are sometimes especially aggregated in the region of the ear.

The Aquatic Salamanders (Newts) lay fertilized eggs on plants. The Land Salamanders, on the contrary, are viviparous and deposit their offspring, which pass through their metamorphosis more or less completely in the uterus of the mother, in water. The spotted Land Salamander produces thirty to forty larvæ, each twelve to fifteen mm. long, with four legs and external gill tufts, while the black Land Salamander of the higher Alpine regions bears only one [two ?] completely-developed offspring. In the latter case, of the numerous eggs which enter the two uteri, only the lowest on each side develops into an embryo, which derives its nourishment from the rest of the eggs which run together so as to form a common mass, and is able to undergo all the stages of development within the uterus.

Fam. **Tritonidæ.** Aquatic Salamanders or Newts. Of slender form, with laterally compressed swimming tail. *Triton cristatus* Laur., large Newt. *Tr. alpestris* Laur. (*igneus* Bechst.), (Bergsalamander). *Tr. tæniatus*, Schn., small Newt.

Fam. **Salamandrinæ.** Land-Salamanders. Clumsy body, with cylindrical tail. *Salamandra maculosa* Laur., the Spotted Salamander: distributed over almost all Europe to North Africa. *S. atra* Laur., Black Salamander, in the high mountains of South Germany, France and Switzerland.

* Rusconi, "Amours des Salamandres aquatiques," Milano. 1821.

Rusconi, "Histoire naturelle, développement et métamorphose de la Salamandre terrestre." Paris, 1854.

v. Siebold, "Ueber das Receptaculum seminis der weiblichen Urodelen," *Zeitschr. für wiss. Zool.*, 1858.

Fr. Leydig, "Ueber die Molche der würtembergischen Fauna," *Archiv für Naturgesch.*, 1867.

R. Wiedersheim, "Salamandrina perspicillata und Geotriton fuscus, etc., Genua. 1875.

Order 3.—BATRACHIA* = ANURA.

Amphibia of stout form, with naked skin, without tail, with procœlous vertebræ and well-developed extremities.

The body is short and stout and is without a tail. On the head are the wide mouth and the large eyes, the iris of which has usually a golden lustre. The eye-lids are well developed, and the lower, which is transparent, can be drawn as a nictitating membrane completely over the eye. The nasal apertures are placed far forward on the extremity of the snout, and can be closed by membranous valves. In the auditory organ there is generally a tympanic cavity, which communicates with the buccal cavity by means of a short wide Eustachian tube and is bound externally by a large tympanic membrane, which is sometimes free and is sometimes concealed beneath the skin.

Only a few of the Batrachia are without teeth (*Pipa*, *Bufo*); as a rule there are small hooked teeth arranged in simple rows at least on the vomer, in the Frogs and *Pelobatidæ*, on the maxillaries and premaxillaries also. The tongue is absent only in a small group of exotic forms; it is usually attached between the rami of the lower jaw in such a way that its posterior part is completely free, and can be protruded as a prehensile organ from the wide mouth.

Ribs are, as a rule, absent, but the transverse processes of the dorsal vertebræ attain a considerable length. A pectoral and pelvic girdle is in all cases present. The former is distinguished by firm connection with the sternum, the latter by the styliform elongation of the ilium. The hyoid bone in its definitive form is considerably simplified; the body, is supported by large anterior horns, while the branchial arches on each side are reduced to a single posterior horn.

In the skin, which is usually naked, glands with an acrid milky secretion are often aggregated in many places, especially in the region of the ear, where they form large glandular projections (*parotid*). Glandular aggregations occur also on the middle division of the hind legs (*Bufo calamita*) and on the sides of the body.

* Rösel von Rosenhof, "Historia naturalis ranarum nostratium," Nürnberg 1758.

Daudin, "Histoire naturelle des Rainettes, des Grenouilles et des Crapauds," Paris, 1802.

Rusconi, "Développement de la grenouille commune." Milano. 1826.

C. Bruch, "Beiträge zur Naturgeschichte und Classification der nackten Amphibien." *Würzb. naturwiss. Zeitschr.*, 1862.

C. Bruch, "Neue Beobachtungen zur Naturgeschichte der einheimischen Batrachier." *Würzb. naturwis. Zeitshr.*, 1863.

A. Ecker, "Die Anatomie des Frosches," Braunschweig, 1864.

Reproduction takes place in the spring. Copulation is confined to an external approximation of the two sexes, and almost always takes place in the water. The male, which sometimes has a wart-like elevation on the thumb (*Rana*) or gland on the arm (*Pelobates*) embraces the female from the back, usually behind the front limbs, and pours out the seminal fluid over the spawn as it issues in strings or in clumps. The individual eggs are surrounded by a viscous layer of albumen which swells up in the water.

The upper half of the ovum is of a darker colour than the lower. The process of segmentation begins in the upper part, and the constrictions which lead to the formation of the segmentation spheres proceed more rapidly in this region than at the lower pole. With the end of segmentation a cavity—the segmentation cavity—appears in the mass of cells: it is placed nearer to the upper pole than to the specifically heavier lower pole. The germ [blastoderm], with medullary plate and folds, arises on the upper half; it quickly, even before the closure of the medullary canal to form the medullary tube, grows round the yolk. After development of the branchial arches and before the mouth is formed, the embryos which have a short tail leave their egg membranes as tadpoles at a stage of development which varies with the species. They then attach themselves by means of two suckers to the gelatinous remains of the spawn (similar suckers are present on the throat of the Triton-larvæ, where however they are stalked). Most larvæ leave the egg membranes with more or less developed rudiments of three pairs of branched external gills (vol. i., fig. 111). The body gradually increases in length and the fin-like tail developes. Later the mouth is formed and the larva begins to feed. Soon the external branchial appendages disappear, while the skin grows over the gill slits like an operculum in such a manner that only one gill aperture is left, through which the water flows out of the branchial chambers on either side.

During these processes fresh lancet-shaped gill-plates are developed in double rows along each branchial arch. The mouth is armed with a horny beak, which is used in gnawing vegetable and also animal substances. The intestine has become very long and much coiled, and the lungs have grown out of the pharynx in the form of long sacs. As development proceeds the hinder extremities first make their appearance on the body of the Tadpole close to the attachment of the strongly-developed swimming tail. As the pulmonary respiration increases, the branchial apparatus becomes more and more reduced, and the animal undergoes an ecdysis, with which is connected not only the loss of the internal gills, but also the appearance of the anterior extremities which have been long concealed beneath the skin. The horny beak is now cast off, and the eyes, which have hitherto been concealed beneath the skin, appear on the surface, and are of considerable size. The larva has now become an exclusively air-breathing, four-legged Frog, which has only to lose its swimming tail in order to acquire its definitive form and be fitted for its terrestrial life (vol. i., fig. 112).

Some *Batrachia* are true land animals (Toads and Tree-Frogs), which especially love dark and damp hiding places: others live indifferently on land or in water. In the first case the five toes of the hind feet are either entirely without a connecting membrane or only have an incomplete one; exceptionally (*Pelobates*), however, they are completely webbed. In the second case, on the contrary, the hind feet are, as a rule, completely webbed. The land Frogs usually seek the water only at spawning time; they crawl, run, and hop on the land, or dig passages and holes in the earth (*Pelobates*, *Alytes*), or they are able to climb up shrubs and trees by means of suctorial discs on the ends of their toes (*Dendrobates*, *Hyla*).

Fig. 630.—*Dactylethra capensis.*

Tribe 1. **Aglossa.** Batrachia without tongue. The tympanic membrane is not exposed. The eyes are placed anteriorly near the angles of the mouth. The hind feet have entire webs. They live in hot localities, especially of the New World.

Fam. **Pipidæ.** Body toad-like, flat, without teeth on jaws and palate. *Pipa dorsigera* Schn., Surinam Toad.

Fam. **Dactylethridæ.** The body is more frog-like, with teeth on the maxillaries and præmaxillaries. *Xenopus* (*Dactylethra*) *capensis* Cuv. (Krallenfrosch), (fig. 630); *Myobatrachus paradoxus* Schleg.

Tribe 2. **Oxydactylia.** *Batrachia* with freely movable tongue and pointed fingers and toes.

Fam. **Ranidæ.** Water-Frogs. Batrachians with long hind limbs, which are adapted for jumping, and the toes of which are usually connected by entire swimming membranes. There are small hooked teeth on the maxillaries, præ.

maxillaries, and usually also on the vomer. *Rana esculenta* L., the green Frog. Green with dark spots and yellow longitudinal streaks on the back. The male has two vocal sacs. Leaves its place of concealment at the end of April, and spawns at the end of May or the beginning of June. On the banks of stagnant water. *R. temporaria* L., the brown frog, with dark spots on the head in the auditory region. It appears very early, and copulates in March; but only remains in the water to spawn, and then frequents meadows and fields. Steenstrup has divided this frog, which is widely divided over Europe, into two species (*R. oxyrhina, platyrhina*). *R. mugiens* Daud., Bull-frog. North America; *Pseudis paradoxa* L., South America, distinguished by the size of its larvæ.

Fam. **Pelobatidæ.** Land-frogs. Toad-frogs. With more or less warty, rough, and richly glandular skin, and clumsy toad-like form; with teeth on the maxillaries. *Alytes obstetricans* Laur. (fig. 626); *Pelobates fuscus* Laur.; *Bombinator igneus* Rös. (Unke, Feuerkröte).

Fam. **Bufonidæ.** Toads. Of clumsy build, with warty glandular skin (ear-glands) and toothless jaws. The posterior feet have five digits, and are but little longer than the anterior, so that the animal is unable to spring with the same agility as the Frogs; but they can in many cases run with great speed. *Bufo vulgaris* Laur., the common Toad; *B. viridis* Laur. (*variabilis*), the green Toad; *B. calamita* Laur. (Kreuzkröte).

Tribe 3. **Discodactylia.** Batrachians with tongue and with broad digits, the points of which are provided with suctorial discs.

Fam. **Hylidæ.** Tree-frogs. With maxillary teeth and without parotids. *Hyla arborea* L., Tree-frog, cosmopolitan; *Nototdelphys ovifera* Weinl., Mexico. The female has a brood-pouch on the posterior part of the back. The larvæ have bell-shaped external branchial vesicles. *Phyllomedusa bicolor* Bodd., South America; *Dendrobates tinctorius* Schn., Cayenne.

CHAPTER VII.

Class III.—REPTILIA.*

Scaly or armoured cold-blooded animals with exclusively pulmonary respiration and two ventricles incompletely separated from one another. Embryos with an amnion and an allantois.

The body-form of the *Reptilia* varies far more than does that of the *Amphibia*, but repeats on the whole the types described for the latter class. The trunk still plays the principal part in locomotion, and accordingly the vertebral column presents a uniform segmentation adapted for serpentine movements. The body, except in the

* J. G. Schneider, "Historia Amphibiorum naturalis et literaria." 1799 to 1801.
A. Günther, "The Reptiles of British India," London. 1864.
E. Schreiber, "Herpetologia europaea," Braunschweig. 1875.

Tortoises, is elongated and more or less cylindrical, and is either altogether apodal, as in the Snakes, or is provided with two or four extremities, which as a rule serve only to support and push on the body which glides along the ground on its belly. In correspondence with this mode of locomotion a cervical region is scarcely at all marked, and even when more developed is always relatively rigid; the tail on the other hand is long and movable.

The skin, as opposed to the predominating soft and naked skin of the *Amphibia*, is tough and firm, in consequence of ossifications of the cutis as well as of a cornification of the epidermis. The former may give rise to bony scutes, overlapping one another in a tectiform manner (*Scincoidea*), or to larger bony plates, which constitute a hard, more or less continuous, dermal armour (*Crocodiles, Tortoises*).

In general pigments are present in the dermis as well as in the deeper layer of the epidermis; they determine the diverse colouring of the skin, and sometimes cause a true change of colour (green Tree Snake, *Chamæleon*). Cutaneous glands are also widely distributed among the *Reptilia*. Many Lizards in particular possess rows of glands on the inside of the femur and in the anal region, which open by distinct pores sometimes on wart-like protuberances (femoral pores, anal pores). In the Crocodiles, too, larger groups of glands are placed beneath the dermal armour both at the sides of the anus and on the sides of the rami of the lower jaw.

The **skeleton** only exceptionally presents the embryonic form of a cartilaginous cranial base and persistent notochord. The vertebral column is more distinctly divided into regions than is that of the *Amphibia*, although the thoracic and lumbar regions still allow of no sharp limitation. In the cervical region the first vertebra becomes the atlas and the second the axis. While fossil Hydrosaurians and the *Ascalabota* possess biconcave vertebræ, the vertebral bodies of other Reptiles are always bony and generally procœlous.

Ribs are very generally present, often along the whole length of the trunk. In the Snakes and the snake-like Lizards, in which a sternum is absent, all the vertebræ of the trunk with the exception of the atlas bear ribs, which, to compensate for the absence of limbs, are capable of free movements. In the Lizards and Crocodiles (fig. 573) there are short cervical ribs. The thoracic ribs are joined to a sternum by means of special sternocostal pieces. In the Crocodiles there is in addition an *abdominal sternum*, which extends over the belly to the pelvic region, and is composed of a number of ventral ribs (without dorsal part). The sacral vertebræ, of which there are usually two, have very large transverse processes and ribs.

The skull (fig. 631) articulates with the atlas by means of an unpaired, often trifid condyle of the occipital bone, and presents a complete ossification of nearly all its parts, the primordial cranium being almost completely replaced. In the occipital region all four elements are present as bones; but the basi-occipital (Tortoises) and the supra-occipital (Crocodiles and Snakes) may be excluded from the boundary of the *foramen magnum*. In the periotic capsule there is a *fenestra rotunda*, as well as the *fenestra ovalis* with the columella. The *opisthotic*, which usually fuses with the exoccipital, takes part in bounding the *fenestra ovalis* (in the Tortoises the exoccipital and the opisthotic are separate).

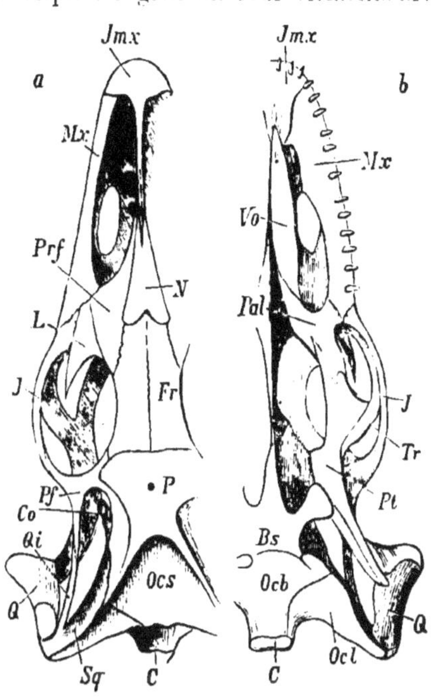

FIG. 631.- Skull of *Monitor* (after Gegenbaur). *a*, from above. *b*, from below. *C*, occipital condyle; *Ocs*, supra-occipital; *Ocl*, exoccipital; *Ocb*, basi-occipital; *P*, parietal; *Fr*, frontal; *Pf*, post-frontal; *Prf*, præfrontal; *L*, lacrymal; *N*, nasal; *Sq*, squamosal; *Q*, quadrate; *Qi*, quadratojugal; *J*, jugal; *Mx*, maxillary; *Jmx*, præmaxillary; *Co*, columella; *Bs*, basi-sphenoid; *Pt*, pterygoid; *Pal*, palatine; *Vo*, vomer; *Tr*, os transversum.

The *prootic*, on the other hand, is separate in all Reptiles; at its anterior there is in front of the lateral parts of the occipital region a separate *prootic*, at the front margin of which is the foramen for the third branch of the trigeminus. The *epiotic* is fused with the supra-occipital. The anterior expansion of the cranial capsule, and the development of the sphenoidal region, present considerable differences. At the base of the skull there is a *basisphenoid* in place of the *parasphenoid*. The *alisphenoids* and the *orbitosphenoids* are as a rule wanting, and are often replaced by processes of the parietals (*Chelonia*) or of the fronto-parietals (*Ophidia*). In the *Chelonia* and Lizards there is a large membranous interorbital septum, which may also contain ossifications. The bones of the roof of the cranium are always very large—sometimes paired, sometimes unpaired. The frontal bone in many cases takes no part in the formation of the roof of the cranial cavity, and only lies on the interorbital septum. Behind the lateral parts of the frontal in the temporal region are the *postfrontals* (*Pf*). In the ethmoidal region the median part remains in part cartilaginous, and is covered above by the paired *nasal* bones (*N*), and at the base by the *vomer* (*Vo*), which in the Snakes and Lizards is paired. The lateral parts are always separate from the median, and are known as the lateral ethmoids or praefrontals (*Prf*). In the Lizards and Crocodiles *lachrymals* (*L*) are present on the outer side of the praefrontals, bounding the anterior margin of the orbits.

The *squamosal* (*Sq*) is more intimately applied to the cranium, and the *quadrate* (*Q*) is always a strongly developed bone. In *Chelonia* and *Crocodilia* the quadrate and maxillo-palatine apparatus are immovably united with the wall of the skull; in Snakes and Lizards, on the other hand, they are more or less freely movable. In the first case not only are the large pterygoid and palatine bones fused with the sphenoid, but the quadrate bone is very firmly connected with the superior maxillary (*i.e.* jugal) arcade. In the Crocodiles a transverse bone (*os transversum*) is developed between the pterygoid and maxillary, and also a superior temporal arcade by which the squamosal is connected with the postfrontal on either side. In the Lizards, in which the maxillo-palatine apparatus and quadrate are movably articulated to the skull, the jugal arch is completely absent [*i.e.*, the jugal is not connected with the quadrate by bone]. On the other hand, these animals possess not only a transverse bone (os transversum) (fig. 631, *Tr*), already mentioned for the Crocodiles, but also a column-like bone—the *columella*—which extends between the parietal and pterygoid. The facial bones are, however, most movable upon one another in the *Ophidia*, which are without the jugal arcade, but present a large os transversum. The two rami of the mandible, which in all *Reptilia* and lower Vertebrates are composed of several pieces, are in these animals connected at the symphysis by an elastic band, an arrangement which permits of considerable extension towards the sides.

The visceral skeleton is reduced to the hyoid bone, from the anterior arch of which the dorsal element (*hyomandibular*) is separated off, enters into relation with the auditory apparatus, and is known as the *columella*. The hyoid bone is most reduced in the Snakes.

The limbs and their girdles are completely absent in most Snakes. In the *Peropoda* and *Tortricidae*, however, traces of hind limbs are found in the anal region, but they are hidden beneath the skin, except the terminal part, which bears a claw. In the *Lacertilia* the extremities present very various grades of development; while the pectoral and pelvic girdles are without exception present, though they are sometimes very rudimentary, the anterior and posterior limbs may be completely absent (Blindworms), or the one pair may be present without the other as small rudiments. In most cases, however, both pairs of extremities are completely developed, and provided with five digits. Sometimes the digits are connected by swimming membranes (Crocodiles), or the extremities are modified to form flat swimming fins (fossil Hydrosaurians and Turtles).

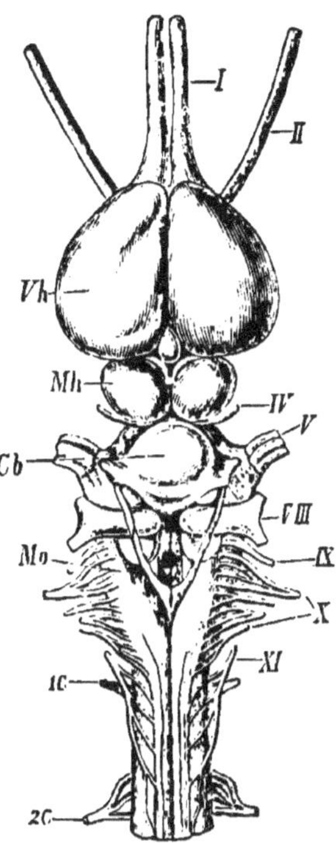

FIG. 632.—Brain of the *Alligator* seen from above (after Rabl-Rückhard). *Vh*, prosencephalon (cerebral hemispheres); *Mh*, mesencephalon (corpora bigemina); *Cb*, cerebellum; *Mo*, medulla oblongata; *I*, olfactory nerve; *II*, optic; *IV*, trochlear (fourth); *V*, trigeminus (fifth); *VIII*, auditory nerve; *IX*, glossopharyngeal (ninth); *X*, vagus (tenth); *XI*, spinal accessory (eleventh); 1*C*, first spinal nerve; 2*C*, second spinal nerve.

The **nervous system** (fig. 632) is decidedly higher than that of the *Amphibia*. The hemispheres are distinguished by their considerable size, and begin to cover the mesencephalon. The cerebellum shows various grades of development progressing from the Snakes to the Crocodiles, and in the latter recalls that of Birds by the contrast of its large median lobe and its small lateral appendages. Of the cranial nerves the facial is no longer united with the trigeminal, and the glossopharyngeal appears as an independent nerve, which has, however, several connections with the vagus. The spinal accessory also arises independently except in the Snakes. Finally the hypoglossal, which passes out through a single or double opening in the skull, enters the category of the cranial nerves.

The eyes are without lids in the Snakes, Geckos, and Amphisbaenas, but are protected in these animals by a transparent capsule, which is separated from the cornea by a space filled with lachrymal fluid. In all other cases there is an upper and lower eyelid. An independent nictitating membrane at the inner angle of the eye is always accompanied by the appearance of a special gland (*Harderian gland*). Peculiar folds of the choroid, which correspond to the processus falciformis of the eyes of Fishes and to the so-called pecten of the eye of Birds, are present in the eye of Lizards.

The auditory organ has a simple tubular cochlea and a corresponding fenestra (*fenestra rotunda*). A tympanic cavity with *Eustachian tube* and tympanic membrane is wanting only in the Snakes and apodal Lizards. In these cases the operculum, which covers the *fenestra ovalis*, and the columella which is attached to the operculum, are buried among the muscles, as in numerous *Amphibia*. When a tympanic cavity is present, the columella is applied by its cartilaginous end to the tympanic membrane, which in many Lizards is still concealed beneath the skin, while a wide Eustachian tube leads into the pharynx. A cutaneous fold above the tympanic membrane of the Crocodiles may be regarded as the first rudiment of an external ear.

The olfactory organ of the *Reptilia* shows, principally in the *Chelonia* and *Crocodilia*, a considerable augmentation of the surface of the mucous membrane, the folds of which are supported by cartilaginous turbinals. The external nares can be closed only in the Water-Snakes and the Crocodiles by an arrangement of valves. In the Crocodiles and Chelonians the internal nares open far back on the palatal part of the mouth. In the Snakes and Saurians there is also a second olfactory organ embedded between the turbinals and the vomer (*nasal glands*, Rathke, *Jacobson's organ*, Leydig), the nerve of which arises at the end of the olfactory lobe, and is spread out like a cup around a cartilaginous papilla.

The sense of taste is by no means always located in the tongue, since in Snakes and many Lizards this organ serves for feeling, and in other cases—*e.g.*, the Chamæleon—is used as a prehensile organ. Leydig* has recently discovered cup-shaped sense-organs in the buccal cavity of Snakes and Saurians. In the Snakes they are arranged alongside the rows of maxillary teeth, in Saurians they are embedded in small pits of the connective tissue.

Alimentary canal.—Excepting in the *Chelonia*, whose jaws possess a horny cutting investment, which constitutes a kind of beak, the jaws are provided with conical or hooked prehensile teeth, which hold fast the prey, but cannot masticate it. As a rule, the teeth are confined to the jaws, and are always arranged

* Fr. Leydig, "Zur Kenntniss der Sinnesorgane der Schlangen," *Arch. für mikr. Anatomie*, Bonn, 1872.

in a single row; sometimes they are fastened to the upper edge (*acrodont*), sometimes to an external, strongly projecting ledge of the flat dental groove (*pleurodont*), rarely, as in Crocodiles, they are wedged into special alveoli. Hooked teeth may also be present on the palatine and pterygoid bones, and in this case they frequently (*e.g.*, in non-poisonous Snakes) form an inner arched row on the roof of the mouth. In the poisonous Snakes special teeth of the upper jaw are traversed by a groove or canal, and enter into close relation with the ducts of poison glands, the secretion of which passes into the wound through the groove or canal in the poison teeth. Salivary glands are found in Snakes and Lizards, both in the lips and on the lower jaw, and a sublingual gland may also be present. The possession of the latter is characteristic of the *Chelonia*.

The œsophagus is very long, and is capable of an extraordinary degree of dilatation. Its walls are usually folded longitudinally, but they may also be beset with large papillæ, as in the Turtles. The stomach is usually arranged longitudinally, except in the *Chelonia*, which possess like the Frogs a transversely-placed stomach. The stomach of the Crocodiles resembles that of Birds, both by its rounded form and by the strength of its muscular walls. The small intestine is but little coiled, and remains relatively short: in the Land-Tortoises alone, which live on vegetable matter, is the intestine more than six or eight times longer than the body. The broad large intestine (rectum) usually begins with an annular valve, and often with a cæcum, and leads into the cloaca, which opens beneath the root of the tail by a round opening, or as in the Snakes and Lizards by a transverse slit (*Plagiotrema*). Liver and pancreas are never absent.

The *Reptilia* breathe exclusively by lungs, which have the form of spacious sacs, with alveolar projections of the walls, or with wide spongy cavities (Tortoises and Crocodiles). In the Snakes and snake-like Lizards the lung on one side is more or less reduced, while the other obtains a correspondingly greater size. The posterior end of the latter loses not only the cellular alveolar spaces, but also the respiratory vessels, and has the form of an air-reservoir (foreshadowing the air-sacs of Birds), which renders respiration possible during the slow process of swallowing. The afferent air-passages are always differentiated into a larynx beginning with a slit-like glottis, and into a long trachea and bronchial tubes, supported by cartilaginous and often bony rings. A membranous or cartilaginous epiglottis is present in many Tortoises, Snakes, and Lizards. The Geckos and Chamaeleons alone have a vocal apparatus. The renewal of the air necessary for respiration is, except in the Chelonia, always effected by aid of the ribs.

The **circulatory organs** (fig. 60) present various grades of development, even to the complete division of the heart, and to the separation of the venous and arterial blood. Not only are there two auricles which are distinct even externally, but the ventricle also is divided into a right and left chamber. The partition wall of the ventricles is indeed perforated in Snakes, Lizards, and Chelonians; but in the Crocodiles it is complete, and effects the separation into a right and left ventricle. In the first cases the pulmonary arteries and the aortic trunks arise from the wide thin-walled right division of the ventricle. In the Crocodiles, on the other hand, the pulmonary arteries and the aortic trunks have a separate origin (fig. 633). The complete number of aortic arches is present only during the fœtal life; in later life their number becomes much reduced. Originally five pairs of vascular arches—as also in Birds and Mammals—are present; they embrace the gullet and join to form the two roots of the aorta. Most of them, however, undergo reduction, losing their connections with each other, so that finally each aortic root (Saurians) arises from two vascular arches, or is the prolongation of a single aortic arch. The aorta, which passes out from the heart, is divided into three trunks—a right and left aorta and a pulmonary artery—each with a separate opening into the ventricle [*i.e.*, three distinct arterial trunks leave the

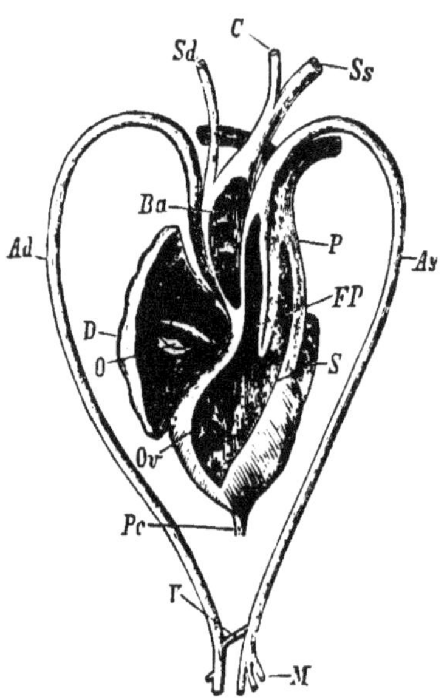

FIG. 633.—Heart and large vascular trunks of *Alligator lucius* seen from the ventral side and partly opened (after Gegenbaur). *D*, right auricle; *S*, left auricle; *O*, ostium venosum of the right auricle; *Ov*, auriculo-ventricular aperture; *Ba*, bulbus arteriosus; *C*, carotis primaria; *Sd*, *Ss*, subclavian arteries; *Ad*, right aortic arch; *As*, left aortic arch; *P*, pulmonary artery; Γ, connection of right with left aortic arch; *M*, mesenteric artery; *Pc*, connection of the heart with the pericardium; *FP*, position of the foramen Panizzæ.

ventricle or ventricles, and not one only as in the Ichthyopsida]. In the Snakes and Lizards the left arterial trunk is prolonged into the left aortic root without giving off vessels (fig. 60, *Aos*), while the right and larger before being prolonged into the right aortic root gives off a common stem for the two carotids (fig. 60, *l*), between which and the corresponding aortic roots a connecting vessel (ductus Botalli), constituting a second persistent aortic arch, may be retained (many Lizards). In the *Chelonia* the right aortic arch likewise gives off the carotids and subclavians, while the left gives off the visceral arteries. In consequence of the very small size of the aortic root of the latter, the aorta appears to be mainly a prolongation of the right aortic arch. Crocodiles present the same arrangements, but in them the right arterial trunk arises from the left ventricle and receives arterial blood from the latter. In this case also, in spite of the complete division of the heart, the mixture of venous and arterial blood is not wholly avoided, since there is a communication—the *foramen Panizzæ*—between the right and left aortic arches. When the separation of the two ventricles is incomplete, mixture of the two kinds of blood takes place in part in the heart, although the entrance into the pulmonary vessels can by special valvular arrangements be separated from the ostia of the arterial trunks in such a manner that the arterial blood principally flows into the latter, and the venous into the former (Brücke). In the venous system there is, as in the *Amphibia*, a renal-portal as well as an hepatic-portal circulation. In the *Chelonia* and *Crocodilia*, however, the renal-portal system is more and more reduced, for the greater part of the blood of the iliac veins passes to the liver. The system of lymphatic vessels presents extraordinarily numerous and wide lymph spaces, and is arranged exactly like that of the *Amphibia*. Contractile lymph hearts have only been discovered in the posterior part of the body at the junction of the trunk and tail. They are paired and situated on the transverse processes or ribs.

The **kidneys** of Reptiles belong, as in Birds and Mammals, to the hinder region of the trunk, and correspond accordingly only to the posterior broad part of the Amphibian kidney. In Lizards and Chelonians a urinary bladder projects on the anterior wall of the cloaca. The urine is not by any means always fluid, but is often a whitish mass of firm consistency, and contains uric acid.

The **generative organs** (fig 634) resemble those of Birds. The morphological relations of the generative organs of the higher vertebrates are attained, inasmuch as the anterior region of the kidney (primordial kidney and Wolffian duct), which in the *Amphibia* still functions as a urinary organ, is here transformed into the efferent apparatus of the testis (epididymis and vas deferens), and in the female sex vanishes or rarely persists as a rudiment (*Rosenmüller's* organ, canal of *Gärtner*), while in the female the Müllerian

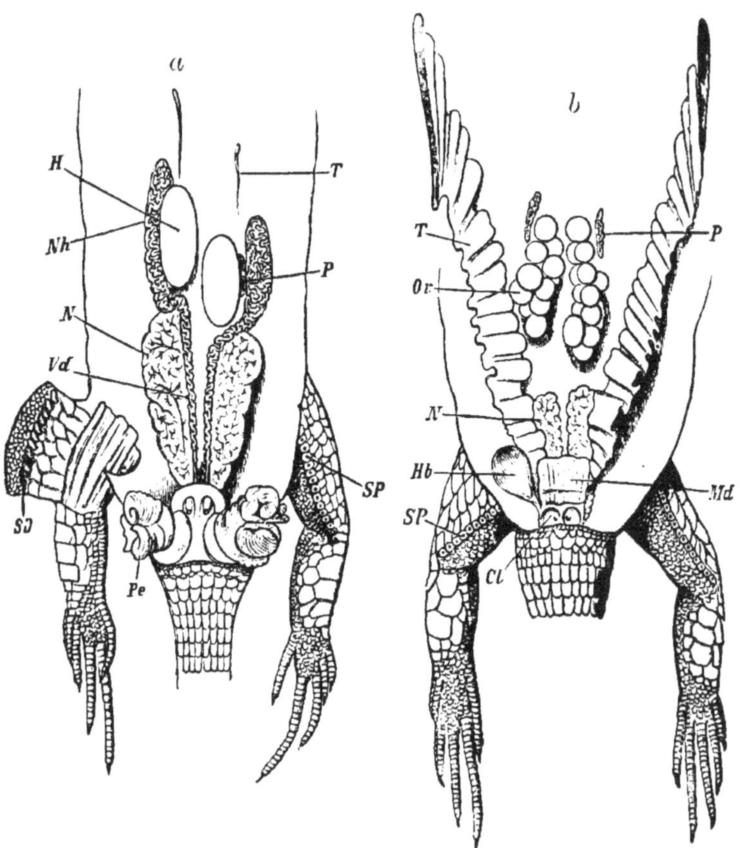

FIG. 634.—Urogenital apparatus of *Lacerta agilis* (after C. Heider). *a*, of the male. *N*, kidney; *H*, testis; *Nh*, epididymis; *Vd*, vas deferens; *P*, remains of the primordial kidney (Wolffian body); *T*, remains of the Müllerian duct; *Pe*, penis; *SP*, pores of femoral glands; *SD*, femoral glands.—*b*, Of the female. *Hb*, urinary bladder; *Md*, rectum (cut); *Cl*, cloaca *Ov*, ovary; *T*, Müllerian duct developed into oviduct.

duct becomes the oviduct. The oviducts as well as the vasa deferentia open separately into the cloaca. The oviducts begin with a wide peritoneal ostium, have a sinuous course, and secrete the calcareous and usually membranous egg shells. The eggs in many cases remain a long time in the terminal part of the oviduct (which may then be termed the uterus), sometimes till the embryonic development is completed.

The males always possess external organs of copulation, to which in the females similarly arranged rudiments (*clitoris*) correspond In Snakes and Lizards (*Plagiotrema*) these organs consist of two protrusible hollow tubes, which are either smooth or covered with spines and lie retracted in a pouch-like cavity behind the cloaca. When protuded their surface is traversed by a groove which conveys the sperm from the genital openings of the cloaca. In the *Chelonia* and *Crocodilia*, on the other hand, an erectile penis supported by fibrous bodies projects on the anterior wall of the cloaca. This penis also has a groove in which the semen is received and passed on, but it cannot be invaginated like the two penises of Snakes and Lizards. Copulation always leads to the fertilization of the ova within the body of the mother. But few Reptiles, e.g., *Pelias berus* amongst the Snakes, and the Blindworm amongst the Lizards, are viviparous. Most forms are oviparous, and bury their eggs in damp earth in sheltered warm spots, and take no further trouble about their fate. Some of the Pythons, however, are an exception to this; inasmuch as they coil their body together over the eggs which they have laid, and afford warmth and shelter to the developing brood.

The developmental history * of the Reptiles is very similar in its general features to that of Birds. The ovum is relatively large, and is sometimes surrounded by a layer of albumen within the shell. The segmentation is partial and leads to the formation of a discoidal blastoderm, with primitive groove and medullary folds. Before the medullary folds have closed, a transverse depression appears at the dilated anterior end of the medullary groove: this depression is the head fold, which leads to the origin of the cranial flexure, a feature always found in the higher vertebrates. [The cranial flexure is found in all vertebrates except Amphioxus.] The embryo which at first lies flat on the yolk, becomes gradually more and more sharply marked off from the latter, for the ventral walls of the boat-shaped body grow together, and leave only a small opening (*umbilicus*).

* C. E. v. Baer, "Ueber Entwickelungsgeschichte der Thiere," II., Königsberg, 1837.
H. Rathke. "Entwickelungsgeschichte der Natter." Königsberg, 1839.
H. Rathke, "Ueber die Entwickelung der Schildkröten," Braunschweig, 1848.
H. Rathke. "Untersuchungen über die Entwickelung und den Körperbau der Crocodile." Braunschweig, 1866.
L. Agassiz. "Embryology of the Turtle." *Contributions to the Nat. Hist., etc.* II., Boston, 1857.

Thus it happens that the central digestive canal, which at first has the form of a shallow groove, becomes converted into a tube which remains for some time connected with the yolk at the umbilicus by a narrow duct.

The appearance of a membrane enclosing the embryo and known as the *amnion* (fig. 635) is characteristic. The amnion arises in the following way. The outer layer (somatopleure) of the blastoderm is raised at the anterior and posterior end of the embryo, and forms two folds covering the head and tail end. These folds soon extend over the lateral portions, and fuse over the body of the embryo, so as to form a closed sac filled with fluid. Another organ which is characteristic of the higher vertebrates is the *allantois*. This arises at the posterior end of the body as a vesicular evagination of the ventral wall of the alimentary canal, and grows out to a sac of considerable size. The walls of this sac, which is filled with fluid, are, unlike those of the amnion which is entirely without vessels, extraordinarily vascular and represent an embryonic respiratory organ, which in the long duration and complicated developmental processes of embryonic life is of great importance. The appearance of the allantois is correlated, not only with the disappearance of branchial respiration, but also with the complete absence of a metamorphosis; the young animal being completely organised when it leaves the egg.

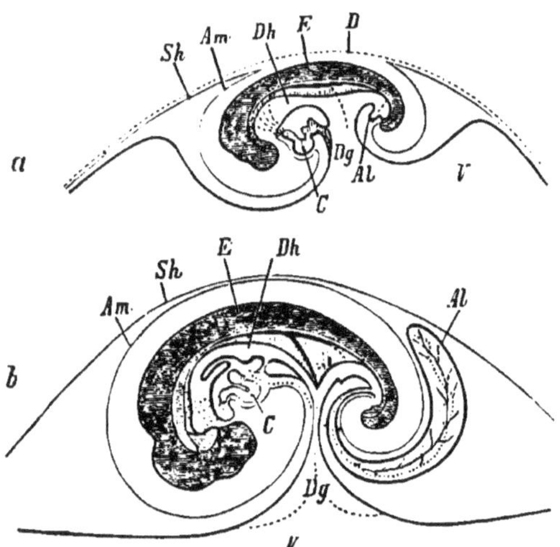

FIG. 635.—Two stages in the development of the chick (after v. Baer) to show the development of the amnion and the allantois.—*a*, The two folds of the amnion are still widely separate from one another; only the first rudiments of the allantois are visible.—*b*, Later stage with closed amnion. *E*, embryo; *D*, vitelline membrane; *Am*, amnion; *Sh*, Serous membrane; *Dh*, alimentary cavity; *Dg*, umbilical passage; *V*, Yolk; *C*, heart; *Al*, allantois.

Some Snakes and Lizards extend far north, while the *Crocodilia* are confined to the torrid zone, and only isolated examples of the *Chelonia* belong to the torrid zone.

The *Reptilia* of the cold and temperate regions fall into a sort of winter sleep, and in the hot climates there is a summer sleep which comes to an end with the beginning of the rainy season.

Most Reptiles are very tenacious of life, and can exist a long time without food and with limited respiration, and are capable, though in a less degree than the *Amphibia*, of reproducing injured or lost parts of the body.

The oldest fossil remains of Reptiles belong to the Primary period, but appear only very sparingly in this period, being confined to the Kupferschiefer formation (*Proterosaurus Speneri*). The Secondary period (Trias and Jura) can show a far greater variety of forms. At this time the Saurians and Hydrosaurians were predominant. The scaly Lizards first appear in the upper strata of the Jura, and are most abundant in the Tertiary period, which also presents a few remains of Snakes. *Chelonia* first appear—excepting the doubtful footprints of the Trias—in the Jura. Land-Tortoises are first met with in the Tertiary formations.

Sub-class 1.—PLAGIOTREMATA (LEPIDOSAURIA).

Reptiles with scales and dermal shields, either apodal or provided with extremities. They have a transverse anal slit and a double penis.

Order 1.—Ophidia * (Snakes).

Apodal Plagiostremata without pectoral girdle; with bifid protrusible tongue; usually with freely movable, always displaceable, maxillary and palatine bones; without urinary bladder.

The Snakes are chiefly characterized by the absence of extremities, and by the distensibility, sometimes extraordinary, of the mouth and pharynx. They cannot, however, be sharply separated from the Lizards. Formerly the limitations of this order rested entirely on the absence of extremities, and thus not only were the *Cæciliadæ* amongst the *Amphibia*, but also the Blindworms and other genera of apodal Lizards, included in it. The *Amphisbænidæ* also were

* Gray, "Catalogue of Reptiles in the collection of the British Museum," Part III., Snakes. London, 1849.
Günther, "Catalogue of Colubrine Snakes in the collection of the British Museum," London, 1858.
Jan, "Iconographie générale des Ophidiens," Livr. I.—XXVII. Paris, 1860—1868.
Lenz, "Schlangenkunde," 2 Auflage, Gotha. 1870.

formerly regarded as Ophidia. Moreover, many Snakes have the rudiments of posterior extremities, which are placed at the root of the tail and have a conical claw projecting at the side of the anus. No Snake has a pectoral girdle or any trace of an anterior pair of extremities.

In the skull of the Snakes (fig. 636) the temporal arcades are absent [*i.e.*, the postfrontal is not directly connected with the squamosal, and there is no jugal or quadrato-jugal connecting the maxilla with the quadrate]. The cranial cavity is very long. The anterior and middle parts of its lateral walls are formed by descending wing-like processes of the parietal and frontal bones. The maxilla is connected with the palato-pterygoid arcade by an os transversum, and these bones are so completely movable upon one another and the cranium, that the mouth is capable of being considerably dilated and laterally extended. The quadrate bone is very movably articulated with the squamosal, which is also movably attached on the occipital region. The two rami of the lower jaw are as movable as are the parts of the maxillo-palatine apparatus. They are connected at the symphysis by an elastic ligament, which permits of a considerable amount of lateral movement.

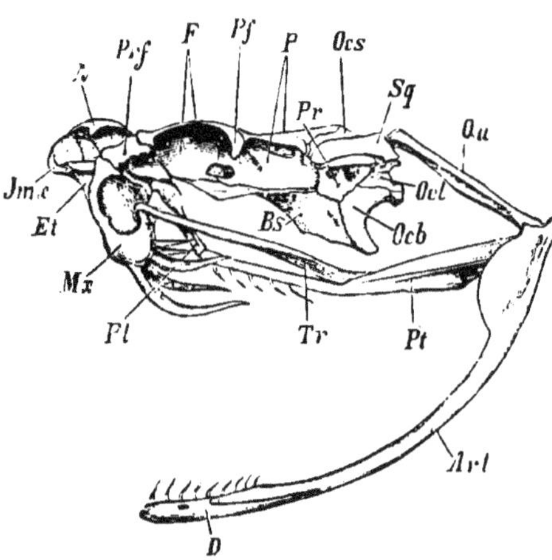

Fig. 636.—Skull of *Crotalus horridus*. *Ocb*, Basioccipital; *Ocl*, exoccipital; *Ocs*, supraoccipital; *Pr*, prootic; *Bs*, Basi-sphenoid; *Sq*, squamosal; *P*, parietal; *F*, frontal; *Pf*, post-frontal; *Prf*, præ-frontal; *Et*, median ethmoid; *N*, nasal; *Qu*, quadrate; *Pt*, pterygoid; *Pl*, palatine; *Mx*, maxillary; *Jmc*, præ-maxillary; *Tr*, transverse; *D*, dentary; *Art*, articulare of the lower jaw.

The armature of the jaws consists of a number of recurved prehensile teeth, which are arranged in a single row on the lower jaw, and usually in a double row on the maxillo-palatine apparatus; they chiefly serve to hold the prey fast while it is being swallowed. Hooked teeth may also be present on the præmaxillæ (*Python*). Only in the *Opoterodonta* are the teeth confined to the upper jaw or to the lower jaw. Besides these solid hooked teeth many snakes possess in the upper jaw grooved teeth, or hollow poison teeth, which are traversed by a canal; the canal is connected with the duct of a poison gland, and through it the secretion of the latter is poured. Frequently the maxilla is much reduced, and contains on each side only one large perforated poison-tooth, near which, however, other larger and smaller supplementary teeth are always placed (*Solenoglypha*). The grooved teeth are rarely more numerous, and are attached to the maxillaries either quite in front (*Proteroglypha*), or behind a row of hooked teeth (*Opisthoglypha*). In both cases the maxilla is larger than that of the *Solenoglypha*. In the *Aglyphodonta*, however, where there are no grooved teeth, the maxilla attains the greatest size and the richest dentition. While the grooved teeth are immovably fixed, the tubular poison teeth are erected, with the maxillaries to which they are attached, when the mouth is opened, and are, when the snake strikes, driven into the flesh of the prey. Simultaneously the secretion of the poison gland, which is forced out by the pressure caused by the contraction of the temporal muscles, is injected into the wound; it is thus mixed with the blood, and quickly causes the death of the victim.

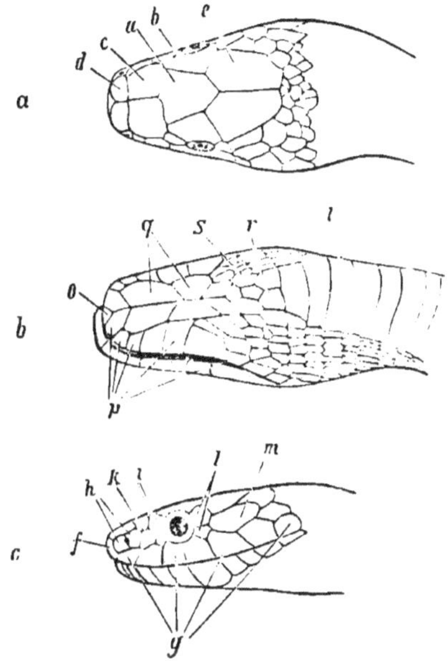

FIG. 637.—Head of *Calopeltis Aesculapii*. *a*, Dorsal view, *b*, ventral, *c*, Lateral view of head of *Tropidonotus viperinus* (after E. Schreiber). *a*, Frontal scute; *b*, supra-ciliary scutes; *c*, posterior nasal scute; *d*, anterior nasal scute; *e*, parietal scute; *f*, rostral scute; *g*, upper labial scutes; *h*, nasal scute; *i*, praeorbital scutes; *k*, loreal scute; *l*, postorbital; *m*, temporal scutes; *o*, chin scute; *p*, lower labial scutes; *q*, mental scute; *r*, cervical scutes; *s*, cervical scales; *t*, ventral scutes.

The hard structures of the integument, which have the form of scales, scutes, and splints, vary much in form, number, and arrangement. While the dorsal surface of the trunk is always covered with smooth or keeled scales; the head is covered with scales as well as with scutes and plates, which, like those of the Lizards, are distinguished according to their special position as frontal, parietal, and occipital scutes; also as rostral, nasal, temporal, and labial scutes, etc.

(fig. 637). The mental scutes—*i.e.*, the scutes in the mental groove on the ventral surface between the rami of the lower jaw (fig. 637, *q*) —may be mentioned as peculiar to most snakes; in front of these two accessory labial scutes on either side form with the median labial scute (*o*) the anterior boundary of the mental groove. The scutes on the abdomen are for the most part broad, and invest the trunk like transverse bands (fig. 637 *b, t*); but scales and small median scutes may also be present here. The ventral surface of the tail, on the other hand, is, as a rule, covered by a double, or rarely by a single, row of scutes. Snakes moult several times in the course of the year: they strip off the whole of the epidermis on which the sculpture of the cutis is repeated.

The internal organisation corresponds with the requirements of the elongated form of the body, as well as with the mode of locomotion and nourishment. A long and extensible gullet with thin walls leads into the dilated saccular stomach, which is followed by a relatively short small intestine. The larynx is placed extraordinarily far forward, and can be projected into the mouth during the long and difficult act of swallowing. The trachea is extremely long, and often contains respiratory air-cells in its course. The left lung is usually entirely rudimentary, while the right lung is correspondingly large, and is transformed at its posterior end into a vesicular air-reservoir.

The auditory organ is without an apparatus for conducting sound. and the eyes have no movable lids. The eye-ball, with its usually vertical pupil, is covered by the skin, which is here transparent, and behind which it is bathed by the lacrymal fluid. The nasal apertures are usually placed quite at the apex or on the lateral margins of the snout. The forked horny tongue serves not as an organ of taste, but as a tactile organ, and is enclosed in a sheath, from which it can be protruded through an indentation of the extremity of the snout, even when the mouth is closed.

The Snakes move principally by means of lateral flexions of the vertebral column. The vertebræ are very numerous, and almost always bear ribs in the region of the trunk. The centra are concave in front and convex behind; they are connected with one another by free ball and socket joints, and by horizontal articular surfaces of the transverse processes in such a manner that dorso-ventral movements are impossible. The ribs are also freely articulated with the vertebral bodies, and can be moved backwards and forwards, movements which are of great use in assisting locomotion. The Snakes run in a certain sense on the extreme points of their ribs, which are attached to dermal scutes; for they move by alternately pushing the ribs forward, and drawing after them the ventral scutes, which are attached to one another and to the ribs by muscles.

The Snakes feed exclusively on living animals, both warm-blooded and cold-blooded, which they attack suddenly, kill and swallow whole without mastication. Swallowing is effected thus: the teeth on the jaws are alternately hooked further and further forwards into the body of the prey, as a result of which the mouth and pharynx of the snake are in a sense gradually drawn over the animal, whose surface is at the same time made slippery by the abundant secretion of the salivary glands. During this process the larynx is projected forward between the rami of the jaws, so that respiration can be maintained. After the completion of this laborious operation of swallowing, the animal appears entirely prostrated, and passes a long period in inactivity, during which the very slow but complete digestion takes place.

Snakes copulate, and are for the most part oviparous. They lay a small number of large eggs, in which the embryonic development may be already far advanced. There are, however, also viviparous Snakes: for example, the sea-snakes (*Hydrophidæ*) and the vipers (*Pelias berus*).

Most of the species distinguished by size and beauty of colour belong to the warmer zones, only the smaller forms extend into northern temperate climates. Many Snakes are fond of the water and are truly amphibious. Others live for the most part on trees and shrubs, or on sandy ground; others exclusively in the sea. In the temperate countries they fall into a kind of winter sleep, in the hot countries they undergo a summer sleep in the dry season.

Sub-order 1. **Opoterodonta.** With narrow, non-distensible, slit-like mouth. and immovably connected facial bones, without or with only a short tail. They have solid hooked teeth only in the upper jaw or in the lower jaw. Posterior limbs present as rudiments. They live beneath stones, or in passages in the earth, and feed on insects.

Fam. **Typhlopidæ.** *Typhlops lumbricalis* Merr. (fig. 638), Antilles. *T. vermicularis* L., Greece. *Stenostoma nigricans* Dum. Bibr., South Africa.

Sub-order 2. **Colubriformia.** Both jaws armed with solid hooked teeth. In the upper jaw the last tooth may be grooved, and then may be either without poison glands, or may be connected with the duct of a small poison gland. This sub-order includes the **Aglyphodonta** and the **Opisthoglypha**.

Fam. **Uropeltidæ.** With short pointed head, mouth not distensible, but with teeth in both jaws. *Uropeltis philippinus* Cuv.

Fam. **Tortricidæ.** With small hardly discernible head and short conical tail. The teeth are small, and there are teeth on the palatine bones. They have a rudiment of the pelvis with small anal claws. *Tortrix scytale* Hmpr., South America: *Cylindrophis rufa* Gray, Java.

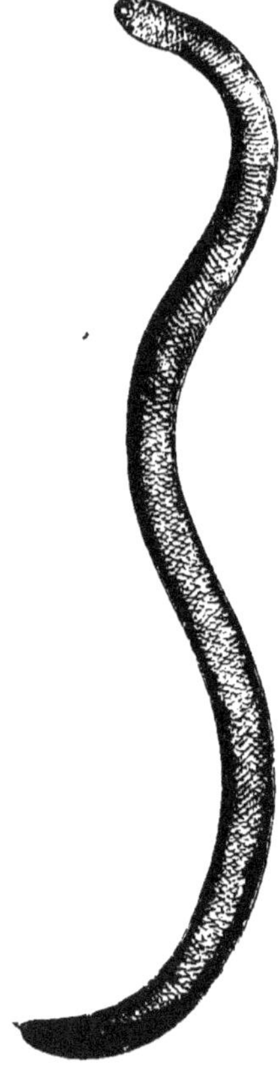

FIG. 638.—*Typhlops lumbricalis* (règne animal).

Fam. **Pythonidæ.** With long oval heads covered with scutes or scales. with rudiments of hind limbs which terminate with an anal claw at the sides of the cloaca. *Eryx jaculus* Wagl., South Europe; *Boa constrictor* L., Brazil: *Python reticulatus* Schn., Sumatra.

Fam. **Colubridæ.** The head is not very broad, and is distinct; it is covered with scutes. The dentition is complete. The tail has a double row of scutes on the under surface. *Coronella austriaca* Laur.=*C. lævis* Lac., widely distributed in Europe; *Liophis cobella* L. Brazil; *Tropidonotus natrix* Gesn., Ringed snake. With obliquely keeled scutes. The species is widely distributed in Europe. *Tr. tesselatus* Meyr.; *Coluber* (*Calopeltis*) *Æsculapii* Gesn.=*C. flavescens* Gm., the snake of Æsculapius. South Europe. Schlangenbad, Austria; *Zamenis atrovirens* Shaw, South Europe; *Herpetodryas carinatus* L., Brazil.

Fam. **Dendrophidæ.** Tree snakes. Body thin and slender, head usually long, flat and distinct from the neck. The ventral scutes usually with two keels. Ventral caudal scutes in two rows. *Dendrophis picta* Gm., East Indies; *Ahætulla smaragdina* Boie, West Africa.

Fam. **Dryophidæ.** Body very long and slender, as is the head; snout thin and sometimes prolonged into a flexible appendage. *Dryophis argentea* Daud., Cayenne.

Fam. **Psammophidæ.** Sand snakes. The posterior tooth of the upper jaw is grooved. *Psammophis lineatus* Dum. Bibr., Mexico; *Cœlopeltis lacertina* Wagl., Egypt.

Fam. **Dipsadidæ.** The body tolerably slender, strongly compressed; with short tail, broad at the end. and very distinct. There are usually posterior grooved teeth. *Dipsas dendrophila* Reinw., East Indies; *D. fasciata* Fisch., West Africa.

Fam. **Scytalidæ.** The hindmost tooth in the upper jaw is the longest, and is grooved. *Scytale coronatum* Dum. Bibr., Brazil. *Oxyrhopus plumbeus* Wied., South America.

Sub-order 3. **Proteroglypha.** Poisonous snakes with large grooved teeth, which are placed anteriorly in the upper jaw, and behind which there are usually solid hooked teeth. The palatine and pterygoid bones, as well as the lower jaw, are armed with hooked teeth.

Fam. **Elapidæ.** Resemble the *Colubridæ*. Head covered with scales; usually with two rows of sub-caudal scutes. *Naja tripudians* Merr., the Cobra, Bengal; *N. haje* L. Cleopatra's Snake, Egypt; *Elaps corallinus* L., the coral Snake, South America. (Fig. 639.)

Fam. **Hydrophidæ.** Sea-Snakes. With scarcely distinct head which is covered with scutes, and compressed body which is prolonged into a strongly compressed

FIG. 639.—*Elaps corallinus* (règne animal).

swimming-tail. They are viviparous. *Platurus fasciatus* Daud., Indian Ocean *Hydrophis* (*Pelamis*) *bicolor* Daud. (fig. 640), Indian Ocean.

Sub order 4. **Solenoglypha.** Snakes with triangular head and relatively short tail. The small upper jaw has a hollow poison tooth on either side, and one or more reserve teeth. Small solid hooked teeth are also present on the palate and in the under jaw.

Fam. **Viperidæ** (Vipers). Head strongly marked off and broad, without pits between the nares and eyes. There are usually two rows of scales on the under side of the short tail. *Vipera aspis* Merr. In wooded mountain regions of South Europe. *V. ammodytes* Dum. Bibr. The sand viper, with a soft horny prominence on the tip of the snout. Italy and Dalmatia. *Pelias berus*,

(Kreuzotter). Common Viper, distinguished by the black-brown zigzag band on the back. Found in the mountain forests of Europe.

Fam. **Crotalidæ.** With a pit between the eyes and nose. *Crotalus durissus* L., Rattlesnake of south-east of North America; *C. horridus* L. South America *Bothrops atrox* L. Brazil.

FIG. 640.—*Hydrophis bicolor* (règne animal).

FIG. 641.—*Pygopus (Bipes) lepidopus* (règne animal).

Order 2.—SAURII * (LACERTILIA)—LIZARDS.

Plagiotrema with pectoral girdle and sternum, usually with tympanic cavity and movable eyelids; with a non-extensible mouth and with urinary bladder.

The Lizards always have an elongated and sometimes a snake-like body. As a rule there are four extremities, which, however, scarcely

* Tiedemann, "Anatomie und Naturgeschichte der Drachen." Nürnberg, 1811.
J. E. Gray, "Catalogue of the specimens of Lizards in the Collection of the British Museum." London. 1845.
Fr. Leydig, "Die in Deutschland lebenden Arten der Saurier." Tübingen, 1872.

carry the body raised from the ground; but in locomotion are used principally for pushing the body forward; they may also be used for clinging (Chamælion), climbing (Geckos), and digging. They usually end with five clawed digits. They are sometimes so short and rudimentary, that they have the appearance of stumps applied to the serpent-like body, and are without separated digits (*Chamæsaura*). In other cases rudiments of the posterior limbs alone exist (*Pseudopus*) (fig. 641), or anterior limbs only are present (*Chirotes*); or finally external limbs may be entirely absent (*Anguis, Acontias, Ophisaurus*). The pectoral and pelvic girdles are however present, and in all Lizards except *Amphisbæna* there is at least a rudiment of the sternum, which increases in size as the anterior limbs become more developed, and then serves for the attachment of a correspondingly greater number of ribs. The ribs are only wanting on the most anterior cervical vertebræ, and sometimes on some of the lumbar as well as on the caudal vertebræ. The anterior ribs present a peculiar modification in *Draco*, being extremely long and serving to support lateral expansions of the skin, which can be used as wings.

The cranial capsule (fig. 631) does not usually extend into the orbital region, behind which it is imperfectly closed by membranous structures (membranous *interorbital septum*). The squamosal is firmly attached to a strongly projecting process (*parotic process*) of the posterior temporal region. The hinder end of the maxilla is frequently connected with the postfrontral (*Pf*) by a bony bridge, the jugal (fig. 631 *J*), which encloses the orbit; while a bone (quadrato-jugal) passes from the jugal to the quadrate, bridging over the temporal region.

An important character of the Lizards as opposed to the Snakes consists in the fact that the bones of the jaws are not movable upon one another. Parts of the maxillo-palatine apparatus are indeed movably connected with the skull (*Hatteria* = *Sphenodon* excepted), especially the pterygoids, which are applied to the articular processes of the basisphenoid, and usually articulate with the quadrate; but the individual bones of the maxillo-palatine apparatus are firmly connected with one another, and with the anterior part of the skull. The pterygoids are firmly attached to the maxillaries by a transverse bone, and serve to support the parietal bones by a rod-shaped *columella* [a bone which extends from the parietal to the pterygoid on each side]. On the top of the skull the parietal bones and the occipital segment are connected by fibrous tissue, and in consequence are slightly movable upon one another. The quadrate bone is movably articulated with the parotic process of the temporal region and supports the lower jaw, the rami of which are firmly connected at the symphysis.

The dentition of the Lizards in form, structure and mode of fixture of the teeth, presents far greater diversity than does that of the Snakes; it is however not so complete since the palate has never an inner row of teeth, but only small lateral groups on the pterygoids. The teeth are almost always attached directly to the bone, either on the edge of the jaw (*Acrodont*), or on the inner side of the jaw (*Pleurodont*). This distinction corresponds to the geographical distribution of the *Iguanas*, those found in the eastern hemisphere being Acrodonts, and those in the western Pleurodonts. The shape of the tongue seems important, and the principal groups are distinguished and named according to this characteristic.

Most Lizards have eyelids, an exposed tympanic membrane and a tympanic cavity. Only the *Amphisbænas* and *Geckos* are without eyelids, and have the same arrangement for covering the eyes as have the Snakes. In the *Scincoideæ* the lower eyelid can be raised like a transparent curtain without hindering the sight. In the *Chamælionidæ* the single eyelid is a muscular cutaneous ring of skin with circular opening.

The integument of Lizards resembles in its general features that of Snakes, but presents much greater variety. Sometimes there are flat or keeled scales, sometimes scutes and larger plates, for the distribution of which on the head the terminology already described for Snakes is used. In addition, more irregular hardenings of the skin may occur—warty protuberances which give the skin an appearance similar to that of the Toads (*Geckonidæ*). On the other hand there are often cutaneous lobes on the throat, crests on the back and on the top of the head, also folds of skin on the sides of the trunk, on the neck, etc. Although the skin of Lizards is in general poor in glands, yet in many forms cutaneous glands and corresponding rows of pores along the inner side of the thigh (fig. 634, *SP*) and in front of the anus are constantly present.

As a rule, the females after copulation (which in temperate regions takes place in summer) lay a small number of eggs; some genera are viviparous (*Anguis*, *Seps*). Most are harmless, and are useful by destroying Insects and Worms; larger species, as the *Iguana*, are hunted for the sake of their flesh. By far the greater number, and all the larger and more beautifully coloured species inhabit the warmer and hot countries.

Fossil remains of Lizards have been found in great numbers, the oldest from the uppermost strata of the Jura. The Lizards of the chalk (*Mosasaurus*, etc.), which are most nearly related to the Monitors, were of gigantic size.

Sub-order 1. **Annulata.** Body snake-like, with hard scaleless skin, which is divided into rings by transverse furrows (fig. 642). These rings are again crossed by longitudinal furrows in such a way that the surface has an elegantly plated, mosaic-like appearance. There are large scutes only on the head and throat. There is no sternum, and the pectoral girdle, except in *Chirotes*, remains very rudimentary.

Rudiments of a pelvis are in all cases present. As a rule extremities are wanting, but small front feet (*Chirotes*) may be present. Eye-lids and tympanic membrane are absent; the small eyes are covered by the integument. A columella is also absent. The tongue is short and thick, without sheath, and the dentition, as in the scaly Lizards, is either acrodont or pleurodont. They are harmless animals, and live for the most part in America, like the *Cæciliadæ*, beneath the ground, usually in ant-hills, and feed on Insects and Worms.

Fam. **Amphisbænidæ.** *Amphisbæna alba* L., Brazil; *A. fuliginosa* L., South America (fig. 642). *Chirotes lumbricoides* Flem. Mexico.

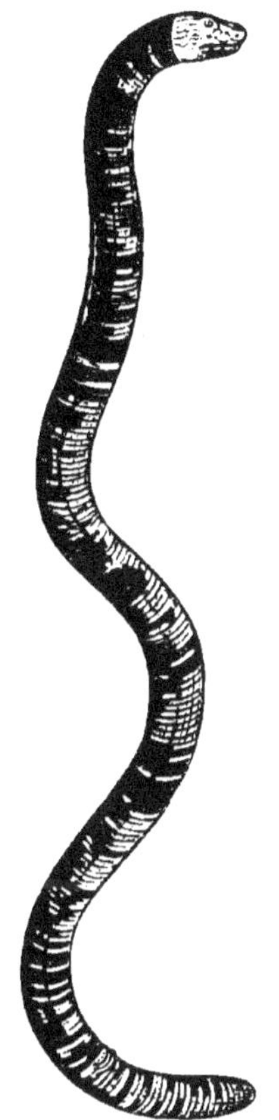
FIG. 642.—*Amphisbæna fuliginosa* (règne animal).

Sub-order 2. **Vermilinguia.** Lizards of the Old World, with vermiform tongue, which can be protruded with great rapidity to a great distance, and deep laterally compressed body, which is covered with a shagreen-like skin. The structure of the skull differs considerably from that of the other Lizards, in that the parietal bones are not movable on the occipital, but are firmly united to the latter and to the occipital crest, which is continued over the parietals.

Fam. **Chamæleonidæ.** The feet are prehensile, and end with five digits, which are arranged in bundles of two and three. The digits of each bundle are connected together as far as the claws, and the two bundles work on one another like the arms of a pair of pincers. The long and slender tail is prehensile, coiling round twigs and branches to attach the animal. They are all *acrodont.* The tympanic membrane is hidden, being covered by the integument. The Chamaeleons possess the remarkable power of changing the colour of their skin; the change is dependent on the light-stimulus of surrounding objects, and is subordinated to the will of the animal. Recent researches, especially those of Brücke.* have contributed to the explanation of this phenomenon. Two different layers of pigment are placed beneath the thin epidermis—a superficial layer of clear, yellow pigment, and a deeper layer of dark-brown to black, the mutual extension and position of which varies. *Chamæleon vulgaris* Cuv., South Spain and Africa.

Sub-order 3. **Crassilinguia.** With thick and short fleshy tongue, which is hardly indented at the point; as a rule it is rather rounded, and cannot be protruded. Eyelids are usually present. The tympanic membrane is usually exposed. In all cases there are four limbs, with digits directed forwards. They live exclusively in the hotter regions of the Old and New Worlds. The eastern and western hemispheres contain types surprisingly alike, which, however (with the exception of the Geckos) can be sharply distinguished by their dentition; all those found in America are pleurodont, those of the Old World are acrodont.

FIG. 613.—*Platydactylus mauritanicus.*

Fam. **Ascalabotæ.** Geckos. Lizards of clumsy Salamander-like form, and of small size; with viscous lobes on the digits for attachment, and with biconcave vertebræ. They are all pleurodont, and without palatal teeth. They are shy, nocturnal animals, and their eyes are large and without lids. They climb and run very skilfully on smooth and steep walls, with the help of their retractile claws and the lobes on their digits. They live for the most part in hot countries: only a few are found in South Europe. They are harmless, but are erroneously considered as poisonous on account of the acrid fluid of their clinging digits. At night they make a loud cry, sounding like the word "Gecko." *Platydactylus mauritanicus* L. (fig. 613); *Pl. muralis* Dum. Bibr. Mediterranean coasts; *Hemidactylus verruculatus* Cuv. Mediterranean coasts; *Ptychozoon homalocephalum* Kuhl., Java.

Fam **Iguanidæ.** Leguana. The body, which is somewhat laterally compressed, is supported by long slender legs, which are pre-eminently adapted for climbing. The head is more or less pyramidal, and often raised like a helmet, and of a peculiar shape, in consequence of the possession of a membranous jugular sac.

* E. Brücke, "Untersuchungen über den Farbenwechsel des afrikanischen Chamaeleons," *Denkschr. der k. Akad. der Wissensch., Wien*, 1852.

Tympanic membrane usually exposed. Many of them have a spiny dorsal crest, and change their colour like the Chamaeleons.

The following Iguanas belong to the western hemisphere, and are pleurodont: *Polychrus marmoratus* Cuv., Brazil; *Iguana tuberculata* Laur.=*sapidissima* Merr., West Indies; *I. delicatissima* Laur., tropical America; *Cyclura carinata* Gray, Cuba; *Basiliscus mitratus* Daud., South America.

The following belong to the eastern hemisphere, and are acrodont: *Calotes ophiomachus* Merr., East Indies; *Draco volans* L., Java; *Lophiura amboinensis* Schloss.

The New Zealand genus, *Hatteria* = *Sphenodon*, which was formerly reckoned among the *Iguanidæ*, shows such considerable differences in its organization that Günther established for it a third order of scaly Reptiles under the name of **Rhynchocephalia**,* which Huxley holds to be allied to the extinct Triassic Lacertilian genera *Hyperodapedon* and *Rhynchosaurus*.

Fam. **Humivagæ**. Lizards with broad flat body, supported by shorter limbs; of almost toad-like aspect. The skin is not unfrequently covered with spiny scales. They live on the ground in stony and sandy places, where they hide themselves in pits and holes.

To the *Humivagæ* of America, which are all pleurodont, belong *Phrynosoma orbiculare* Wiegm., Tapayaxin, Mexico; *Tropidurus cyclurus* Wied., Brazil.

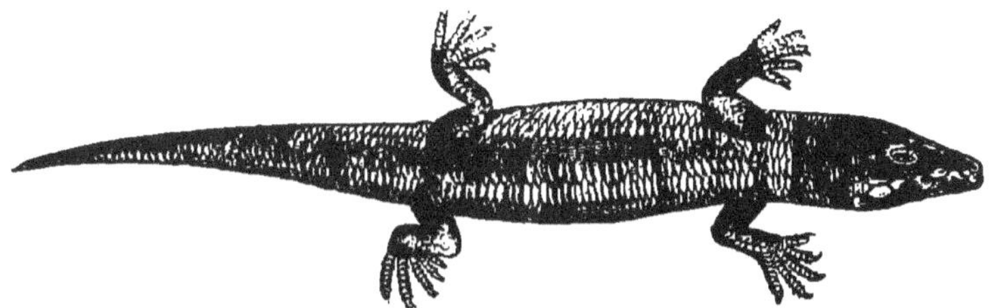

FIG. 644.—*Scincus officinalis* (règne animal).

To the *Humivagæ* of East India and Africa, which are acrodont, and possess canine teeth, belong *Phrynocephalus helioscopus* Kp., Siberia; *Uromastix spinipes* Merr., Egypt; *Agama colonorum* Daud., Egypt; *Stellio vulgaris* Latr., Hardun, Egypt.

Sub-order 4. **Brevilinguia**. Scaly Lizards, with elongated, often snake-like body. The limbs are very diversely developed. The tongue is short and thick, without sheath, more or less indented at the thinner anterior end, and but slightly protrusible. Eyelids are, as a rule, present. The tympanic membrane is often concealed beneath the skin.

Fam. **Scincoideæ**. Sand-Lizards. The more or less snake-like body is covered with smooth bony scales. The crown of the head is invested with larger scutes. *Anguis fragilis* L., Blindworm, Europe; *Scincus officinalis* Laur. (fig. 644) Egypt; *Seps chalcidica* Merr., Dalmatia; *Acontias meleagris* Cuv., Cape.

* A. Günther. "Contribution of the Anatomy of *Hatteria* (*Rhynchocephalus*) Gray." *Phil. Trans. Roy. Soc.*, London, vol. 157. ii., 1867.

Fam. **Ptychopleuræ.** The body is provided with two lateral folds of skin, covered with small scales. These folds extend from the region of the ear to near the anus, and form the boundary between the dorsal and ventral surfaces. *Zonurus Cordylus* Merr. = *griseus* Cuv., South Africa; *Pseudopus Pallasii*, Cuv., South-east Europe, and in lower Austria; *Pygopus (Bipes) lepidopus* Lacep., New Holland (fig. 641); *Chamæsaura anguina* Schn., Cape; *Ophisaurus ventralis* Daud., North America.

Sub-order 5. **Fissilinguia.** *Pleurodonta*, with long and thin, protrusible bifid tongue, usually with complete eyelids, and always with exposed tympanic membrane. The scales of the trunk are small and imbricated, those of the long tail mostly lozenge-shaped.

Fam. **Lacertidæ.** Mostly brightly coloured, very agile Lizards, with long tail and head covered with scutes. The ventral surface is covered with usually rhomboidal scales, arranged in oblique rows. *Lacerta vivipara* L., Germany and South Europe, viviparous; *L. ocellata* Daud.: *L. viridis*, green with black spots in front, Dalmatia; *L. agilis* L. = *stirpium* Daud., common Lizard: *L. muralis* Merr., South Europe; *Heloderma horridum* Wiegm., Mexico.

Fam. **Ameividæ.** Lizards of the New World whose head is covered with scutes, as in the *Lacertidæ*, while the abdomen is covered with rhomboidal scutes, arranged in transverse rows. *Tejus monitor* Merr. = *T. Tejuexin* L., Brazil, live in holes in the earth and in hollow trunks of trees. Feed on Mice, Insects, and Worms, and are, including the long tail, four or five feet long. They are hunted and eaten. *Ameiva vulgaris* Licht., West Indies.

Fam. **Monitoridæ.** Elongated Lizards of large size, without femoral pores. The crown of the head, the back and the abdomen, are covered with small plate-like scales. The separation of the ventricles of the heart is the most complete in the whole order. *Psammosaurus scincus* Merr. = *Varanus arenarius* Dum. Bibr., Egypt; the Land Crocodile of Herodotus; *Monitor niloticus* Hassl., eats the eggs of the Crocodile.

The **Proterosauria** and **Thecodontia** are fossil groups of Saurians. The former represent the oldest Lizards, and are distinguished by the possession of biconcave vertebral bodies and bifid spinous processes. They are found in the Kupferschiefer. The *Thecodontia*, also with biconcave vertebræ, possessed compressed teeth wedged into alveoli, with their crowns covered with finely serrated striæ; they belonged to the Triassic period.

The fossil **Dinosauria** must be mentioned as a special order of Reptiles. These were the colossal terrestrial inhabitants of the Jura, Weald, and lower chalk; in several features of their structure they recall Mammals, especially the Pachydermata.

Other orders of fossil Saurians, as the **Ornithoscelida**, present modifications which point in various ways to the organisation of birds. Characterised by the præacetabular extension of the ilium and the downward direction of the elongated pubis and ischium, they possessed, at least in the group which includes the Jurassic genus

Compsognathus, very long cervical vertebræ, an almost bird-like head. a very long neck, short anterior and very long posterior ribs. The astragalus is fused with the long tibia, as in birds.

The **Pterosauria** or **Pterodactyls**, which likewise lived principally in the Jurassic period, were flying Saurians. The external finger of the hand was elongated in the form of a sabre, and of considerable strength (fig. 118); it probably supported an expansion of the integument, which enabled the animal to float along in the air, or even to fly.

Rhamphorhynchus Gemmingii H. v. M., lithographic slate. *Pterodactylus longirostris* Cuv., Jura.

Sub-class 2.—Hydrosauria.*

Aquatic Reptiles of considerable size, with teeth wedged into the jaws, and leathery or armoured skin, with swimming fins or powerful feet, the digits of which are connected by webs.

The Hydrosaurians, represented at the present time by the Crocodiles, are characterised by their usually gigantic size, by an organisation corresponding to their aquatic habits, and by their high development.

The numerous fossil forms were exclusively inhabitants of the sea, and had swimming fins resembling those of Whales; the bones of the arm were short, the carpal bones and the phalanges were numerous, and the digits were connected. The vertebral column, which was movable in its individual regions, and still composed of broad amphicœlous vertebræ, was prolonged into a tail of considerable size, which was probably surrounded by a membranous fin. At a higher grade of development the vertebral column consists of opisthocœlous reptilian vertebræ, and ends with a swimming-tail, surrounded by a cutaneous fold. The extremities become more and more like feet, and the distinctly separated digits are still webbed. Such forms no longer inhabit the high sea, but are found on the coast in lagoons, and near the mouths of rivers. They go up on to the land, and move quickly upon it, but they are unable to turn quickly and easily.

The dentition shows that the *Hydrosaurians* are powerful preda-

* R. Owen, "Palæontology." London, 1860.

Huxley, "On the dermal armour of Jacare and Caiman, etc." *Journ. Proceed. Linn. Soc.*, vol. iv., 1860.

Rathke, "Untersuchungen über die Entwickelung und den Körperbau der Crocodile." Braunschweig, 1866.

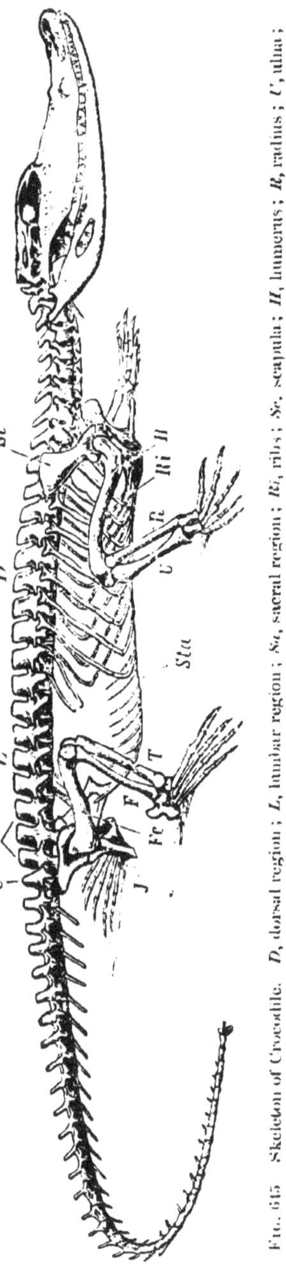

FIG. 645. Skeleton of Crocodile. *D*, dorsal region; *L*, lumbar region; *Sa*, sacral region; *Ri*, ribs; *Sc*, scapula; *H*, humerus; *R*, radius; *U*, ulna; *Sta*, sternum abdominale; *Fe*, femur; *T*, tibia; *J*, ischium; *C*, caudal vertebræ.

tory animals. The flat head is prolonged into a long snout; the long jaws are armed with sharp conical prehensile teeth, which are wedged into deep alveoli; the crowns of the teeth are sometimes smooth, sometimes striated or superficially folded, and are gradually replaced by succeeding supplementary teeth. Ribs are present in great numbers, not only in the very long thoracic region, but also in the cervical and abdominal regions.

In the Crocodiles there is, in the abdominal region, a narrow *sternum abdominale*, which is prolonged to the pelvic girdle, and bears on its sides, a number of abdominal ribs, the upper ends of which do not reach the vertebral column (fig. 645). The internal organisation probably presented different grades of perfection in the various groups, of which only the highest—viz., that found in living Crocodiles—can be known to us.

Order I.—ENALIOSAURIA.

Hydrosauria with naked leathery skin, biconcave vertebræ and swimming fins (confined to the secondary period).

The remains of these gigantic inhabitants of the sea, which lived through the secondary period from its beginning to its end, show that they were the most powerful marine animals of that time. They were of extreme length, and possessed a usually elongated, flat snout with numerous conical prehensile teeth, a very long mobile trunk, and fin-like extremities as in the Whales.

Fam. **Nothosaurii** (*Sauropterygii* Owen). With elongated bones of the upper jaw which reach to the point of the long snout: without superior temporal bones: with simple conical teeth. Belong to the Trias. *Nothosaurus mirabilis* Münst.; *Simosaurus* H. v. M., and others.

Fam. **Plesiosaurii** (*Sauropterygii* Owen). With long snake-like neck, short head and tail, and elongated swimming fins. They lived in the Jura and the chalk. *Plesiosaurus* Conyb.

Fam. **Ichthyosaurii** (*Ichthyopterygii* Owen). With very short neck, thick elongated body, short swimming fins, and long tail probably surrounded by a fin. The snout, pointed and elongated like a beak, is principally formed by the præmaxillary bones. The teeth present a striated and folded surface, and are closely crowded together. They are found principally in the Jura, rarely in the chalk. *Ichthyosaurus communis* De la Beche, etc.

Order 2.—Crocodilia (Loricata).

Hydrosauria, with bony dermal plates and teeth wedged into the bones of the jaws, to which they are confined; with four partly clawed feet and long, keeled swimming tail.

The extremities no longer have the form of swimming fins, but of freely articulated legs and feet with separated digits. The integument is granular and leathery, and contains, especially on the dorsal surface, large and in part keeled, osseous plates. On the tail these plates form a dentated crest, paired in front, but in its hinder part simple.

The broad flat skull is distinguished by the corroded appearance of the surface of the bones, and possesses separated alisphenoids, and above the maxillo-jugal arcade a supra-temporal arcade, which is separated from the orbit by a bony bridge (process of the post-frontal and jugal). The roof of the skull is formed by an unpaired parietal and frontal, to which are joined the paired nasal bones. The upper jaws are firmly united with the skull and are elongated so as to form a long snout, at the end of which the paired præmaxillary bones are wedged in. The sides of the snout are formed by the maxillary bones which are very large. The præmaxillaries, which bound the nasal apertures, and the maxillaries develop horizontal palatal plates, which meet in the middle line and form the anterior part of the hard palate. The lacrymal is always of considerable size. Behind, the palatine and pterygoid develop palatal plates which unite suturally in the median line, and constitute a completely closed roof for the buccal cavity. The posterior nares which are surrounded by the paired vomers open at the posterior margin of the buccal cavity. The conical teeth, which are completely confined to the bones of the jaws [præmaxillæ, maxillæ, and mandible], are deeply wedged into alveoli, and they present slightly compressed striated crowns. The fourth tooth of the mandible is usually distinguished by its great size as a prehensile tooth, and, when the jaws are shut, fits into a gap or an excavation in the upper jaw. In the *Teleosauria* the vertebræ are amphicœlous; in the *Steneosauria*, which are also extinct, they are opisthocœlous, and in the Crocodiles of the present day procœlous.

The internal organisation of the living Crocodiles is the highest amongst all Reptiles. The eyes have vertical pupils and two lids as well as a nictitating membrane. The nasal openings lie far forward on the point of the snout, and, as well as the ears which are placed far back, can be closed by cutaneous valves. The buccal cavity, to the floor of which is attached a flat non-protractile tongue, is without salivary glands, and leads by a wide œsophagus into the rounded muscular stomach, which resembles that of Birds in form and structure, and specially in the aponeurotic discs of its internal lining. The stomach is followed by a thin-walled duodenum, which is beset with papillæ, and passes into the small intestine, which is folded in a zigzag fashion. There is no cæcal appendage to the short wide large intestine. The latter becomes narrow and almost funnel-shaped, before it opens into the cloaca, from the anterior wall of which arises the erectile copulatory organ. The structure of the heart is the most perfect found in all Reptiles, and, in the complete separation of a right venous and a left arterial portion, affords a direct transition to that of the warm-blooded animals. Finally, the free communication of the body cavity by openings of the so-called peritoneal canals, which recall the abdominal pores of the Ganoids and Selachians, deserves to be mentioned as peculiarities of the Crocodilia.

Three groups of Crocodiles are to be distinguished: two of these the *Teleosauria* (*Amphicœlia*) and *Steneosauria* (*Opisthocœlia*) are extinct. The former with the genera *Mystriosaurus* Kp. and *Teleosaurus* Geoffr. are confined to the Jurassic formation, the latter with *Steneosaurus* Geoffr. *Cetiosaurus* Owen, etc., occur in the Jura and in the chalk. Only the third group of the Crocodiles or *Procœlia* has persisted from the chalk onwards through the tertiary period to our own time.

Sub-order **Procœlia = Crocodilia**, *s.str.*

With procœlous vertebræ and long compressed swimming tail, the dorsal side of which bears a double cutaneous crest, which becomes single at the posterior end. The anterior feet with five free digits; the posterior with four digits, which are more or less united by webs. Live in the mouths and lagoons of great rivers in the warmer climates of the Old and New Worlds, and seek their prey by night. The hard-shelled eggs are laid in the sand and in holes on the banks.

Fam. **Crocodilidæ.** The so-called canine teeth (fourth tooth of the lower jaw) fit into a notch of the margin of the upper jaw. Hind feet with complete swimming membrane. *Crocodilus vulgaris* Cuv., Nile. *C. rhombifer* Cuv., Cuba.

Fam. **Alligatoridæ.** The snout is broad and without notch for the so-called canines of the mandible. Swimming membranes only partially developed or rudimentary. [Found only in America.] *Alligator lucius* Cuv.; *Caiman (Jacare) sclerops* Schn.

Fam. **Gavialidæ.** *Rhamphostoma gangeticum* Geoffr., East Indies; *Rhynchosuchus Schlegelii* Gray, Australia.

Sub-class 3.—Chelonia.*

Reptilia of short, stout form of body, with an upper and lower osseous shield which covers the dorsal and ventral surfaces. There are four feet, and the jaws are without teeth.

No other group of Reptiles is so clearly defined and characterised to the same extent by peculiarities of form and organisation as is that of the Chelonia. The investment of the body by an upper, more or less arched and usually osseous dorsal shield (*carapace*), and by a lower ventral shield (*plastron*), joined to the former by lateral arches, forms a character as distinctive of the Chelonia as is the possession of wings and feathers of the class Aves.

The shield-like, dermal armour (fig. 646) beneath which the head extremities and tail can often be retracted, owes its origin partly to osseous parts of the vertebral column and partly to the accessory dermal bones, which are intimately connected with the former. The flat plastron contains nine more or less developed osseous pieces, an anterior unpaired *interclavicular*, and four pairs of lateral pieces (the anterior being distinguished as *clavicularia*) between which there

* H. Rathke. "Ueber die Entwickelung der Schildkröten," Braunschweig. 1848.

Gray, "Catalogue of the Shield Reptiles in the Collection of the British Museum. Part I., London 1855, Suppl., 1870, Append., 1872.

L. Agassiz. "Embryology of the Turtle." Natural History of the United States, vol. III., part III., 1857.

may be left a median space, closed by skin or cartilage (*Trionyx*, *Chelonia*, etc.). The spinous processes and ribs of the thoracic vertebræ take part in the formation of the large carapace, as well as a number of paired and unpaired osseous dermal plates, which are placed partly

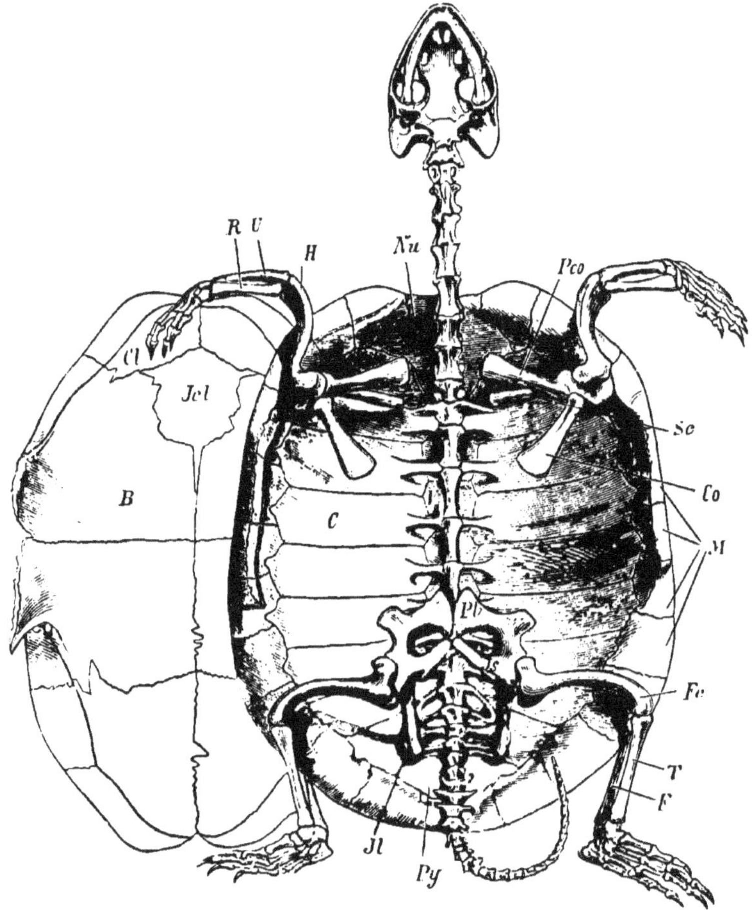

FIG. 616.—Skeleton of *Cistudo* (*Emys*) *europaea*. *V*, vertebral (neural) plates; *C*, costa plates; *M*, marginal plates; *Nu*, nuchal plate; *Py*, pygal plate; *B*, plastron (ventral shield); *Cl*, clavicle; *Jcl*, inter-clavicle; *Sc*, scapula; *Co*, coracoid; *Pco*, acromial process (pro-coracoid); *Pb*, pubis; *Js*, ischium; *Jl*, ilium; *H*, humerus; *R*, radius; *U*, ulna; *Fe*, femur; *T*, tibia; *F*, fibula.

in the median line in the neck (*nuchal plate*), and in the sacral region (*pygal plate*), and partly laterally at the edge of the shield (22 marginal plates).

While the spinous processes of eight of the thoracic vertebræ (2nd to 9th) appear in the median line as horizontal plates [*neural plates*], the ribs of the same vertebræ (2nd to 9th, these ribs are distinguished from the first and last ribs by their greater length) are transformed into broad transverse plates [*costal plates*], which are joined with one another by indented sutures, and present the special peculiarity of giving off broad processes, which arch over the muscles of the back, and are connected with the neural plates (expanded spinous processes). In addition, larger plates, which owe their origin to cornifications of the epidermis, are usually present. They are applied to the outer surface of both the dorsal and ventral shields and are used, in the case of some of the larger species, as tortoise-shell. They by no means correspond with the subjacent bony pieces, but are very regularly arranged in such a manner, that in the dorsal shield a median and two lateral rows of plates can be distinguished, and round the periphery a circle of marginal plates. On the ventral surface, on the other hand, there is a double row of such plates.

Unlike the middle (thoracic) region of the vertebral column, the vertebræ of which are firmly connected with the dorsal shield, the cervical and caudal vertebræ are always movable upon one another. The cervical region is exceedingly flexible, and can be more or less completely retracted within the shell; it consists of eight long vertebræ, which are without ribs. The ten rib-bearing vertebræ are followed by two or three sacral vertebræ, which project beneath the carapace, and by a considerable number of very movable caudal vertebræ.

The head is tolerably arched: the bones of the skull are firmly united to one another by sutures, and form a broad roof, which is prolonged into a strongly developed occipital crest. The skull is characterised by the possession of a pair of parietal bones and of large anterior frontals. Descending lamellar processes of the parietal bones extend along the sides of the cartilaginous cranial capsule as far as the short basisphenoid. The temporal fossa is most completely roofed in in the marine *Chelonia* by broad osseous plates which are formed by the postfrontal, jugal, quadrato-jugal, and the squamosal. The opisthotic remains as an independent bone behind the prootic, which forms the lateral walls of the cranial cavity. All the parts of the maxillo-palatine apparatus as well as the quadrate are firmly connected with the bones of the skull, and are marked off from one another by serrated sutures. The facial parts of the skull are strikingly short, and the nasal bones are absent. The bony palate is formed by the broad palatine and the unpaired vomer, behind the palatine plates of which the posterior nares open. The pterygoids are very broad and lamellar. Teeth are completely absent, both on the palatal bones and on the high, relatively short bones of the jaws, but the edges of the latter are covered, like the beak of a bird, by sharp cutting, serrated horny plates, which enable certain species to bite with great vigour and to inflict sensible wounds.

The four limbs enable the *Chelonia* to creep and run on land: in the aquatic forms, however, they are swimming feet or fins. The position of the pectoral and pelvic girdles, and of the corresponding muscles, between the dorsal and ventral shields, is remarkable; but is fully explained developmentally by the growth of the anterior and posterior ribs. The scapula is formed of an ascending rod-like bone, the upper end of which is attached to the transverse process of the anterior thoracic vertebra by a ligamentous or cartilaginous connection. A strong acromial process (procoracoid) reaches from the scapula to the unpaired portion of the ventral shield, to which it is likewise attached by a ligamentous or cartilaginous connection. The pelvis closely resembles that of the Saurians, and except in the Land Tortoises is not firmly connected with the carapace.

In the organs of digestion and reproduction Chelonians partly resemble Crocodiles and partly Birds. They especially resemble the former in the structure of the male generative organs, and in the possession of peritoneal canals, which are, however, closed. The opening of the genital ducts and the ureters into the neck of the urinary bladder, which accordingly functions as a urogenital sinus, is worthy of remark. The eyes are placed in closed orbits, and have lids and a nictitating membrane. There is always a tympanic cavity with a wide Eustachian tube, a long columella and a tympanic membrane, which is visible externally. The tongue is attached to the floor of the buccal cavity, and is not protrusible; in the Land Tortoises it is beset with long papillæ.

The copulation lasts a day, and during that time the male is carried on the back of the female. The eggs are laid in small number, except in the marine Chelonia, in which they are more numerous. They contain within the shell a layer of albumen, surrounding the yolk, and are buried in the earth, in the aquatic Chelonians near the shore. According to Agassiz the North American Marsh Tortoises lay eggs only once in the year, while they copulate twice (in the spring and autumn). The first copulation, according to this investigator, takes place in *Emys picta* in the seventh year, the first deposition of eggs in the eleventh year of the animal's life.

These facts agree with the slow growth of the body of the Tortoises, and the great age which they attain.

The Chelonians belong mainly to warmer climates, and live principally on vegetables; many of them, however, also live on Mollusca, Crustacea, and Fishes.

Fossil remains are first found, but rarely, in the upper white Jura. More numerous remains are found in the Tertiary period.

Fam. **Cheloniadæ.** Turtles. With flat dorsal and often cartilaginous ventral shield, between which the head and extremities cannot be retracted. The latter are fin-like feet, with immovably connected digits, which are usually without nails, and are covered by a common skin. The anterior limbs are much longer than the posterior. *Chelonia esculenta* Merr.: *Ch.* (*Caretta*) *imbricata*

Fig. 647.—*Thalassochelys caretta* (règne animal).

L., Atlantic and Indian Ocean; *Thalassochelys caretta* L. = *corticata* Rond. (fig. 647), Atlantic Ocean and Mediterranean; *Sphargis coriacea* Gray. Rare in the Mediterranean, more common in the Atlantic Ocean and South Sea.

Fam. **Trionycidæ.** Soft or Mud Tortoises. With flat, oval, incompletely ossified dorsal shield, and long retractile neck. Jaws with cutting edges, surrounded by fleshy lips. The head and feet are not retractile. The nasal openings are on the long snout. *Trionyx ferox* Merr. A fierce animal. Found in the rivers of Georgia and Carolina. Good to eat.

Fam. **Chelydæ.** Head and feet not retractile. Latter end with free digits, which are webbed and furnished with claws. *Chelys fimbriata* Schweig., Matamata. South America.

Fam. **Emydæ.** Freshwater Tortoises. Dorsal shield flat, plastron usually small. Feet thick, with freely movable digits, which are connected by a web. They swim excellently, and move also with great facility on land. They principally inhabit sluggish rivers and ponds. *Cistudo europæa* Schneid. *lutaria* Gesn., the common Tortoise of South Europe and East Germany: *Emys caspica*, on the Caspian Sea, in Dalmatia and Greece; *Chelydra serpentina* L., with very sharp jaws, in North America.

Fam. **Chersidæ.** Land Tortoises. With high, arched, ossified carapace: head and feet retractile. The digits are immovably connected as far as the nails to thick club-feet, with indurated soles. They live in damp and shady localities in warm and hot climates, and feed on plants. *Testudo græca* L., *nemoralis* Aldr., = *marginata*, South Italy: *T. tabulata* Daud., in America.

CHAPTER VIII.

Class IV.—AVES,* BIRDS.

Warm-blooded oviparous animals, covered with feathers. The chambers of the heart are completely separated. The right aortic arch persists. There is a single occipital condyle, and the anterior limbs are transformed into wings.

As opposed to the poikilothermic Vertebrates (*i.e.*, Vertebrates whose temperature varies with that of the external medium) the blood of *Aves* and *Mammalia* possesses a high temperature, which remains tolerably constant in spite of the changing temperature of the external medium. This maintenance of a constant temperature demands above everything a great energy of metabolism. The surface of all the vegetative organs, especially of the lungs, kidneys, and alimentary canal, has a relatively greater extension in the warm-blooded than in the cold-blooded animals. The operations of digestion, preparation of blood, circulation and respiration are carried on with much greater energy. With the need of a richer nourishment, the processes of vegetative life take a disproportionately more rapid course, and as the high and uniform temperature of the blood is a

* Joh. F. Naumann, "Naturgeschichte der Vögel Deutschlands," 13 Bde., Stuttgart, 1822-1860.
"Naumannia. Archiv für Ornithologie," Herausgegeben von Ed. Baldamus. Köthen, 1849.
"Journal für Ornithologie." Herausgegeben von J. Cabanis. Cassel, 1853—.
"The Ibis," 1859.
Tiedemann. "Anatomie und Naturgeschichte der Vögel." Heidelberg, 1810-1814.
C. E. v. Baer. "Entwickelungsgeschichte der Thiere." I. und II., 1828-1837.
Remak, "Untersuch. über die Entwick. der Wirbelthiere," Berlin, 1850-55.
Huxley, "On the Classification of Birds." Proceed. Zool. Soc., 1867.
Gray and Mitchell, "The Genera of Birds." London, 1844-49.
C. Sundevall. "Tentamen." Stockholm, 1872-73.

condition necessary to their very maintenance, they seem to be the principal source of warmth produced. Since the loss of heat is greater when the temperature of the external medium is lowered, the activity of the vegetative organs must considerably increase in the colder season of the year, and in the northern climates.

In addition to the continual addition of new quantities of heat, a second cause contributes to the maintenance of the constant temperature of the warm-blooded animals. This is the protection afforded by the special nature of the covering of the body. While the Vertebrates with a variable temperature have a naked or armoured skin, Birds and Mammals have a more or less close covering of hairs or feathers, which limits to a great extent the loss of heat by radiation. The large aquatic animals, on the other hand, have a scanty covering of hair, but they develop thick layers of fat beneath the cutis, which serve for the retention of heat, and at the same time for hydrostatic purposes.

There is in all cases a mutual relation of a complicated kind between the factors which favour the withdrawal of heat and the conditions of the retention and the formation of heat, a relation which in spite of many variations in its individual factors results in the equalization of the heat generated and the heat lost. Some Mammals are able to maintain their proper temperature only within certain limits of the external temperature; these animals are to a certain extent incompletely homothermic, and when the temperature sinks below a certain point they fall into the so-called winter-sleep (*hibernation*), *i.e.*, a state of rest characterised by an almost complete absence of movement, and by a diminution in the energy of all the vital processes. In the class of Birds, whose higher temperature permits of no interruption or limitation of the vital functions, there is no example of hibernation. But these animals have numerous means of heat adjustment at their disposal; in particular, the swiftness of their flight enables them to leave their homes at the approach of the cold season, and to betake themselves to warmer climates, where food is abundant. The common migrations of the migratory birds, migrations which sometimes extend over great distances, to a certain extent take the place of the winter-sleep of the hibernating animals; in the Mammalia whose organisation permits of hibernation migrations like those of Birds are very rare.

The most essential peculiarity of Birds, and one with which many characteristics both of external appearance and of internal organisation are correlated, is their power of flight. This peculiarity in connection with these characters determines the sharp definition as well as the relatively great uniformity of the class, which, indeed, is descended from the Saurians, but exists at the present day without any forms transitional to other groups. On the other hand, the remains of a group of Saurians (*Archæopteryx lithographica*) have been discovered in the Solhenhofen lithographic slate which combine characters of the Pterodactyls with those of the Birds.

The entire structure of the body of Birds corresponds with the two principal modes of locomotion—on the one hand flight, and on the other walking and hopping on the earth. The trunk, which is oval, is supported in an obliquely horizontal position on the two hind legs, the pedal surface of which stretches over a relatively large area. Posteriorly the body is prolonged into a short rudimentary tail, the last vertebra of which serves for the support of a group of stiff steering, or tail feathers (*rectrices*). In front it is prolonged into a movable neck, on which is balanced a light, rounded head, with a projecting, horny beak. The anterior extremities, which are transformed into wings, lie folded together at the sides of the body.

Arrangements for lessening the weight of the body are discernible in the special structure of all the systems of organs; these are especially noticeable in the structure of the osseous skeleton. The bones contain air-spaces (*pneumaticity*), which communicate with the air-sacs of the body through openings in the dense and firm osseous substance, which is however confined to a relatively thin layer. This pneumaticity is most highly developed in those birds which combine a quick and enduring power of flight, with a considerable size of body (Albatross, Hornbill, Pelican). In these cases all the bones except the quadrato-jugal and the scapula appear to be pneumatic, while on the other hand in the *Ratitæ* (Ostrich), which have lost the power of flight, all the bones except some of the cranial bones are filled with marrow.

The Skeleton.—Except in the Ostrich-like birds, the cranial bones very early fuse together to form a light and firm skull, which articulates with the atlas by means of a single condyle. The squamosal and periotic bones (prootic, epiotic and opisthotic) fuse to form a single bone which is united with the occipital and with which the quadrate articulates. The large frontal bones take the principal part in the formation of the cranial roof. Almost the whole of the upper edge of the large orbit, which in the Parrots is closed by a lower ring, is formed by the frontal bones. An independent lachrymal is present at the anterior margin of the orbit. The ethmoid region and the cranial

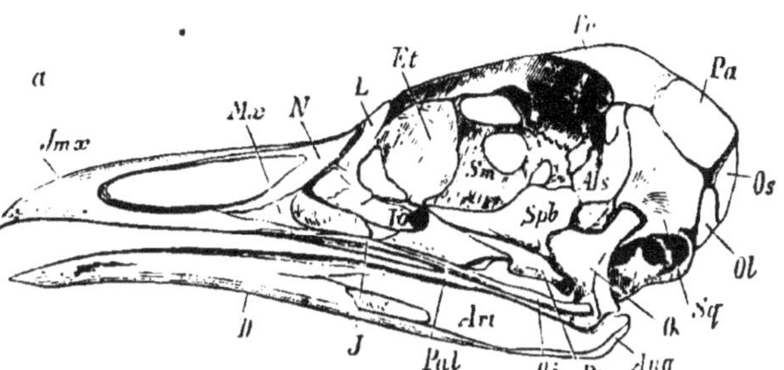

capsule are widely separated by an interobital septum of considerable size. The latter, sometimes together with the remains of the fused orbitosphenoids, frequently remains membranous and unossified in its median part, and rests on an elongated bony rod corresponding to the basisphenoid. At the base of the temporal region there are two bones—the basitemporals (Parker)—which are ankylosed with one another, and are probably to be referred to a parasphenoid. In all cases independent alisphenoids are present. The ethmoid region is composed of a vertically placed, unpaired ethmoid, situated in the anterior prolongation of the interorbital septum, and of lateral ethmoids which separate the eyes and nasal cavities, and through which the olfactory nerves pass into the nasal cavities. The lateral ethmoids may be swollen and contain ethmoidal cells. In front of them are developed the two nasal cavities with their bony or cartilaginous septum, which is the prolongation of the unpaired ethmoid, and affords a support to the rolled-up turbinal bones, which are sometimes also attached to the vomer. The facial bones unite to form a projecting beak, the margins of which are covered with horny substance, and which is often movably connected with the skull. The suspensorium

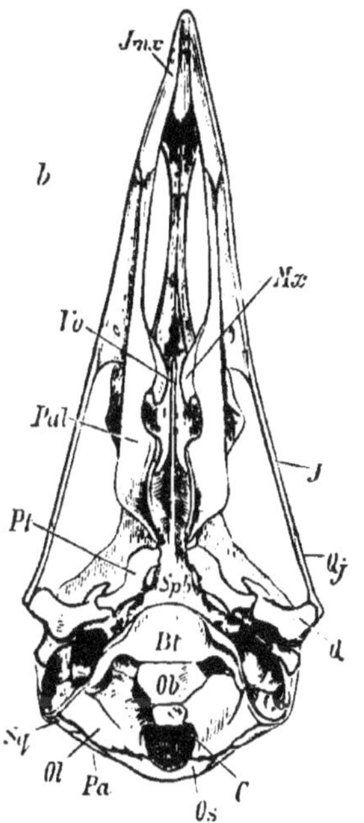

FIG. 148.—Skull of *Otis tarda* (bustard). *a*, From the side; *b*, from beneath. *Ob*, basi-occipital; *C*, condyle; *Ol*, ex-occipital; *Os*, supra-occipital; *Sq*, squamosal; *Bt*, basi-temporal (parasphenoid); *Spb*, basi-sphenoid; *Als*, alisphenoid; *Sm*, inter-orbital septum; *Et*, median ethmoid; *Pa*, parietal; *Fr*, frontal; *Mx*, maxillary; *Jmx*, præmaxillary; *N*, nasal; *L*, lachrymal; *J*, jugal; *Qj*, quadrato-jugal; *Q*, quadrate; *Pt*, pterygoid; *Pal*, palatine; *Vo*, vomer; *D*, dentary; *Art*, articulare; *Ang*, angulare.

of the lower jaw and the maxillo-palatine apparatus are enabled by special articular arrangements to move on the temporal bone and on corresponding processes of the basisphenoid. The quadrate, which is articulated to the temporal bone, has, besides the articular surface of the lower jaw, movable connections with the long rod-like quadrato-jugal, and with the usually styliform pterygoid which runs obliquely inwards, while the base of the upper beak presents a thin elastic place below the frontal bone, or is separated from the frontal bone by a transverse movable suture. When the beak is opened, and the lower jaw is moved downwards, the pressure on the quadrate bone is transferred to the rod-like quadrato-jugal and the pterygoid bones, and from these is transmitted partly directly and partly by means of the palatine bones to the upper beak, so that the latter must be more or less raised at that point. Therefore, when the mouth is opened, the end of the beak is raised. The greater part of the upper beak is formed by the unpaired præmaxilla, with the sides of which the maxillæ are fused, while an upper median process ascends between the nares and unites with the frontal on the inside of the nasal bones.

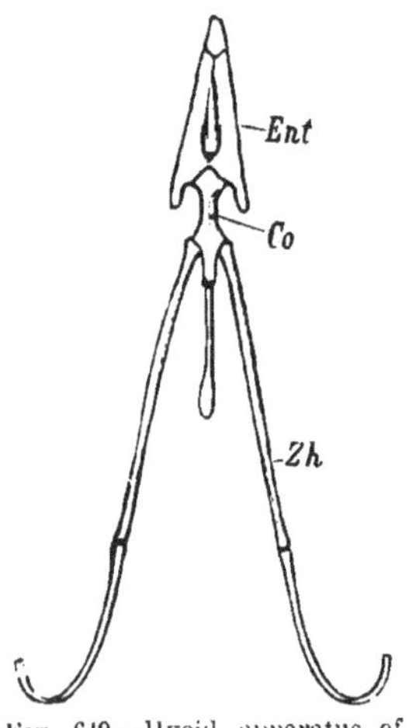

Fig. 649.—Hyoid apparatus of *Corvus cornix*. *Co*, body of hyoid; *Zh*, cornua; *Ent*, entoglossal bone.

The hyoid bone (fig. 649) is prolonged into a posterior rod; its anterior * cornua are usually two-jointed and are not connected with the skull, but in some cases they are much elongated and arch over the skull as far as the forehead (Woodpecker). They then constitute in connection with the muscles of their sheath a mechanism for the protrusion of the tongue.

In the **vertebral column** (fig. 650), a very long movable cervical region, a rigid dorsal and pelvic region and a rudimentary, only slightly movable caudal region can be distinguished. In Birds there is no separation of thoracic and lumbar regions as in Mammals, since all the dorsal vertebræ bear ribs and the region corresponding with the lumbar region takes a share in the formation of the sacrum. The cervical and dorsal regions also are not sharply distinct from

* Usually described as the posterior cornua. Ed.

one another, since the cervical vertebræ, as in the Crocodiles, bear ribs, which unite with the transverse processes to form a foramen transversarium. The neck is long and always freely movable, and contains 9 to 23 vertebræ (Swan). The shorter dorsal vertebræ are always less numerous; they have superior and inferior spinous processes and all bear ribs, to the ventral ends of which sterno-costal bones are articulated at an angle which projects backwards (fig. 650, *Stc*). The sternocostals also articulate with the margin of the sternum, and serve when they are extended to increase the distance between the latter and the vertebral column. But, since the ribs are firmly applied to one another by means of posterior processes (*processus uncinati*), the movement of the sternocostal ribs must necessarily affect the thorax as a whole, and dilate it (inspiration). The **sternum** is a broad flat bone which covers not only the thorax but a great part of the abdomen, and bears a projecting keel-like crest which serves for the attachment of the muscles of flight (*Carinatæ*). The sternal crest is reduced or obsolete only when the power of flight is feeble or entirely absent (*Ratitæ*).

The rib-bearing dorsal vertebræ are followed by a tolerably extensive division of the vertebral column, which corresponds to the lumbar and sacral regions, and which, by the fusion of a number of vertebræ with each other, and with the long iliac bones of the pelvic girdle, presents the characters of the sacrum. The **sacrum** is much elongated, and includes sixteen to twenty and more vertebræ; of these a certain number can be shown to be lumbar (præsacral), and are almost always preceded by two to three rib-bearing dorsal vertebræ. Then follows the true sacrum; it consists of two vertebræ, which are equivalent to the sacral vertebræ of Lizards and Crocodiles, and constitute by means of their transverse processes (with fused ribs) the main support of the pelvis near the cup of the hip-joint (*acetabular vertebræ*). Finally the true sacrum is followed by a postsacral region, which is composed of from three to seven of the anterior caudal vertebræ. The short caudal region, which succeeds the postsacral, consists, as a rule, of from seven to eight movable vertebræ, of which the last is represented by a vertical, laterally compressed plate—the *pygostyle*—to which the muscles for the movement of the steering feathers (*rectrices*) of the tail are attached. This deep ploughshare-shaped terminal body is composed of from four to six vertebræ, so that the reduction of the number of the caudal vertebræ, as compared with that of the *Saururæ* (*Archæopteryx*), is by no means so considerable.

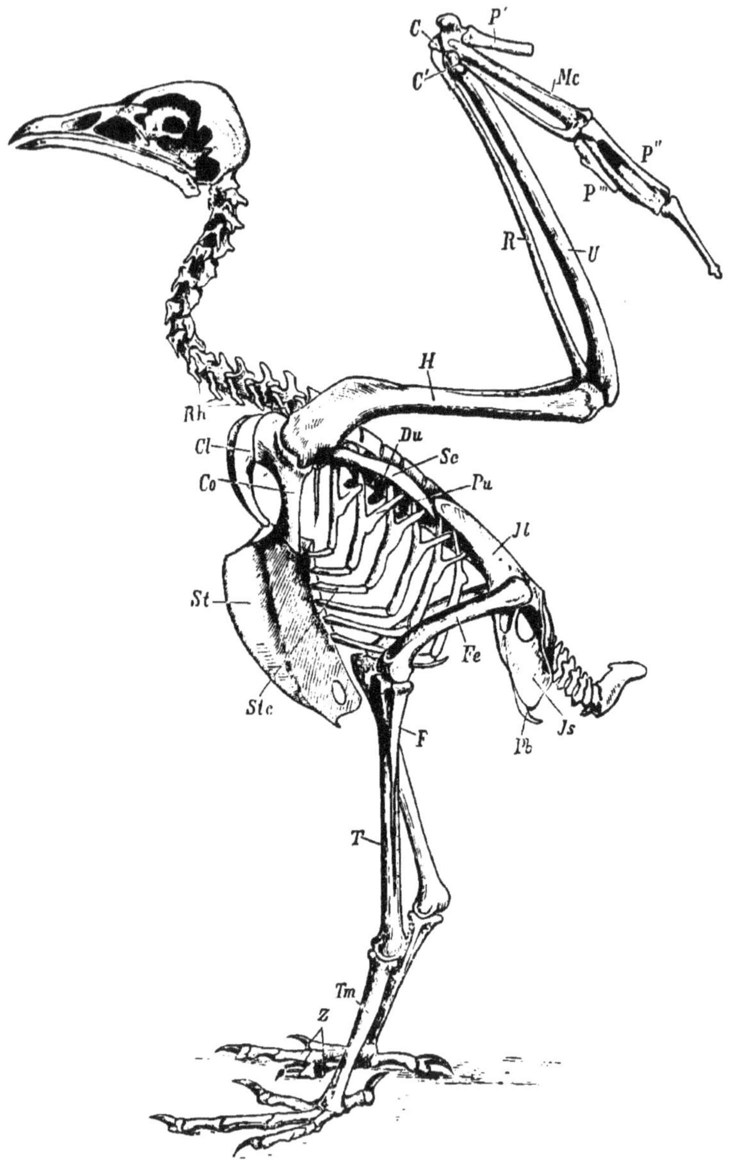

FIG. 650.—Skeleton of *Neophron percnopterus.* *Rh*, cervical ribs; *Du*, inferior spinous process of the thoracic vertebræ; *Cl*, clavicle; *Co*, coracoid; *Sc*, scapula; *St*, sternum; *Ste*, sternocostal bones (sternal ribs); *Pu*, uncinate process of the thoracic ribs; *Jl*, ilium; *Js*, ischium; *Pb*, pubis; *H*, humerus; *R*, radius; *U*, ulna; *C C'*, carpus; *Mc*, metacarpus; *P' P'' P'''*, phalanges of the three fingers; *Fe*, femur; *T*, tibia; *F*, fibula; *Tm*, tarso-metatarsus; *Z*, toes.

Shoulder girdle and wings.—The peculiarities of the anterior extremities are connected with their transformation into wings. Their connection with the thorax is always a firm one, since flying organs, whose movement pre-supposes a great expenditure of muscular power, require the necessary support on the trunk. While the scapula, which is a long, sabre-shaped bone, lies along the dorsal side of the thoracic framework, the clavicle and coracoid, as pillar-like supports for the shoulder-joint, are attached to the sternum. The two clavicles are fused so as to form a fork (*furcula*). The anterior extremity consists of a humerus, a longer forearm composed of radius and ulna, and the reduced hand. The latter contains only two carpal bones, an elongated metacarpus and three fingers—viz., the pollex or thumb, bearing the so-called bastard wing (*alula*), a middle finger, and a little finger. The humerus, the forearm, and the hand are so placed when at rest, that the humerus is directed backwards, the long forearm lies tolerably parallel to it and is directed forwards, while the hand is again bent backwards.

Pelvic girdle and legs.—The girdle of the hind limbs has the form of an elongated pelvis connected with a great number of lumbar and sacral vertebræ, and except in the ostrich (*Struthio camelus*) is without a symphysis pubis. The short and powerful femur is directed obliquely horizontally forwards, and concealed beneath the flesh and feathers of the abdomen, in such a manner that the knee-joint is not visible externally. The crus, which is much longer and more extensive, is chiefly composed of the *tibia*, the *fibula* being quite rudimentary and represented by a styliform bone on the outer side of the tibia. The crus is always followed by a long forwardly directed bone—the *tarso-metatarsus*—which is composed of the fused tarsal bones of the distal row (*intertarsal joint*), and of the metatarsal bones. The tarso-metatarsus varies much in length, and is the cause of the differences in the length of leg. At its lower end it divides into three processes, which are provided with articulating heads, for the attachment of the same number of toes. When a fourth toe is present, there is always a small bone on the inner side of the metatarsus, and to this bone the fourth inner toe is attached. The three or four toes (only in one case is the number reduced to two) are composed of several phalanges, the number of which increases from within outwards in such a manner that the first toe has two, the fourth (external) five joints.

[The digits present in the Avian pes are Nos. 1, 2, 3, 4, or 2, 3, 4, or in the Ostrich 3, 4. Digit No. 5 is never present.]

The muscles of the thorax are powerfully developed in connection with the power of flight, and a peculiar muscular arrangement, in consequence of which the toes are mechanically bent when the animal sits, deserves mention.

Exoskeleton.—The most important character in the external appearance of Birds is their covering of feathers. The skin is only naked in a few places—as on the beak, the toes, usually on the tarso-metatarsus, and sometimes on the neck (Vulture), or even on the abdomen (Ostrich); and also on the cutaneous outgrowths of the head and neck (Gallinaceous birds and Vulture). While the naked skin at the base of the beak is soft (so-called *cere*), the edges of the beak are usually cornified, and are only exceptionally soft (Ducks, Snipe), and then are richly innervated and serve as a fine tactile organ. The skin on the toes and tarso-metatarsus is also cornified, so as to form a firm horny covering, which is sometimes granular, more often divided into scales, and which may afford important systematic characters. When this integument forms a long continuous horny sheath on the anterior and lateral surfaces of the metatarsus, the latter is termed *laminiplantar* (Thrushes and singing-birds). The following special horny structures may be mentioned—the claws on the toes, the so-called spurs on the posterior and internal edge of the metatarsus in the male *Gallinaceæ*, and sometimes on the thumb-joint of the wing (*Parra*).

The **feathers** of Birds correspond to the hairs of *Mammalia*, and like them arise in pits of the dermis lined by the epidermis. At the bottom of the pit there is a vascular papilla, the epidermal investment of which gives rise by its rapid growth to the first rudiment of the hair or feather, around which the epidermal lining of the pit lies like a sheath. In the feather the axial part or stem with the *calamus* and shaft (*rhachis*) is to be distinguished from the vane (*vexillum*). The cylindrical hollow calamus is partly embedded in the skin, and encloses the dried-up papilla (pith); the rhachis is the projecting part of the stem, and bears a number of lateral processes—the *barbs*—which with their attached parts (*barbules*) constitute the vane (*vexillum*). The lower slightly concave side of the rhachis presents along its whole length, from the end of the calamus to the point of the feathers, a deep longitudinal groove, at the base of which a second feather—the *aftershaft* (*hyporhachis*)—arises, which, as well as the main-shaft, gives off two rows of barbs, but only in rare cases (Cassowary) reaches the length of the main shaft, and is sometimes (wing and tail quills) completely absent. The barbs *(rami)* send off barbules *(radii)* arranged in two rows, and the barbules (at any rate of the front row) bear barbicels and hooklets, which by their mutual interlocking effect the firm connection of the whole *vexillum.*

According to the nature of the stem and barbs, the following kinds of feather can be distinguished—*contour feathers (pennæ)*, with stiff shaft and firm *vexillum*; *down feathers (plumæ)*, with soft shaft and vane, the barbs of which bear round, or knotty barbules without hooklets; and finally *filoplumes (filoplumæ)*, with slender bristle-like shaft, the vane of which is reduced or absent. The *pennæ* determine the external outline of the plumage, and attain their greatest size as *remiges* in the wings and as *rectrices* in the tail. The *plumæ* form the deep layer of the plumage, and are covered by the contour feathers; they serve for the retention of warmth. The filoplumes, on the other hand, are distributed more among the *pennæ*, and at the angle of the mouth have the appearance of stiff bristles (*vibrissæ*). There are, moreover, many forms of feathers intermediate between these principal forms. In the autumn there is a complete change of feathers (*autumnal moult*), whereas the *spring moult* by which the bird acquires its breeding plumage is only rarely connected with a complete new formation of the plumage. As a rule, the spring moult consists in a colouring of the feathers (probably by chemical change in the pigments already present), and sometimes in a mechanical breaking off of certain parts of the feathers.

Birds have no sebaceous or sweat-glands, but there is often a bilobed gland above the last caudal vertebra. This gland (the *uropygial* gland, or oil-gland) has a simple duct, and its oily secretion serves to anoint the feathers.

The plumage is only rarely distributed evenly over the whole of the body (*Aptenodytes*). Usually the contour feathers are arranged in rows—the so-called *pterylæ*—between which there are spaces—the *apteria*—which are naked or only covered with down (fig. 651). The form and distribution of these spaces present modifications which can be used in classification.

The grouping of the feathers on the anterior limbs and on the tail determines the utility of these organs as wings and steering apparatus respectively. The wing has to a certain extent the form of a fan, which can be folded at two points—viz., the elbow joint and the carpal joint; its surface is formed by the large *remiges* on the under surface of the hand and forearm, but partly also by special folds of the skin, which stretch between the body and humerus, and between the humerus and forearm. The posterior of these folds is of importance in connecting the wing with the body; the anterior has a relation to the mechanism by which the wing is unfolded, inasmuch as it contains an elastic band which extends along its outer edge from the humerus to the articulation of the hand, and which, when the forearm is extended, exercises a traction on the thumb side of the carpal joint, and so causes the simultaneous extension of the hand.

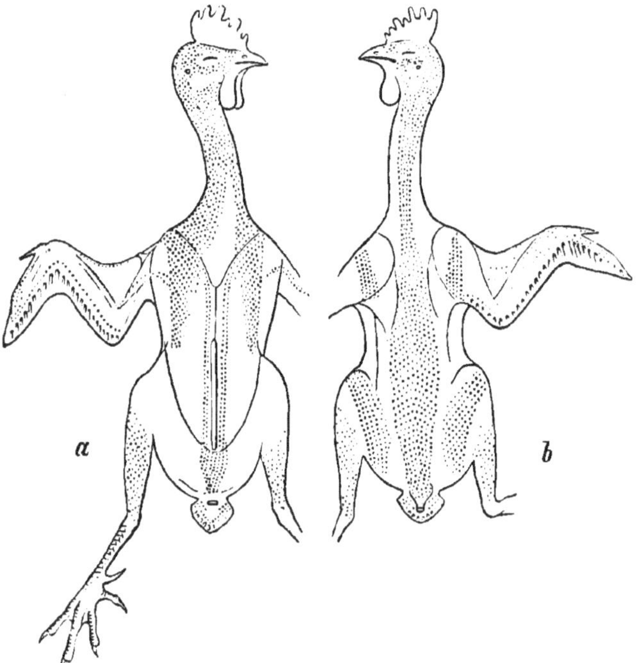

FIG. 651.—*Pterylæ* and *apteria* of *Gallus Bankiva* (after Nitzsch). *a*, ventral side; *b*, dorsal side.

The large wing-feathers (*remiges*) are attached along the lower edge of the hand and forearm. Those which are attached to the hand, from the extremity of the wing to the carpal joint, are known as *primary* remiges (fig. 652, *HS*), while those attached to the forearm as far as the elbow joint are called *secondary* remiges (*AS*). There are usually ten primaries, and a greater but variable number of smaller secondaries.

A number of feathers (*coverts*) attached to the upper end of the humerus are called *scapulars* (*parapterum*), and some feathers fastened to the thumb-joint (sometimes replaced by a spur) are called the bastard-wing (*alula*). All the remiges are covered at their base by shorter feathers, which are arranged in overlapping rows, and are known as coverts (*tectrices*, fig. 652, *T*). In certain cases the wings may become so much reduced that the power of flight is almost or quite lost, a condition which is met with in some running and land birds (*Dinornithidae*, Kiwi and Ostrich), and also in certain water (Penguin).

The great contour-feathers of the tail are called *rectrices* (*Rt*), because during flight they are used for altering the direction and for steering. There are, as a rule, twelve (sometimes ten or twenty and more) rectrices attached to the last caudal vertebra in such a way that they can be moved singly, and unfolded laterally like a fan, as well as be all raised or depressed together. The roots of the tail quills are covered by a number of coverts, which in some cases attain an extraordinary size and shape, and constitute an ornament to the Bird (Peacock). When the power of flight is absent the tail loses its significance as a steering apparatus, and the tail quills are reduced or completely absent. In such cases, however, some of the coverts may attain a considerable size as ornamental feathers.

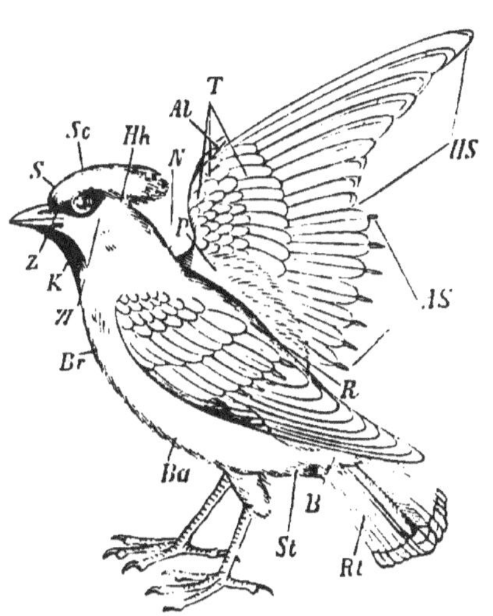

Fig. 652.—Nomenclature of the plumage and regions of *Bombycilla garrula* (wax-wing)(slightly modified after Reichenbach). *S*, forehead; *Sc*, occiput; *Hh*, hind-head; *Z*, lore; *W*, cheek; *N*, nape; *R*, back; *K*, throat; *Br*, breast; *Ba*, belly; *St*, vent; *B*, tail coverts; *Rt*, tail, with tail quills (rectrices); *Hs*, primaries; *As*, secondaries; *T*, coverts (tectrices); *P*, scapulars (parapterum); *Al*, bastard wing (alula).

The hind limbs, which are principally used in the locomotion of the animal on firm ground present numerous varieties, according to the mode of locomotion of the Bird. In the first place, walking legs, or *pedes gradarii*, and wading legs, or *pedes vadantes*, are to be distinguished (fig. 653). The former are much more completely feathered, being covered at least as far as the articulation of the heel: but they vary considerably. The following varieties may be distinguished (fig. 653):—*pedes adhamantes*, with four toes directed forwards, *Cypselus* (*a*); *pedes scansorii*, with two toes directed forwards and two backwards, *Picus* (*b*); *pedes fissi*, with three toes directed forwards and one back, the anterior toes being free to their

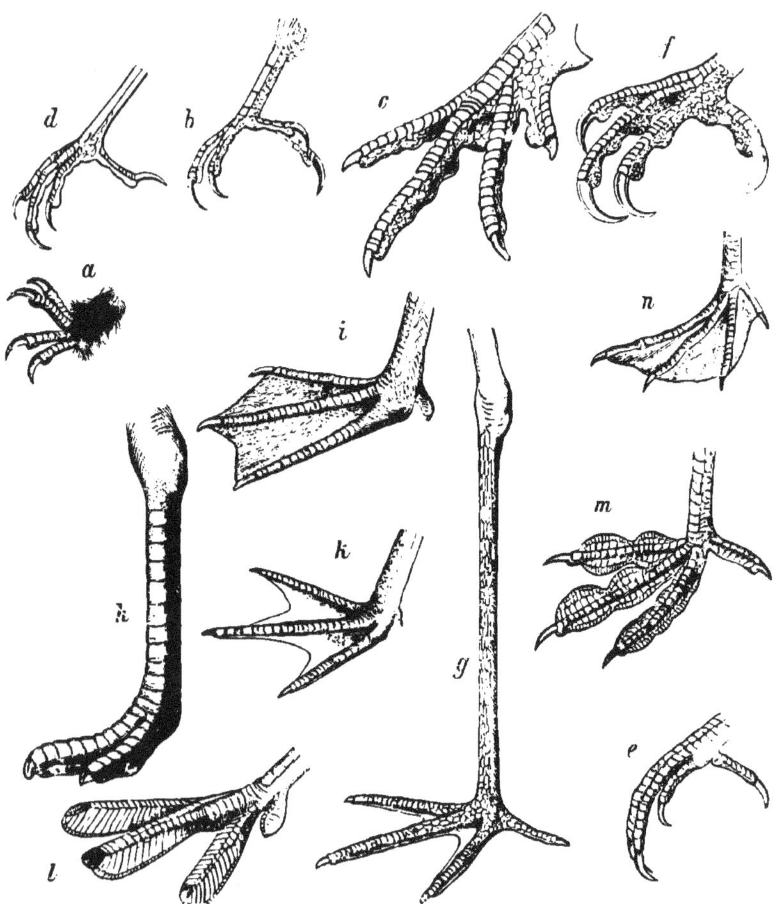

FIG. 653.—The most important forms of Birds' feet (*b*, *c*, *d*, *f*, *n*, from the règne animal). *a*, pes adhamans of *Cypselus apus*; *b*, P. scansorius of *Picus capensis*; *c*, P. ambulatorius of *Phasianus colchicus*; *d*, P. fissus of *Turdus torquatus*; *e*, P. gressorius of *Alcedo ispida*; *f*, P. insidens of *Falco biarmicus*; *g*, P. colligatus of *Mycteria senegalensis*; *h*, P. cursorius of *Struthio camelus*; *i*, P. palmatus of *Mergus merganser*; *k*, P. semi-palmatus of *Recurvirostra avocetta*; *l*, P. fissi-palmatus of *Podiceps cristatus*; *m*, P. lobatus of *Fulica atra*; *n*, P. steganus of *Phaëton æthereus*.

roots, *Turdus* (*d*); *pedes ambulatorii*, with three toes directed forward, the inner toe backwards, the middle and outer toes united at their roots, *Phasianus* (*c*); *pedes gressorii*, the inner toe is placed behind; of the three anteriorly directed toes, the middle and outer are fused as far as the middle, *Alcedo* (*e*); *pedes insidentes* (*f*), the inner toe is behind, the three anteriorly directed toes are united by a short membrane, *Falco* (*f*). Sometimes the outer toes of the *pes scansorius* (*Cuculus*), and the inner toe of the *pes adhamans* (*Colius*) can be turned both forwards and backwards. The wading legs (*p. vadantes*) as opposed to the walking legs (*p. gradarii*) are characterised by the partly or completely naked, unfeathered tibial region; they are found principally in Water Birds, amongst which the *Grallatores* have wading legs with a very long metatarsus—the so-called *pedes grallarii*. The *p. grallarii* may be distinguished into *p. colligati*, in which the anterior toes are united at their roots by a short membrane, *Ciconia* (*g*); and the *p. semicolligati*, in which this membranous connection is confined to the middle and outer toes, *Limosa*. The running legs (*p. cursorii*) are powerful *pedes grallarii* without hind toes, and with three (*Rhea*) or two *Struthio* (*h*) strong front toes. The short wading legs of the swimming birds, as well as the longer legs of the *Grallatores*, present with regard to the structure of their feet the following types:—Swimming feet, or *pedes palmati*, when the three anteriorly directed toes are connected as far as their extremities by an undivided swimming membrane or web, *Anas* (*i*): half swimming feet, or *p. semipalmati*, when the web only reaches to the middle of the toes, *Recurvirostra* (*k*); split swimming feet, or *p. fissipalmati*, when the toes have an entire cutaneous border, *Podiceps* (*l*); lobed feet, or *p. lobati*, when the border is lobed on each joint, *Fulica* (*m*). When the hind toe is also included in the web membrane the feet are termed *p. stegani*, *Halieus* [*Phalacrocorax*] (*n*). Finally, the hind toe may be reduced or completely absent in the *Natatores* and *Grallatores*.

The **Brain** of Birds (fig. 79) is much more highly developed than that of Reptiles, and completely fills the roomy cranial cavity. The hemispheres are, indeed, still without superficial convolutions, but already possess a rudimentary corpus callosum (Meckel). They cover not only the thalamencephalon, but also the two large, laterally displaced *corpora bigemina*. The differentiation of the *cerebellum* is still further advanced, since there is a median part corresponding to the so-called *vermis* of Mammalia, and small lateral appendages.

In consequence of the cervical flexure of the embryo the medulla oblongata forms an angle with the spinal cord, the posterior columns of which diverge from one another in the posterior enlargement of the lumbar region so as to form a second sinus rhomboidalis. The cranial nerves are all separate and their distribution is essentially the same as in the *Mammalia*. The spinal cord reaches almost to the end of the neural canal of the vertebral column.

Sense Organs.—The eyes always attain a considerable size and a high development. The eyelids are always movable, especially the lower lid and the transparent nictitating membrane, which is drawn over the eye by a peculiar muscular apparatus. The eyeball (fig. 654) of the Bird has an unusual form, in that the hind part on which the retina is spread is a segment of a much larger sphere than is the small anterior part. The two parts are connected by a median portion, which has the shape of a short truncated cone, with the smallest end directed forwards. This form of the eyeball is most marked in the nocturnal birds of prey, and least in the aquatic Birds in which the axis of the eye is shorter. There is always a bony sclerotic ring behind the edge of the cornea. The cornea, except in the swimming Birds, is strongly arched, while the anterior surface of the lens in the nocturnal Birds alone possesses a considerable convexity. The *pecten* (wanting only in *Apteryx*) is a peculiar structure of the avian eye. It consists of a process of the choroid, which traverses the retina and passes obliquely through the vitreous humour to the lens. It corresponds to the falciform process of the piscine and reptilian eye. The avian eye is characterised not only by the sharpness of vision consequent on the large size and complicated structure of the retina, but also by the highly-developed power of accommodation, which is principally due to the muscle of the so-called *ciliary ligament* (Krampton's muscle), and also to the great mobility of the muscular iris (dilatation and contraction of the pupil).

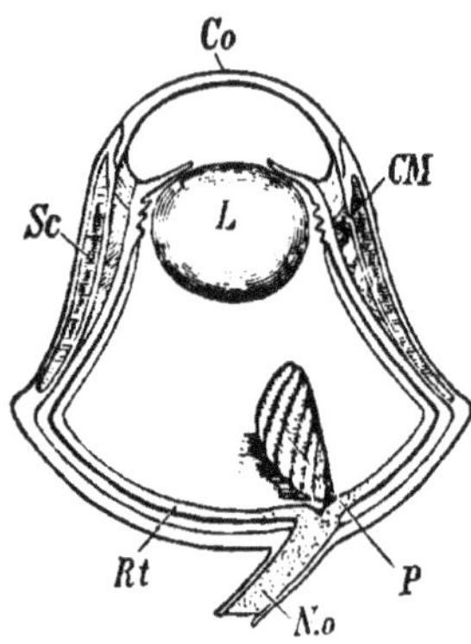

Fig. 654.—Eye of a nocturnal bird of prey (after Wiedersheim). *Co*, cornea; *L*, lens; *Rt*, retina; *P*, pecten; *N.o*, optic nerve; *Sc*, ossifications of the sclerotic; *CM*, ciliary muscle.

The **auditory organ** (fig. 578 *II.*) is enclosed by spongy masses of bone. It possesses three large semicircular canals and a dilated cochlea (*lagena*). The vestibule, which on account of its small size may also be regarded as the lower dilated part of the cochlea, has two openings: the *foramen ovale* which is closed by the terminal piece (*operculum*) of the columella and looks into the tympanic cavity, and a second more rounded opening, the *foramen rotundum*, which is closed by membrane. In addition to the labyrinth there is always a tympanic cavity which communicates with the air-spaces in the neighbouring bones of the skull, and with the pharynx close behind the posterior nares by means of the Eustachian tubes. Towards the exterior the tympanic cavity is closed by a tympanic membrane, to which the long rod-shaped auditory ossicle (*columella*), corresponding to the stapes of Mammalia, is fastened. On the outer side of the tympanic membrane there is a short external auditory meatus, the opening of which is often surrounded by a circle of larger feathers, and in the Owls is overlapped by a cutaneous valve which is likewise beset with feathers, and constitutes a rudimentary pinna.

The **olfactory organ** has three pairs of turbinal bones in the spacious nasal cavities, which are separated by an incomplete septum (*nares perviæ*). The two nasal apertures, except in *Apteryx*, lie more or less near the root of the upper beak; sometimes (Crows) they are covered and protected by stiff hairs: in the *Procellaridæ* they are elongated into a tube and join one another. A so-called nasal gland usually lies on the frontal bone, more rarely beneath the nasal bone or at the inner corner of the eye; it opens by a simple duct into the nasal cavity.

The sense of **taste** is connected with the soft base of the tongue which is rich in papillæ. The tongue is soft throughout its whole extent only in the Parrots. In most other cases it has a firmer covering, and in many cases lends important aid in mastication. In general the **tongue** as well as the beak may be regarded as a tactile organ. In rare cases (Snipe, Duck) the beak is the seat of a finer tactile sensibility, owing to the possession of a soft skin rich in nerves and in the end-corpuscles of Vater.

Alimentary canal.—In spite of great differences in the mode of nourishment the avian digestive organs present a fairly uniform structure; their peculiarities have relation to the power of flight. The jaws are covered by a hard horny sheath and transformed into the beak. True teeth are entirely absent, at least in living Birds as opposed to the fossil *Odontornithes* (*Ichthyornis, Hesperornis*); dental papillæ were however discovered by Etienne Geoffroy St. Hilaire in the jaws of the embryo Parrot. While the upper beak is formed by the fused præmaxillæ, the maxillæ and the nasal bones, the lower corresponds to the two rami of the lower jaw, the fused extremities of which are known as the *myxa*. The lower edge reaching from the angle of the chin to the extremity is termed the *gonys*, the edge of the upper beak is the *culmen*, the region between the eye and the base of the beak which is covered by the cere (*ceroma*) is the lore. The form and development of the beak vary extremely according to the special mode of subsistence (fig. 655).

The tongue, which is always movable, lies on the floor of the

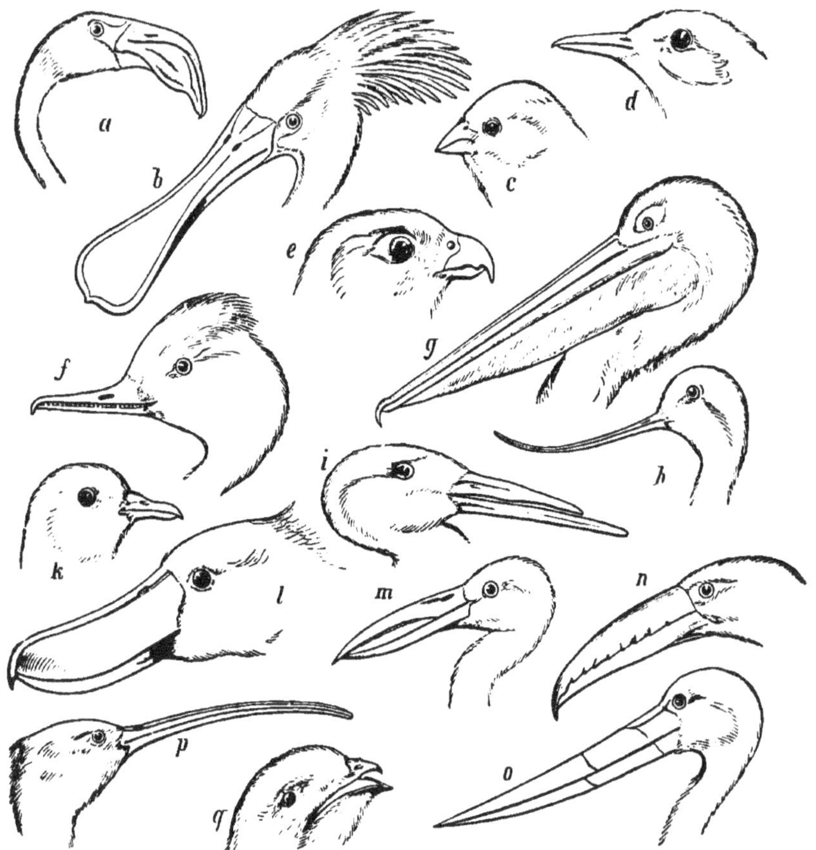

FIG. 655.—Forms of beaks (*a*, *b*, *c*, *d*, *k*, after Naumann; *g*, *i*, *m*, *o*, règne animal; *l*, from Brehm). *a*, *Phœnicopterus antiquorum*; *b*, *Platalea leucorodia*; *c*, *Emberiza citrinella*; *d*, *Turdus cyanus*; *e*, *Falco candicans*; *f*, *Mergus merganser*; *g*, *Pelecanus perspicillatus*; *h*, *Recurvirostra avocetta*; *i*, *Rhynchops nigra*; *k*, *Columba livia*; *l*, *Balæniceps rex*; *m*, *Anastomos coromandelianus*; *n*, *Pteroglossus discolor*; *o*, *Mycteria senegalensis*; *p*, *Falcinellus igneus*; *q*, *Cypselus apus*.

buccal cavity. It consists of the horny or fleshy covering of two cartilages attached to the anterior end of the hyoid bone, and serves for deglutition, and frequently for seizing food. The buccal cavity, which in the Pelicans is dilated into a large cervical sac supported by the rami of the lower jaw, receives the secretion of a number of salivary glands. There is no velum palati. The muscular, longitudinally folded œsophagus, the length of which in general depends on that of the neck, frequently possesses—especially in the birds of prey, but also in the larger granivorous birds (Pigeons, Fowls, Parrots)—a crop-like dilatation, in which the food is softened (fig. 656). In the Pigeons the crop bears two small round accessory sacs, the walls of which secrete in the breeding season a cheesy substance used in feeding the young.

The lower end of the œsophagus is dilated into a glandular *proventriculus*, which is followed by the wide muscular stomach (*gizzard*). While the proventriculus has, as a rule, an oval form and is smaller than the gizzard, the latter is provided with muscular walls, which are weaker (birds of prey) or stronger (granivorous birds), according to the kind of food eaten. In the granivorous birds the gizzard is excellently adapted for the mechanical preparation of the softened food material by the possession of two solid plates, which form the horny internal wall, and work against one another. The first loop of the small intestine (corresponding to the duodenum) surrounds the elongated *pancreas*, the ducts of which, as well as the usually double bile ducts, open in this region. The beginning of the short large intestine is marked by a circular valve, and by the origin of two cæca; it presents no distinction into colon and rectum, and passes into the cloaca, into which the urinogenital apparatus also opens. At its entrance into the cloaca it presents a sphincter-like circular fold. A peculiar glandular sac—the *bursa Fabricii*—opens into the dorsal wall of the cloaca.

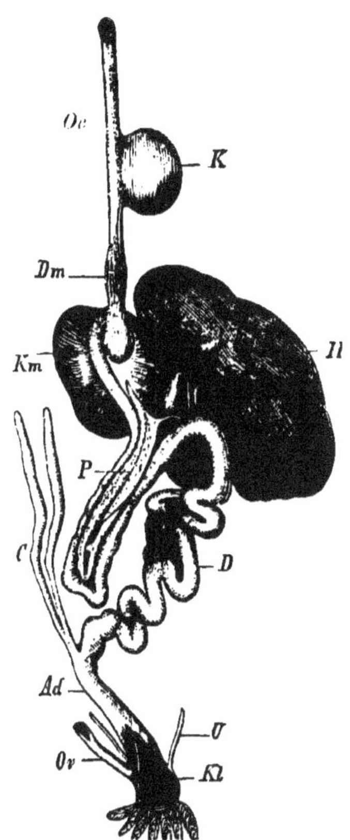

FIG. 656.—Digestive canal of a bird. *Oe*, œsophagus; *K*, crop; *Dm*, proventriculus; *Km*, gizzard; *D*, small intestine; *P*, pancreas (placed in the duodenal loop); *H*, liver; *C*, the two cæca; *Ad*, large intestine; *U*, ureter; *Ov*, oviduct; *Kl*, cloaca.

The large elongated **kidneys** are placed in excavations of the sacrum, and are divided by indentations into a number of lobes. The ureters run behind the rectum and open into the cloaca internally to the genital apertures. The urinary secretion is not fluid, as in the *Mammalia*, but is a white semi-fluid mass, which soon hardens.

The **heart** is completely divided into a right and left half, and lies in the median line, enclosed by the pericardium. As a peculiarity of the heart may be mentioned, the special development of the right atrioventricular valve, which, unlike the tricuspid valve of the Mammalian heart, is a simple strong muscular fold. Since the diaphragm is rudimentary, the thoracic cavity is directly continuous with the abdominal. The pulsations of the heart, in correspondence with the more active respiration, are repeated more rapidly than in *Mammalia*. The right aortic arch persists. The veins open by two superior and one inferior vena cava into the right auricle. The renal-portal circulation still persists in Birds, though it is but slightly developed.

The **lymphatic system** opens by two thoracic ducts (*ductus thoracici*) into the superior venæ cavæ, but also very generally communicates with the veins of the pelvic region. Lymph hearts are only found at the side of the coccygeal bone in the Ostrich and Cassowary, and in some wading and swimming Birds. They are, however, often replaced by vesicular non-contractile dilatations.

Respiratory organs.—The glottis is placed behind the root of the tongue, and leads into a long trachea, which is supported by bony rings. The trachea is not unfrequently longer than the neck, and in such cases, principally in the male sex, is thrown into a number of coils, which either lie beneath the skin (Capercally) or even penetrate into the hollow crest of the sternum (Whooper Swan).

The lower larynx or syrinx.—Except in the Ostrich, the Stork, and some Vultures, the vocal organ is developed at the point where the trachea divides into the bronchi, so that both divisions take part in its formation (fig. 657). The last tracheal rings and the anterior bronchial rings have a modified form, and are often intimately connected with each other; the end of the trachea and the beginning of the bronchi are compressed or dilated into a vesicular form and transformed into the so-called tympanum, which in the males of many Ducks and Divers is dilated into unsymmetrical secondary cavities (tympanic cavity and labyrinth), which serve as resonating apparatuses. The part of the trachea from which the bronchi pass off (*i.e.*, tympanum) is traversed in a horizontal direction by a projecting osseous band—the *pessulus*—which forms a vertical septum between the anterior apertures of the two bronchi. This septum, at its anterior (ventral) and posterior (dorsal) ends, gives off on each side two arched processes, which pass downwards—one along the dorsal, and the other along the ventral edge of the bronchus of its side; and between these cornua the internal wall of each bronchus, which is here membranous, is stretched, and constitutes the *membrana tympaniformis interna*. In the Singing Birds there is in addition a semi-lunar fold (*membrana semilunaris*) on the pessulus, as a prolongation of the membrana tympaniformis interna. In many cases a membranous fold—the *membrana tympaniformis externa*—is developed

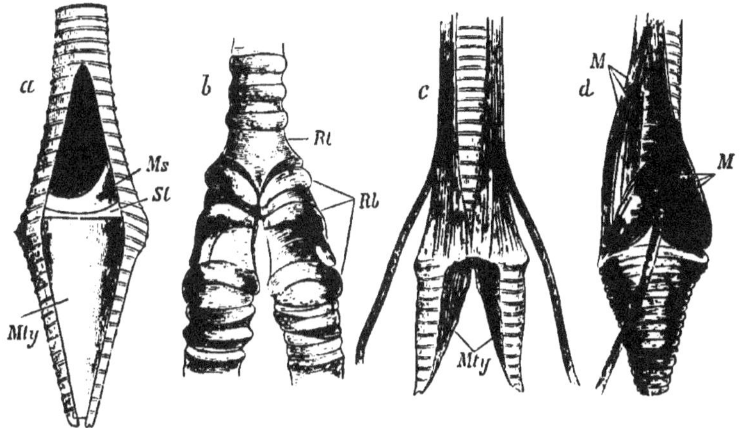

FIG. 657.—Lower larynx of Raven (from Owen). *a*, Side view of larynx laid open.—*b*, Larynx after removal of muscles.—*c*, Larynx with muscles, from the front; *d*, from the side. *St*, pessulus; *Mty*, membrana tympaniformis interna; *Ms*, membrana semilunaris; *Rt*, modified last tracheal ring; *Rb*, the modified three first bronchial rings; *M*, muscles.

on the external side of the tympanum, and forms with the free edge of the internal tympaniform membrane (*i.e.*, with the *membrana semilunaris*), a vocal slit or glottis on either side. The tension of these folds, which function as vocal cords, is regulated by a muscular apparatus, which connects the trachea with the lateral parts of the tympanum, or also with the anterior bronchial rings, and is most highly developed in the singing birds, in which the syrinx may possess five or six pairs of such muscles.

The bronchi are relatively short and lead, at their entrance into the lungs, into a number of wide membranous bronchial tubes, which traverse the pulmonary tissue. The lungs are not, as in Mammals, freely suspended in a closed thoracic cavity and invested by a pleural sac, but are attached to the dorsal wall of the body cavity by cellular tissue, and sunk in the interspaces between the ribs at the sides of the vertebral column. The behaviour of the bronchial tubes and the structure of the finer respiratory air-spaces of the lungs present essential differences from those of the *Mammalia*. The large **air-sacs** are diverticula of the lungs (fig. 658); they have a fairly constant arrangement, extending forwards between the clavicles (*peritracheal* or *interclavicular* air-sac), and also into the anterior and lateral regions of the thorax (*thoracic* air-sacs), and backwards among the viscera, into the pelvic region of the abdominal cavity (*abdominal* air-sacs). The abdominal sacs lead into the cavities of the femora and pelvic bones, while the smaller anterior sacs are prolonged into the air-spaces of the bones of the arm, and into those of the skin, which are sometimes, especially in the large swimming Birds, which fly well (*Sula, Pelecanus*), so numerous that the skin emits a crackling sound when touched (maintenance of temperature, reduction of specific gravity, air reservoirs for respiration). With such arrangements combined with the rudimentary form of the diaphragm already mentioned, and the peculiar structure of the thorax, the mechanism of respiration must be quite unlike that of the Mammalia. The dilatation of the thoracic framework, which also encloses the abdominal cavity, is the result of the extension of the sternocostal bones and the consequent increase in the distance

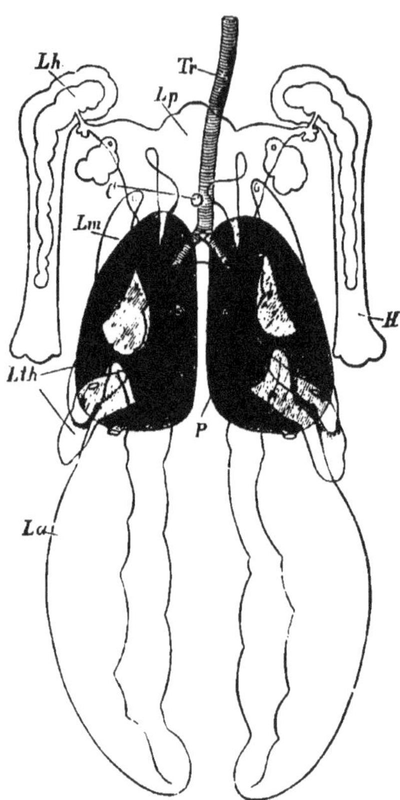

Fig. 658. Lungs and air-sacs of the pigeon (diagrammatic, after C. Heider). *Tr*, trachea; *P*, lungs; *Lp*, peritracheal air-sac with its diverticula (*Lh* and *Lm*) into the humerus (*H*) and between the pectoral muscles; *C*, their connection with the sternal air-spaces; *Lth*, thoracic air-sacs; *La*, abdominal air-sacs.

between the sternum and the vertebral column. The respiratory movements are therefore mainly effected by the sternocostal muscles and the elevators of the ribs, which function as inspiratory muscles.

The **generative organs** closely resemble those of the *Reptilia*. The males are distinguished, not only by their superior strength, but also by the brighter colour of their plumage and the greater variety of their song. There are two oval testes on the anterior side of the kidneys; they become much enlarged at the breeding season, and the left is usually the larger. The epididymis, which is but little developed, leads into the vas deferens, which passes back along the outside of the ureter. The ends of the vasa deferentia are frequently swollen so as to form seminal vesicles, and open on two conical papillæ placed on the hinder (dorsal) wall of the cloaca.

A copulatory organ is, as a rule, wanting; in some of the larger water birds, however (*Ciconia*, *Platalea*, etc.) a rudimentary penis is present as a wart-like process on the front (ventral) wall of the cloaca. It is larger in most of the *Struthionidæ*, the Ducks, Geese, Swans, and in the Curassows and Guans (*Penelope*, *Crax*, *Crax*). In these Birds a curved tube, supported by two fibrous bodies, is attached to the ventral wall of the cloaca. The end of the tube can be retracted by an elastic band. A superficial groove serves to conduct the sperm during copulation. In the two-toed Ostrich, the penis attains a still higher structure, analagous to that of the male copulatory parts of the *Chelonia* and *Crocodilia*. Below the two fibrous bodies, the broad bases of which arise from the front wall of the cloaca, there is a third cavernous body the extremity of which is non-retractile and passes into an erectile bulb—the rudiment of a *glans penis*.

In the female generative organs the ovary and oviduct of the right side are reduced or entirely absent. The generative organs of the left side, however, are correspondingly larger at the breeding season. The ovary is racemose: the oviduct is much coiled, and is divided into three regions: (1) The wide abdominal ostium in front; (2) the coiled glandular part which secretes, from the glands of its longitudinally folded mucous membrane, the albumen which is added in layers and is twisted together at the ends to form the *chalazæ*; (3) a posterior short and wide portion—the so-called uterus—which serves to produce the variously coloured egg-shell, and opens by a short and narrow terminal region into the cloaca on the outer side of the corresponding ureter. When there are copulatory parts in the male, there are also clitoris-like structures at the same place in the female.

Development.—Birds are, without exception, oviparous (relation to power of flight). The egg is remarkable for the large amount of yolk (distinguishable into white and yellow yolk), and its porous calcareous shell (fig. 659). The development requires a high temperature, at least equal to that of the blood. The necessary heat is usually supplied by the bird during incubation.

Fertilization takes place in the upper region of the oviduct before the secretion of the albumen and of the shell membrane, and is at once followed by the partial (discoidal) segmentation which only implicates the clear part of the yolk (formative yolk) around the germinal vesicle—the so-called tread of the cock (*cicatricula*).

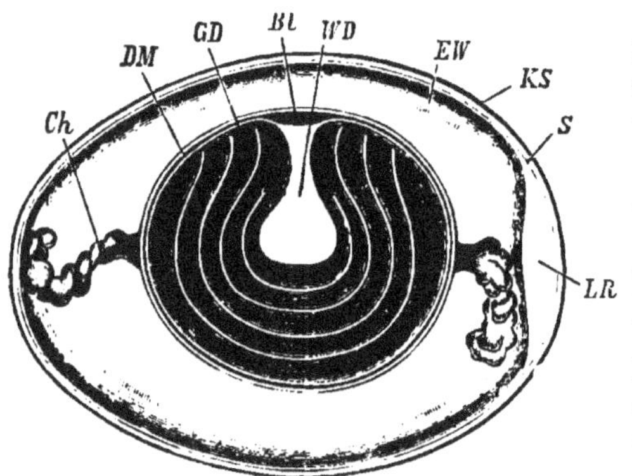

Fig. 659.—Diagrammatic longitudinal section through an undeveloped hen's egg (after Allen Thomson). *Bl*, germinal disc; *GD*, yellow yolk; *WD*, white yolk; *DM*, vitelline membrane; *EW*, albumen; *Ch*, chalazæ; *S*, shell membrane; *KS*, calcareous shell; *LR*, air-chamber.

When the egg is laid, the segmentation is already completed, and the cicatricula has developed into the *germinal disc* or *blastoderm*. The embryo, which later projects from the yolk, developes, as in Reptiles, the characteristic fœtal membranes—the *amnion* and *allantois* (fig. 635). The duration of the embryonic development varies according to the size of the egg and the relative development of the young when hatched. The Bird, when ready to creep out, breaks the blunt end of the shell by means of a sharp tooth placed at the extremity of the upper beak.

The young when hatched have essentially the organisation of the adult animal, although they may still be far inferior to it in the degree of their bodily development. While the *Gallinacei* and the *Cursores*, and most *Grallatores* and *Natatores* have when hatched a complete covering of down, and are so far advanced in development, that they at once follow the mother on land or into water and there seek their own food (**præcoces**); others like the

Passeres, Scansores, Columbinae and *Raptores* leave the egg membranes very early (**altrices**); they are naked, or only covered with down in places, and incapable of free locomotion or of feeding themselves, and remain for some time in the nest, in which they are fed and tended by their parents.

The mental qualities of Birds are incomparably higher than those of Reptiles. The high development of the senses (sight) renders them capable of a sharp discernment, with which is combined a good memory. Under the guidance of its parents the Bird gradually learns to fly and sing; it collects experiences, which it combines so as to arrive at judgments and conclusions; it recognises the surroundings of its nest, distinguishes between friends and foes, and selects the proper means both for the preservation of its existence and for the care of its brood. In some Birds the capacity for profiting by instruction and the faculty of imitation are extraordinarily developed (Starling, Parrot). The emotional side appears no less developed, as may be inferred not only from their general behaviour and the varying expression of their song, but especially from the behaviour of the two sexes at the breeding season. Their instinctive actions are directed to the preservation of the individual, and as in Insects, but in a far higher degree, to the care of their offspring.

In general the manifestations of intelligence as well as of instinct attain their maximum at the time of reproduction, which in the temperate and colder climates usually takes place in the spring (in the Crossbill exceptionally in the middle of winter, *winter plumage, breeding plumage*). The voice* is clearer and richer in the breeding season; the male endeavours to excite the female by his song and the beauty of his plumage. In addition to the changes of plumage and song, the whole behaviour of Birds is modified under the influence of sexual excitement (love-gestures, etc).

With the exception of Fowls, Pheasants, etc., Birds are monogamous; they pair only for the breeding time, and collect together later, and migrate in larger flocks. There are, however, some instances of the migration of single pairs.

Most Birds build nests, and seek for this purpose a suitable place in the district they inhabit. Only a few Birds (Goat-suckers, etc.) are content simply to lay their eggs on the ground, others (Skua, Terns, Ostrich) scoop out a pit or make a depression in moss and grass (*Tetraonidae*). The most skilfully constructed, however, are the

* Cf. A. E. Brehm's "Illustrirtes Thierleben," Tom IV., V. and VI.

nests of those Birds which glue particles of extraneous matter together with their sticky saliva or which weave fine tressworks of moss, wool and grass-stalks (Weavers). As a rule it is the female alone which builds the nest, the male merely helping in collecting the materials. There are, however, instances in which the male takes a share in the construction (Swallows, Weavers); while in other cases (*Gallinacei*, Chaffinch) the male takes no share at all in building the nest. Many sea Birds, as the Auks and Penguins, lay but one egg, the large Birds of prey, Pigeons, Swifts, and Humming-birds, lay two eggs. The number of eggs is larger in the singing Birds and still greater in the swimming Birds of ponds and rivers, and in the Fowls and Ostriches. The duration of the period of incubation is equally various; it depends upon the size of the egg and the degree of development of the young when hatched.

While the Humming-birds and golden-crested Wrens incubate for eleven to twelve days, and the singing Birds fifteen to eighteen days, Fowls require three weeks, Swans double that time, and Ostriches seven to eight weeks. Incubation essentially consists in keeping the eggs at a warm, uniform temperature; this is effected by the body of the sitting bird, and is often facilitated by the presence of naked places on the body. As a rule, the mother alone sits, and the male occupies himself with bringing her food. Not unfrequently, however, as in the Pigeons, Peewits, and many swimming birds, the two parents relieve one another regularly; the male in such cases certainly sits only for a shorter time during the day, while the female sits through the whole night. In the Ostrich the female only sits during the first period of incubation; later the parts are changed, and the male undertakes the chief part of the incubation, especially sitting almost all night. The behaviour of numerous Cuckoos, and especially of our native species, is very remarkable; they leave the building of nests and the care of their brood to other Birds, and lay their small eggs, singly and at intervals of about eight days, amongst the eggs of different singing Birds.

The care and nurture of the young usually falls entirely or mostly on the hen bird. On the other hand, as a rule, both parents take an equal part in the protection of the brood.

Leaving out of consideration the activities which relate to reproduction, the instinct of Birds manifests itself, principally in late summer and autumn, as an impulse to migrate, and still more mysteriously as a true guide on the journey. Few birds of the colder and temperate climates pass the winter in the places where they breed (*Resident birds:* Eagles, Owls, Ravens, Woodpeckers, Magpies, Sparrows, Titmice, Grouse, etc.) Many of them rove over larger and smaller regions in search of food (*Strichvögel:* Thrushes, Bramblings, and Chaffinches, Woodpeckers, Yellow Bunting, Finches, and crested Lark). Others migrate before the beginning of the cold season of the year, when nourishment is deficient, from the northern climates to the temperate, from these to southern regions (*Zugvögel;* Swallows, Storks, Jackdaws, Crows, Starlings, Wildgeese, Cranes, etc.)

There are but scanty materials for the geological history of this class. Leaving out of consideration the feather-tailed *Archæopteryx lithographica** (fig. 119) of the Jura (*Saururæ*), the oldest remains of the swimming and wading birds belong to the chalk. In the tertiary period the remains are indeed more frequent, but are nevertheless insufficient for a more accurate definition. In the Diluvium, on the other hand, numerous types of species still living are found, as well as remarkable gigantic forms which have become extinct within the historical period (*Palæornis, Dinornis, Palapteryx, Didus*).

I. CARINATÆ.

The sternum has a keel (*carina*) for the insertion of the powerfully developed muscles of flight. The remiges of the wing and the rectrices of the tail are usually well developed. Almost all are able to fly.

Order 1.—Natatores (Swimming Birds).

Aquatic Birds with short legs often placed far back; feet either pedes palmati or p. stegani.

The form of the body of the swimming Birds varies extraordinarily, according to the special adaptation to their aquatic habitat. They all have a thick compact plumage, a very rich clothing of down, and a large uropygial gland. The legs are short and are placed far back, and usually feathered as far as the ankle. They end with swimming feet, either *pedes palmati*, or *fissipalmati*, or *stegani*. The Natatores are all excellent swimmers; many are strong flyers, while others are

* This group, which is allied to the reptilian genus *Compsognatus* (*Ornithoscelida*) is especially characterised by the fact that the caudal part of the vertebral column is as long as the body, and was furnished with feathers, arranged in pairs. Since the metatarsal bones are not fused, there is no true avian tarsometatarsus.

incapable of flight, and are almost entirely confined to the water. Most of them also dive with great skill, either shooting down into the water from the air or suddenly diving beneath the water while swimming. The form of the beak varies as much as does the structure of the wings. Sometimes the beak is much arched and armed with cutting edges, and sometimes flat and broad; sometimes elongated and pointed. The form of the beak is correlated with the mode of subsistence: in the first case we have to do with predatory birds, which especially prey on fishes, in the last case with birds, which live on worms and small aquatic animals, but also on fishes. The swimming Birds, with broad soft-skinned beak, search in the mud and feed not only on worms and small aquatic animals, but also on seeds and vegetable matters. The *Natatores* are gregarious, and exist in great flocks on the sea coasts or on inland waters, but some of them are also found on the high seas, far from land. Most of them are migratory. They nest near the water, often in common breeding places, and lay a few eggs either directly on the ground, or in holes, or in simple, rudely-made nests. Many of them are of great importance to man, partly on account of their flesh and eggs, and partly of their down and skins, and partly also on account of their excrements, which are used as manure (guano).

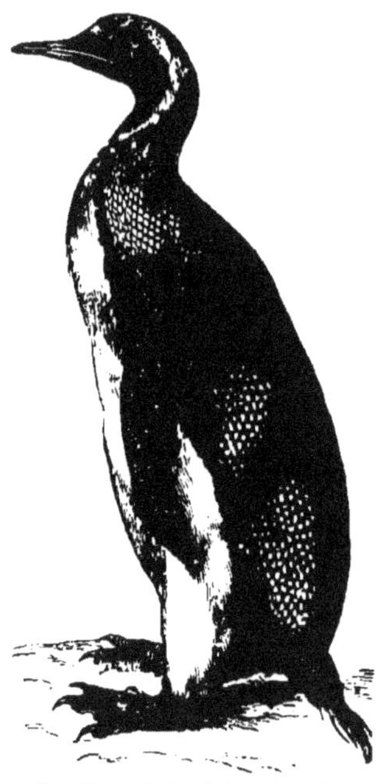

FIG. 660. *Aptenodytes patagonica* (from Brehm).

Fam. **Impennes** (Penguins). The wings are fin-like, without remiges, covered with small scale-like feathers. The tail is short, with stiff feathers. The short swimming-feet have a reduced forwardly-directed hind toe, and are placed so far back that on land the body must be carried almost vertically. They are excellent divers. In the breeding season they stand upright and arranged in long rows—the so-called schools. They lay only one egg in a depression in the ground, and keep it in a vertically upright position during incubation; but they can also carry it about buried in the down between the legs. Both sexes participate in incubation. *Aptenodytes patagonica*, Forst., King-Penguin (fig. 660); *Spheniscus demersus* L., Black-footed Penguin, South Africa and America; *Eudyptes chrysocoma* L., South Sea, Patagonia.

Fam. **Alcidæ** (Auks). The wings are short and ill-adapted for flight. There are, however, small remiges. The swimming-feet with rudimentary or without hind toe. Their common breeding-places are on the coasts (*Vogelberge*), where they lay their eggs singly in holes in the earth or in nests, and bring up their young. *Alca impennis* L., Great Auk; now extirpated. *A. torda* L., Razorbill; *Mormon arcticus* Ill. (*fratercula* Temm.), Puffin; *Uria troile* Lath., Guillemot; *U. grylle* Cuv., Black Guillemot.

Fam. **Colymbidæ** (Divers). The head has a pointed straight beak. The freely projecting metatarsus is strongly laterally compressed. The feet are palmate or fissipalmate. *Podiceps cristatus* L., Great crested Grebe; *P. minor* Gm.; *Colymbus glacialis* L., Great Northern Diver.

Fam. **Lamellirostres.** Beak broad, deep at the base, covered with a soft, richly innervated skin, with transverse lamellæ on its edges (dentated appearance), and ends with a nail-like extremity. The feet are palmate (*p. palmati*), with rudimentary hind toe, which is sometimes naked, sometimes fringed with membrane. *Phœnicopterus antiquorum* L., Flamingo, North Africa: *Cygnus olor* L., Mute Swan; *C. musicus* Bechst., Whooper; *Anser cinereus* Meyer, Gray Goose; *A. hyperboreus* L., Snow Goose; *A. segetum* L., Bean Goose; *Anas boschas* L., Wild Duck, the ancestral species of the various races of domestic ducks; *A. (Tadorna) tadorna* L., Sheldrake: *Mergus merganser* L., Goosander: *M. serrator* L., Redbreasted Merganser; *M. albellus* L., Smew.

Fam. **Steganopodes.** Large swimming birds, with small head, well developed, often long and pointed wings, with swimming feet (*p. stegani*). *Pelecanus onocrotalus* L., Pelican; *Halieus carbo*, Cormorant; *Tachypetes aquila* L., Frigate Bird: *Sula bassana* L., Gannet, North Europe; *Phaëton ætherius* L., Tropicbird.

Fam. **Laridæ** (Gulls). Lightly built Swallow- or Pigeon-like swimming-birds, with long pointed wings and often forked tail, relatively high, three-toed swimming-feet and free hind toe. They dive from the air (Stosstaucher). *Sterna hirundo* L., Tern; *Larus minutus* Pall., Little Gull; *L. ridibundus* L., Blackheaded Gull: *L. canus* L., Common Gull: *Lestris parasitica* L., Skua, North German Coasts; *Rhynchops nigra* L., Skimmer.

Fam. **Procellariidæ** (Sturmvögel). Gull-like birds, with rostrum compositum. Feet palmate, hind toe absent or reduced to a stump. They select rocky and precipitous coasts for their common breeding-places. The female lays one egg and takes turn with the male in incubation. The young are nurtured for a long time. *Diomedea exulans* L., Albatross, South Sea: *Procellaria glacialis* L., Fulmar Petrel, from the Arctic Seas to North German Coasts; *Thalassidroma pelagica* L., Stormy Petrel, Atlantic Ocean.

Order 2.—Grallatores (Waders).

Birds with long thin neck and long beak, with elongated wading legs (p. vadantes).

The *Grallatores* are adapted for an aquatic life, since they have to seek their food in water, but their adaptations are of a different kind to those of the *Natatores*. They live more in swampy places, on the banks of rivers and seas, and wade through shallow water in order to seek snails and worms, or frogs and fishes. They therefore possess, with some exceptions, long wading legs, with usually naked tibiæ projecting freely from the body, and very long metatarsus often covered with scales. But few have running legs (*p. cursorii*) and are land birds (Bustards). Some (Rails) are similar in their mode of life, the shortness of their legs and the structure of their toes, to the swimming birds (*Natatores*), and swim and dive well, but fly badly. Corresponding with the considerable length of the legs, there is a long neck and usually also a long beak. The size and form of the beak varies exceedingly. When small worms, Insect larvæ and Molluscs are sought in mud and loose earth, the beak is long, but relatively weak and soft, and has a sensitive richly innervated extremity; in other cases the beak is very strong, angular, and adapted for the capture of fishes and frogs, and even of small Mammals; and finally in the transitional groups before mentioned it is short and strong, like that of a fowl, with a somewhat arched *culmen*, and adapted for an omnivorous diet. The feet also present great differences in the size and connection of the toes. The wings usually attain a medium size. The tail, on the other hand, is short. The plumage is more uniform and simple, and but rarely presents beautiful and glittering colours. Most *Grallatores* are migratory birds of the temperate regions, and live in pairs, in a monogamous state. They build rude nests on the ground, on the shore, or on trees and houses, more rarely on water. The young are sometimes *altrices*, and sometimes *præcoces*.

Fam. **Charadriidæ**. Plovers. With tolerably thick head, short neck, and hard-edged beak of medium length. *Cursorius europaeus* = *C. isabellinus* M., North Africa and South Europe; *Oedicnemus crepitans* Temm., Steppes in South Europe, Africa and West Asia, also on the great fallow lands in Germany. *Charadrius pluvialis* L., Golden-plover, inhabits the Tundra: *Vanellus cristatus* M., Peewit, Germany and Holland.

Fam. **Scolopacidæ**. Snipe. Head of medium size, strongly arched with a long, thin, usually soft beak, covered with a richly innervated skin. *Totanus hypoleucus* Temm., Sandpiper; *Recurrirostra avocetta* L., Avocet; *Tringa cinerea* Gm., *Machetes pugnax* Cuv., Ruff; *Scolopax rusticola* L., Woodcock; *Gallinago media* Gray, Snipe; *G. gallinula* L., Jack Snipe: *Numenius arquata* L., Curlew.

Fam. **Herodii = Ardeidæ**. Herons and Storks. Large *Grallatores* with powerful body, long neck, and small, partly naked head: beak powerful, without cere, with sharp hard edges, sometimes curved at the point, rarely spoon-shaped. Legs long, and naked far above the intertarsal joint: feet usually *p. colligati*; hind toe rests on the ground. *Ibis rubra* Vieill., the scarlet Ibis of Central America. *I. religiosa* Cuv., the sacred Ibis: *Falcinellus igneus* Gray, Glossy Ibis; *Platalea leucorodia* L., Spoonbill: *Balaeniceps* *rex* Gould; *Ardea cinerea* L.; *A. purpurea* L., South Europe: *Ciconia alba* L., Stork: *Mycteria senegalensis; Leptoptilus argala* Temm.. (Marabu): *Anastomus lamelligerus* Temm., East Indies; *Grus cinerea* Bechst., Common Crane.

Fam. **Rallidæ**. Water-hens, Rails. Intermediate between the *Natatores* and the *Gallinacei*. *Rallus aquaticus* L., Water Rail. Northern and Central Europe to Central Asia; *Crex pratensis* L., Cornerake; *Cr. porzana* L.. Europe, Spotted crake: *Parra jacana* L., America; *Galinula chloropus* L.. Moorhen: *Fulica atra* L., Coot. On the reedy lakes and ponds of Europe.

Fam. **Alectoridæ**. Transitional between the *Grallatores* and *Gallinacei*.

Fig. 661. - *Chauna chavaria* (règne animal).

They resemble the former in the length of their legs, and the latter in their mode of life and in the form of their beak. *Otis tarda* L.. Bustard. Lives as a migratory bird (Strichvogel) in the plains of South-east Europe with one or two females to each male. *O. tetrax* L., more in the South: *Dicholophus cristatus* Ill., Cariama [Seriema], in Brazil, lives on Lizards and Snakes like the Secretary-bird of South Africa. *Psophia crepitans* L., Trumpeter. South America: *Palamedea cornuta* L., with spurs on the wings. *Chauna chavaria* Ill., Screamers (fig. 661). With spurs on the wings. Is domesticated. Receives in German the name of Shepherd's bird from its use as keeper and defender of flocks of hens and geese in South America.

Order 3.—Gallinacei = Rasores.

Terrestrial birds of medium and sometimes considerable size, of stout build, with short, rounded wings, strong beak, usually arched and bent downwards at the point, and with powerful feet adapted for perching (p. insidentes), usually præcoces.

The *Gallinacei* possess in general a stout body with thick plumage, small head and powerful beak, short or moderately long neck, usually short and rounded wings, legs of medium length and well-developed tail, composed of numerous rectrices. There are often naked places on the head, as well as erectile combs and cutaneous folds (wattles), the latter principally as distinctions of the male sex. The beak is usually short, broad and high, and is characterized both by the overlapping cutting edges and by the depressed extremity of the upper beak. Its base is soft and membranous, and covered with feathers, among which a membranous or cartilaginous scale projects over the nasal apertures. The plumage of the *Gallinacei* is close and stiff, and often beautifully marked and ornamented with rich colours and a metallic lustre (male). The tail quills are usually more than twelve in number, and there may be as many as eighteen or twenty. The wings are as a rule short and rounded, with ten primary remiges and twelve to eighteen secondary. The flight, therefore, is clumsy: only the *Pteroclidæ* fly quickly and with skilful turnings. The legs are powerful and short, or of medium length; they are usually feathered as far as the foot-joint, rarely up to the toes. There is often a sharp spur, which serves as a weapon, on the metatarsus of the male above the hind toe, which is articulated high up. The *Gallinacei* live for the most part on the ground, either in forests or in fields, on grassy plains from high mountains down to the sea coast. They are good runners, and seek their food on the ground, feeding specially on berries, buds and seeds, also on insects and worms. They form their rude nests for the most part on the surface of the ground, or in low bushes; more rarely on high trees, and lay a great number of eggs. As a rule the cock lives with a number of hens, and takes no part either in the building of the nest or in the care of the brood. The young are for the most part *præcoces*. The hens are easily domesticated, and on account of their eggs and their well-flavoured flesh, have been made useful as domestic animals from the earliest times.

Fam. **Penelopidæ**. Large, long-legged *Gallinacei*, with well-developed remiges and long, rounded tail, resembling the three-toed Ostrich in the structure of the protrusible penis. *Crax alector* L., Curassow, South America; *Crax pauxi* L.. *C. galeata* Cuv.. Mexico; *Penelope cristata* Gm., Guan, Brazil; *Meleagris mexicana*, Gould., ancestral form of *M. gallopavo*, the Turkey.

Here are allied the **Crypturidæ (Tinamidæ)**, Tinamous, and **Opisthocomidæ**, Hoazin.

Fam. **Megapodiidæ** (Mound-birds). Long-legged *Gallinacei*, of medium size, with short, broad tail and large, strongly-clawed, ambulatory foot (pes ambulatorius), the hind toe of which is articulated at the same level as the front toes. *Megacephalon maleo* Temm., Celebes; *Megapodius tumulus*, North East of Australia.

Fam. **Phasianidæ** (True fowls). The head is partially bare of feathers, especially in the cheek region; it is often adorned with coloured combs, cutaneous lobes or tufts of feathers, and has a strongly-arched beak of medium length, with the point curved downwards. The two sexes are strikingly different, the male being larger and more richly adorned. They are inhabitants of the Old World. *Gallus bankiva* Temm., Island of Sunda; *Lophophorus refulgens* Temm.. Himalayas; *Phasianus colchicus* L., Common Pheasant; *Ph. pictus* L., Golden Pheasant; *Ph.* (*Gallophasis*) *nycthemerus* L., Silver Pheasant, China: *Pavo cristatus* L., Peacock; *Argus giganteus* Temm., Argus Pheasant, Malacca, Borneo: *Numida meleagris* L.. Guinea Fowl, North Africa.

Fam. **Tetraonidæ**. The body is stout, the neck short, the head small and feathered, with at most one naked stripe above the eyes. The legs are short, and are usually feathered down to the toes. *Tetrao urogallus* L., Capercally; *T. tetrix* L., Black Grouse. The hybrid between these two species is called *T. medius*, by Meyer. *T. bonasia* L.. Hazel Grouse; *Lagopus albus* Vieill., Willow Grouse, Scandinavia; *L. alpinus* Nilss.; *Perdrix cinerea*, Briss., Partridge; *P. saxatilis* M.W.; *P. rubra* Temm.. Red-legged Partridge; *Coturnix dactylisonans* Meyer. Quail.

Fam. **Pteroclidæ** (Sand-grouse). Small *Gallinacei*, with small head, short beak, short, weak legs, long-pointed wings and wedge-shaped tail. Feet with short toes: hind toe when present rudimentary and attached high up; it may be absent. *Pterocles alchata* Gray, in Asia Minor and Africa; *Syrrhaptes paradoxus* Pall., in the steppes of Tartary, and lately in North Germany.

Order 4. Columbinæ (Pigeons).

Birds with weak soft beak, swollen round the nasal apertures, with pointed wings of medium size, and short cloven feet (pedes fissi). The young are altrices.

The *Columbinæ* are most nearly allied to the *Pteroclidæ*. They are of medium size, with small head, short neck and short legs. The beak is longer than in the *Gallinacei*, but weaker, and gently arched at the horny, somewhat turned-up extremity. At the base of the beak the scaly cover of the nasal openings is swollen, naked, and membranous. The rather long, pointed wings enable the bird to fly quickly and skilfully. The tail is weak and rounded, and contains usually twelve, rarely fourteen, or sixteen rectrices.

The stiff, beautifully coloured plumage lies smoothly on the body, and presents hardly any difference in the two sexes. The short legs are unfitted for rapid and constant locomotion. The feet are cloven (*p. fissi*) or ambulatory (*p. ambulatorii*), and the well-developed hind toe rests on the ground. The *Columbinæ* have a paired crop, which at the breeding season in both sexes secretes a creamy fluid for the nourishment of the young. They are distributed over all parts of the world. They live in pairs or in flocks in forests, and feed almost exclusively on grain and seeds. The species which live in the north are migratory (Zugvögel); others make short migration (Strichvögel). while others are resident birds. They live in a state of monogamy, and lay two, rarely three, eggs in a rudely-constructed nest. Both

FIG. 662.—*Columba livia* (after Naumann).

sexes take part in hatching and bringing up the young. The young leave the egg almost entirely naked, with closed eyelids, and, as altrices, require the care of the mother for a considerable time.

Fam. **Columbidæ.** The beak, with smooth edges, never dentated. *Columba livia* L., Rock-Dove (fig. 662). Slate-blue, with white wing-coverts and two black bands on the wings and tail. It is the ancestral form of the numerous races of domestic pigeon. It nests on rocks and ruins, and is distributed from the coasts of the Mediterranean over a great part of Europe and Asia. *C.* (*Palumbœnas*) *œnas* L., Stock-Dove; *Palumbus torquatus*, Leach, Ring-dove; *Ectopistes migratorius* L., Passenger Pigeon, North America; *Turtur auritus* Bp., Turtle-Dove; *T. risorius* Sws.; *Goura coronata*, Flem., New Guinea.

Fam. **Didunculidæ.** Beak compressed, lower jaw dentated, with hooked extremity. *Didunculus strigirostris* Gould, Samoa Islands.

The extinct Dodos (**Ineptæ**) were allied to this last family, and have been placed among the pigeon-like birds. They were living in Vasco di Gama's time on small islands on the East Coast of Africa (the Mascarenes), and were still plentiful; they became extinct two hundred years ago. As far as we can judge of the appearance of this bird from the preserved remains (in [London] Oxford and Copenhagen) of skulls, beaks, and legs, and from the old descriptions, and especially from an old oil painting preserved in the British Museum, the *Dodo, Didus ineptus* L., was an unwieldy bird, larger than the Swan, with lax plumage, powerful, four-toed, scraping feet, and strong, deeply-cleft beak.

Order 5.—SCANSORES.

Birds with powerful beak, stiff plumage having but little down, and scansorial feet. The young are altrices.

Within the artificial limits of this order is included a number of groups of very different birds which essentially agree only in the structure of the feet, which are adapted principally for climbing; they present, however, considerable differences in the manner of locomotion, and find their nearest allies in several families of *Passeres*. The beak is always powerful; it is sometimes long, straight, and angular, adapted for hammering and chiseling on trees (Woodpecker); sometimes short and curved like a hook (Parrot), or of colossal size and with dentated edges (Toucan). The legs end with long-toed, scansorial feet, the outer toe of which can in some cases be directed forward. The metatarsus is seldom feathered, more frequently beset with semirings and scutes in front and small scales behind. The wings contain very generally ten primaries. The tail is sometimes used as a support in climbing. Most of the Scansores inhabit forests, nest in hollow trees, and feed on insects, some of them, however, on small birds, and others on fruit and vegetable matters.

Fam. **Ramphastidæ** (Toucans). Raven-like birds, with colossal, marginally serrated beak, and horny, brush-like tongue. *Ramphastus toco* L.; *Pteroglossus Aracari* Ill.

Fam. **Trogonidæ**. Beak short and strong, with usually serrated edges and wide, slit-like mouth, with bristles at the corner of the mouth. The plumage of the male has a metallic lustre. *Trogon curucui* L., Brazil; *Calurus resplendens* Gould, the Quesal, in Central America. Here are allied the Jacamar (*Galbula*) and Puff-bird (*Bucco*).

Fam. **Cuculidæ** (Cuckoos). With gently-curved, deeply-cleft beak, long pointed wings, and wedge-shaped, pointed tail. The feet are scansorial, and the outer toe can be directed forward. *Cuculus canorus* L., European Cuckoo, sparrowhawk-like, with barred plumage; *Coccystes glandarius* L., Great Spotted Cuckoo of South Europe.

Here are allied the **Musophagidæ** (Plaintain-eaters). *Corythaix persa* L., Guinea; *Musophaga violacea* Isert, Plaintain-eater, West Africa. In *Colius* the outer and inner toes can be turned backwards or forwards.

Fam. **Picidæ** (Woodpeckers). Powerfully-built *Scansores*, with strong, chisel-shaped beak, pointed in front, without cere. Metatarsus with transverse scales; feet with strong claws; with firm tail. The tongue is long, flat and horny, and bears at its end arrow-like, short, recurved hooks; it can be rapidly protruded to a considerable distance in consequence of a peculiar mechanism of the hyoid bone. The cornua of the hyoid are bent into wide arches, and extend over the skull to the base of the beak. *Picus martius* L., Black Woodpecker, Europe and Asia; *P. major* L.; *P. medius* L.; *P.* (*Piculus*) *minor* L., Lesser Spotted Woodpecker, Europe; *P. tridactylus* L.; *P. viridis* L., Green Woodpecker; *P. canus* Gm., Greyheaded Green Woodpecker; *Jynx torquilla* L., Wryneck.

Fam. **Psittacidæ** (Parrots). Scansores of the warmer climates, with stout, strongly-bent beak, fleshy tongue, and powerful legs with short metatarsus. The feet, with toes arranged in pairs, are used like a hand to seize the food. The upper beak, which is dentated and covered at its base by a cere, is articulated with the frontal, and its long hooked extremity overlaps the short and broad lower beak. Most of the Parrots belong to America, many also to the Moluccas and Australia. A few are found in Polynesia, New Zealand, and Africa.

Plictolophinæ. Cockatoos. Head usually with movable crest. *Plictolophus leucocephalus* Less.; *Nymphicus Novæ Hollandiæ* Gray; *Calyptorhynchus galeatus* Lath., Van Diemen's Land.

Platycercinæ. Parrakeets. With moderately pointed, rarely rounded wing, and long, graduated, wedge-shaped tail. *Sittace militaris* L., Maccaw, Mexico; *Palæornis Alexandri* L., Ceylon: *Melopsittacus undulatus* Shaw (Wellenpapagei), Australia: *Pezoporus formosus* Lath., Ground-Parrakeet, Australia; *Platycercus Pennantii* Lath., Australia.

Psittacinæ. Tail truncated, or rounded. *Psittacus erithacus* L., Grey Parrot, West Africa: *Psittacula passerina* L., Love-bird, Brazil.

Trichoglossinæ. Lories. The tip of the tongue is pencil-shaped, with feathery, horny papillæ. *Trichoglossus papuensis* L., New Guinea; *Nestor meridionalis* L., New Zealand.

Strigopinæ. Kakapos. Of owl-like appearance, with incomplete feather-disc. *Strigops habroptilus* Gray, New Zealand.

Order 6.—Passeres (Insessores). Passerine Birds.

Birds with horny beak, without cere. Metatarsus covered with laminæ, or scales. The feet are pedes ambulatorii, p. gressorii, or p. adhamantes. The young are altrices. A vocal apparatus with muscles is frequently present.

The birds included in this large order are of small size, and present great differences in the form of their beak; they fly exceedingly well. When on the ground they hop, or more rarely walk, and they remain by preference on trees and in bushes. They are usually divided according to their vocal apparatus into two orders the singing birds or **Oscines**, and the shrieking birds or **Clamatores**; a division which seems the more artificial because the same types of form of beak and of the whole structure of the body are repeated in the two groups. An arrangement based on the form of the beak might lead to less artificial groups. By far the greater number of Passeres live in monogamy, often united in large flocks. Many of them build skilfully-constructed nests, and are migratory.

Tribe 1. **Levirostres.** *Clamatores*, with large, but light beak, short, weak legs, and gressorial or fissate feet, which are adapted for clinging to branches.

Fam. **Buceridæ** (Hornbills). Raven-like birds, of considerable size, with colossal, but always light, dentated, and downwardly-curved beak and horn-like head-dress at the base of the upper beak. *Bucorvus abyssinicus* Gm.; *Buceros rhinoceros* L., Sumatra.

Fam. **Halcyonidæ** (Kingfishers). *Passeres*, with large head and long, keeled, angular beak, relatively short wings and short tail. Metatarsus short; feet gressorial. *Alcedo ispida* L., Europe; *Ceryle rudis* L., Black and white Kingfisher, Africa: *Dacelo gigas* Glog., Australia.

Fam. **Meropidæ.** Bee-eaters. The beak is compressed and gently curved downwards. The plumage is variegated: the legs are weak. The wings are pointed, with long coverts. *Merops apiaster* L., South Europe.

Fam. **Coracidæ.** Rollers. Large, beautifully coloured birds, with deeply-cleft beak with sharp edges and recurved extremity. The wings are long and the feet cloven (*p. fissi*). *Coracias garrula* L., Roller.

Tribe 2. **Tenuirostres.** *Clamatores* and *Oscines* with long, thin beak and ambulatory or cloven feet (*p. ambulatorii* or *fissi*), with long hind toe.

Fam. **Upupidæ.** Hoopoes. Beautifully coloured Clamatores with long, laterally compressed beak: short, triangular tongue and long, strongly rounded wings. *Upupa epops* L., Hoopoe.

Fam. **Trochilidæ.** Humming birds. The smallest of all birds. Variegated plumage with metallic lustre. Slender feet (*p. ambulatorii or fissi*). The long, awl-shaped beak has, in consequence of the projecting edges of the upper beak, the form of a tube, from which the long tongue, which is cleft up to the root, can be rapidly projected. *Rhamphodon nævius* Less., Brazil; *Phaëthornis superciliosus* Sws., Brazil; *Trochilus colubris* L.; *Lophornis magnifica* Vp., Brazil.

Fam. **Meliphagidæ.** Honey suckers. Small, beautifully coloured birds, of stout build, with muscular vocal apparatus, with long gently-curved beak, long metatarsus, wings of medium length and long tail. *Meliphaga auricomis* Sws., Australia; *Nectarinia famosa* Ill.; *N.* (*Cinnyris*) *splendida* Cuv., South Africa.

Fam. **Certhiidæ.** Tree-creepers. *Oscines* with long, slightly-curved beak, pointed, horny tongue, metatarsus covered with scales, and long hind toe with a sharp claw. *Certhia familiaris* L., Common creeper: *Tichodroma muraria* Ill., Wall creeper.

Tribe 3. **Fissirostres.** With short neck, flattened head, and deeply-cleft beak, with long, pointed wings and weak feet (*p. ambulatorii* or *adhamantes*). They all fly with rapidity and dexterity. They catch their food, especially flies, *Neuroptera* and butterflies, during flight with open beak. They live for the most part in warmer climates.

Fam. **Hirundinidæ.** Swallows. Small, delicately-formed *Oscines* with broad, triangular beak, compressed at the point, nine primary rectrices and long, forked tail. They are distributed over all parts of the earth, and construct their nests with skill. The European species pass the winter in Central Africa. *Hirundo* L. Beak short and triangular: metatarsus naked. The first and second remiges of equal length. *H. rustica* L., Swallow. *H.* (*Chelidon* Boie. with feathered metatarsus) *urbica* L., Martin. *H.* (*Cotile* Boie. The nasal apertures free, the tail slightly excavated and moderately long) *riparia* L., Sand Martin. Nests in holes in the earth, which it digs for itself in banks. *H. rupestris* Scop., Crag Swallow, South of France.

Fam. **Cypselidæ.** Swifts. Swallow-like Clamatores, with narrow wings curved in the form of a sabre; short feathered metatarsus and strongly clawed feet (*pedes adhamantes*); sometimes with inwardly directed hind toe *Collocalia esculenta* L., (Salangane), East Indies; *Cypselus apus* L., Swift; *C. melba* L., (*alpinus*), Alpine Swift.

Fam. **Caprimulgidæ.** Goatsuckers. *Clamatores*, with short, uncommonly flat, triangular beak. Their size varies from that of a lark to that of a raven. Plumage soft, owl-like, and of the colour of the bark of trees. The legs are very weak and short. Hind toes half turned inwards, but can also be turned forwards. The middle toe is long, and sometimes has a serrated claw. They live for the most part in forests, and feed especially on moths, which they catch in their open mouth, during their swift, silent flight. As a rule they lay two eggs on the bare ground, without even scraping a hole for their reception. *Caprimulgus* L., the buccal slit extends to close below the eyes. Edge of beak not dentated, is fringed with stiff bristles. *C. europæus* L.; *C. ruficollis* Temm., Spain.

Tribe 4. **Dentirostres.** Principally *Oscines* with variously-shaped, often thin and pointed, sometimes slightly curved beak; upper beak is more or less notched at the point. In the wings, which are of medium length, the first of the ten primary remiges is reduced, and may be entirely absent.

Fam. **Corvidæ.** Beak strong and thick, somewhat curved anteriorly and slightly notched. *Corvus corax* L., Raven; *C. cornix* L., Hooded Crow; *C. corone* L., Carrion Crow; *C. frugilegus* L., Rook; *C. monedula* L., Jackdaw; *Pica caudata* Ray, Magpie; *Garrulus glandarius* L., Jay; *Oriolus galbula* L., Golden Oriole.

Fam. **Paradiseidæ.** Birds of Paradise. With slightly curved, compressed beak. Feet very strong and toes large. The two middle rectrices are often elongated and filiform, with small vane only at the extremity. Male with tufts of lax feathers at the sides of the body, and also on the neck and breast. *Paradisea apoda* L.; *Cicinnurus regius* L., New Guinea, (fig. 663.)

Fam. **Sturnidæ.** Starlings. *Oscines* with straight or slightly curved, strong beak, the point of which is rarely only slightly notched, without rictal vibrissæ. *Sturnus vulgaris* L., Starling; *Pastor roseus* Temm., Rose-coloured Starling; *Buphaga africana* L., Oxpecker.

Here are allied *Pipra aureola* L., Cayenne; *Rupicola crocea* Bp., Cock of the Rock, South America, and the **Cotingidæ.**

Fam. **Laniidæ.** Shrikes. Large, powerful *Oscines* with hooked, strongly serrated beak, strong rictal vibrissæ, and tolerably long, sharply clawed feet. *Lanius excubitor* L., Grey Shrike; *L. minor* L., Lesser Grey Shrike; *L. rufus* Briss., Woodchat Shrike; *L. collurio* L., Red-backed Shrike.

Fam. **Muscicapidæ.** Flycatchers. Beak short, broad and depressed at the

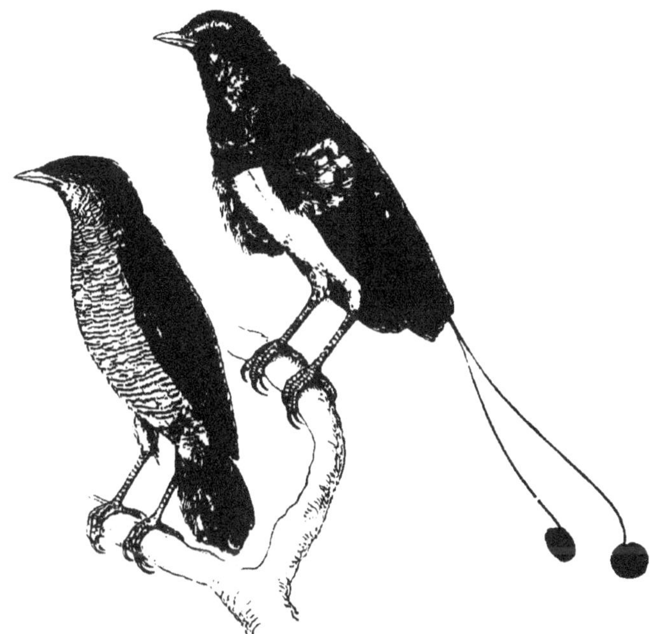

FIG. 663.—*Cicinnurus regius* (male and female).

base, somewhat compressed anteriorly, with hooked curved point. *Muscicapa grisola* L.; *M. atricapilla* L.; *M. collaris* Bechst., (*albicollis*): *Bombycilla garrula* L., Waxwing.

Fam. **Paridæ.** Titmice. Small, beautifully coloured, and very agile *Oscines*, of stout build, with sharp, short, almost conical beak. *Parus major* L., Great Titmouse; *P. ater* L., Coal Titmouse; *P. cæruleus* L., Blue Titmouse; *P. cristatus* L., Crested Titmouse; *P. palustris* L., Marsh Titmouse; *P. caudatus* L., Long-tailed Titmouse; *Aegithalus pendulinus* L., Penduline Titmouse; *Sitta europæa* L., Nuthatch.

Fam. **Motacillidæ.** Wagtails. Slender body; beak tolerably long, and notched at the point. *Anthus pratensis* Bechst., Meadow Pipit; *Motacilla alba* L.; *M. flava* L.; *M. sulphurea* Bechst.; *Accentor alpinus* Bechst., Alpine Accentor.

Fam. **Sylviidæ.** Small *Oscines*, with thin and pointed beak and metatarsus covered with scales in front. *Sylvia nisoria* Bechst., Barred Warbler; *S. atricapilla* Lath., Blackcap; *Phyllopneuste hypolais* Bechst.; *Troglodytes parvulus* Koch, Wren; *Regulus cristatus* Koch: *R. ignicapillus* Naum., Fire-crested Wren.

Fam. **Turdidæ.** Thrushes. The beak is tolerably long, somewhat compressed, and slightly notched before the point, and furnished with vibrissæ at its base. The metatarsus is long, and covered with an anterior and two lateral scales. laminiplantar. *Cinclus aquaticus* Bechst., Dipper; *Luscinia philomela* Bechst., Thrush Nightingale, large nightingale in East Europe: *L. luscinia* L., Nightingale: *L. suecica* L., Blue-throat: *L. rubicula* L., Robin; *Turdus pilaris* L., Fieldfare: *T. musicus* L., Thrush; *T. iliacus* L., Redwing: *T. torquatus* L., Ring-ouzel; *T. merula* L., Blackbird: *T. saxatilis* L., Rock Thrush; *T. migratorius* L., American Robin; *T. cyanus* L., Blue Thrush. The Lyre-bird (*Menura superba*, Dav.)—a large bird found in New Holland—is allied to the Thrushes in the form of its beak.

Tribe 5. **Conirostres.** *Oscines* of small size, with thick head and powerful, conical beak, with short neck, wings of medium length and ambulatory feet (*p. ambulatorii*). The metatarsus is short, and is covered with scales in front. They feed on corn and seeds, berries and fruits, but do not despise insects.

Fam. **Alaudidæ.** Larks. The plumage is earth-coloured: the beak is of medium length, the wings broad and long, and the tail short. *Alauda arvensis* L., Skylark: *A. arborea* L., Woodlark; *A. cristata* L., Crested Lark; *A. alpestris* L., Shore Lark: *A. calandra*, Calandra Lark, South Europe.

Fam. **Fringillidæ.** Finches. With short, thick, conical beak, without notch, but with a basal swelling. *Emberiza citrinella* L., Yellow Bunting; *E. cia* L., Meadow Bunting; *E. nivalis* L., Snow Bunting; *Fringilla cœlebs* L., Chaffinch: *F. spinus* L., Siskin; *F. carduelis* L., Goldfinch; *Passer domesticus* L., House-sparrow: *P. montanus* L., Tree-sparrow: *Coccothraustes vulgaris* Pall., Haw-finch: *Pyrrhula vulgaris* Briss., Bullfinch: *P. canaria* L., Canary: *Loxia curvirostra* Gm., Crossbill.

Fam. **Ploceidæ.** Weaver-birds. Build purse-shaped nests. Live in Africa, East India, and Australia. *Ploceus textor* Gray; *Pl. socius* Gray.

Order 7.—Raptatores (Birds of Prey).

Powerfully-built birds, with curved beak, hooked at the extremity, and strongly clawed feet (p. insidentes). They feed principally on warm-blooded animals.

The *Raptatores* are characterised by their powerful build, by the high development of their sense organs, and by the special development of their beak and of the armature of the feet, by which they are fitted for their peculiar mode of existence. The compressed root of the beak is covered by a soft cere, which surrounds the nasal apertures. The cutting edges, and the hooked and downwardly-curved point of the upper beak are always hard and horny. The strong toes, of which the outer can be turned backwards or forwards, are always armed with powerful claws, which are admirably adapted for the seizure of prey. The feet are *p. insidentes*, and are feathered to the intertarsal joint, rarely to the toes. Before the digestion the food is softened in the crop, from which the feathers and hairs rolled together in balls are ejected as the "castings." As a rule the female alone incubates, but the male assists in procuring food for the helpless young. Some genera of Owls and Falcons are cosmopolitan.

Fam. **Strigidæ.** Owls. With large, anteriorly directed eyes, which are surrounded by a circle of stiff feathers, sometimes in a veil-like manner; with strong, hooked beak, bent downwards from the base. The ear has usually a membranous operculum and external cutaneous fold, on which the feathers may be grouped, so as to give the appearance of a concha. *Strix flammea* L., Barn Owl (fig. 664); *Syrnium aluco* L., Tawny Owl; *Otus vulgaris* L., Long-eared Owl; *O. brachyotus* Gm., Short-eared Owl; *Bubo maximus* Sibb., Eagle Owl; *Ephialtes scops* L., Scops Owl, South Europe; *Surnia passerina* Blas., Sparrow Owl; *Nyctea nivea* Daud., Snowy Owl.

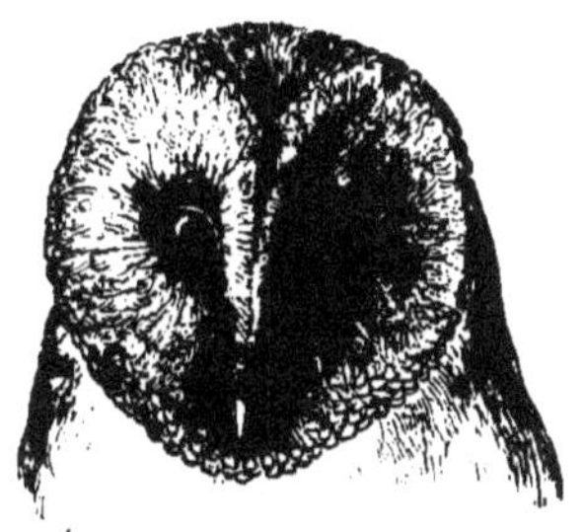

Fig. 664.—Head of *Strix flammea*.

Fam. **Vulturidæ.** Vultures. *Raptatores* of large size, with long, straight beak, only bent downwards at the tip. Nares often pervious (*Cathartinæ*). Head and neck often in great part naked. The head sometimes bears lobed appendages; the neck is often surrounded by a collar of down and feathers. *Sarcorhamphus gryphus* Geoffr., Condor; *S. papa* Dum., King-Vulture, South America; *Cathartes aura* Ill.; *C. atratus* Baird, Turkey Buzzard, South America; *Neophron percnopterus* Sav., Egyptian Vulture; *Vultur cinereus* Gm., South Europe; *Gyps fulvus* Briss.; *Gypaëtus barbatus* Cuv., Lämmergeier, South Europe.

Fam. **Accipitridæ = Falconidæ.** With shorter and usually dentated beak, feathered head (rarely with naked cheeks) and neck. Metatarsus of medium length, and sometimes feathered.

Aquila chrysaëtos L., Golden Eagle, South Germany; *B. imperialis* Kais. Blas., Imperial Eagle, South Europe; *A. fulva* M.W., Golden Eagle, Tyrol; *A. naevia* Briss., Spotted Eagle; *Haliaëtos albicilla* Briss. (*ossifragus* L.), Sea Eagle, Europe, North Africa; *Pandion haliaëtos*, Cuv., Osprey, Northern hemisphere.

Milvus regalis Briss., Red Kite. Seizes its prey from other birds, and only takes small animals—as hamster-rats, moles, and mice; *M. ater* Daud., Black Kite.

Buteo vulgaris L., Buzzard; *B. lagopus* L., Rough-legged Buzzard; *Pernis apivorus* Cuv., Honey Buzzard.

Astur palumbarius L., Goshawk; *Nisus communis* Cuv., Sparrowhawk.

Falco tinnunculus L., Kestrel; *F. peregrinus* L., Peregrine-falcon; *F. candidus* Gm. = *gyrfalco* L., Jer-falcon.

Circus rufus L. (*æruginosus*), Marsh Harrier; *C. cyaneus* L., Hen Harrier.

Fam. **Gypogeranidæ**. Slender body, with long neck, long wings and tail, and much elongated metatarsus. Beak with extended cere, laterally compressed and strongly curved. *Gypogeranus serpentarius* Ill., Secretary bird. Flies badly, but runs well: preys on snakes in Africa.

II. RATITÆ.

Birds incapable of flight, without sternal keel, and without firm remiges or rectrices.

Order 1.—Cursores.

Ratitæ of considerable body size, with three-toed or exceptionally with two-toed cursorial feet.

The Ostriches, which are the largest of living birds, possess a broad and flat, deeply-slit beak with a blunt point, a relatively small, in part naked, head, a long, slightly feathered neck and long powerful cursorial legs. Besides the reduction of the wing-bones, there are other peculiarities of skeletal structure which characterise these birds as being exclusively cursorial. Almost all the bones are heavy and massive, with much reduced pneumaticity. The sternum has the form of a broad, slightly arched plate, without any trace of a keel. The clavicle also is undeveloped, and the uncinate processes of the ribs are rudimentary or entirely absent. The plumage covers the body with tolerable uniformity, except that there are naked places on the head, the neck, the extremities, and the abdomen; but does not present any regular arrangement of *pterylæ*: it approximates in its special structure to the hairy covering of *Mammalia* (Cassowary). While the down is much reduced, the contour-feathers have a more down-like appearance on account of their flexible shaft and lax vane, or they may be stiff and hair-like with setiform barbs, or sometimes, as in the wings of the Cassowary, they are spine-like.

Fam. **Struthionidæ**. Two-toed Ostriches. With naked head and neck, pubic symphysis and long, completely naked, two-toed legs. They inhabit the plains and deserts of Africa. They live in companies, and are polygamous. *Struthio camelus* L., Ostrich.

Fam. **Rheidæ**. Three-toed Ostriches. With partially feathered head and neck, and three-toed feet. They inhabit America. *Rhea americana* Lam., Rhea.

Fam. **Casuariidæ** With high, almost compressed beak, and usually a helmet-shaped, bony knob on the head, with short neck and three-toed legs. *Dromæus Novæ Hollandiæ* Gray, Emeu, Australia; *Casuarius galeatus* Vieill., Cassowary, New Guinea [Ceram].

The reduction of the wings in terrestrial birds is not confined to the Ostriches; but is also characteristic of a number of very strangely organised forms which differ so much from each other that they deserve to be separated into several orders. These birds belong principally to New Zealand; also to Madagascar and the Mascarenes. Some of them are extinct, but have only become so within historic times.

In the uninhabited forest regions of the north island of New Zealand there still lives, though gradually approaching extinction,

FIG. 665.—*Apteryx Owenii.*

an extremely remarkable bird—the Kiwi (**Apteryx Mantelli Australis** Shaw), which is sometimes placed among the ostriches and called the Dwarf Ostrich. A second species of the same genus (*A. Owenii*) belongs to the south island, on which another larger form (*Roaroa*) is said to exist, and has been distinguished as a third species (*A. maxima*, Verr.). These birds (**Apterygia**), which are about the size of a large hen, are entirely covered with long, hair-like feathers, which hang down loosely and completely hide the rudimentary wings. The short, powerful legs are covered with scales; the three anteriorly directed toes are armed with claws adapted for scratching: the hind toe is short and raised from the ground. The head, which is borne on a short neck, is prolonged into a long and rounded, snipe-like beak, at the extreme point of which are the nasal apertures. The Kiwis are nocturnal birds, which by day remain concealed in holes in the earth and go out at night to seek their food. They feed on insect-larvæ and worms, live in pairs, and at the breeding time, which seems to come twice in the year, they lay, in holes scraped in the earth, a strikingly large egg, which according to some is incubated by the female, and according to others by the male and female in turn.

A second group of terrestrial birds of New Zealand, which are incapable of flight, includes a number of forms which are in great part extinct, and some of which attained an enormous size (up to ten feet high). These are the **Dinornithidæ**. Of heavy, unwieldy build, and incapable of raising themselves from the ground, they were unable to resist the pursuit of the natives of New Zealand. The remains of some have been found in the diluvium, and in some cases the bones appear so recent, that it cannot be doubted that they co-existed with man. The traditions of the natives about the gigantic *Moa*, and numerous discoveries of the fragments of eggs in graves, also point to the fact that these gigantic birds have lived in historic times; while, on the other hand, recent discoveries have rendered probable the existence of smaller species at the present day. Recently in the exploration of the mountain chains, between the Rewaki and Tabaka rivers, the footprints of a gigantic bird, the bones of which were already known from the volcanic sand of the north island, have been discovered. The restoration of the skeleton of gigantic species (*Palapteryx ingens, Dinornis giganteus, elephantopus*, etc.) has been partially effected from the bones which have been collected. A skeleton of *Dinornis elephantopus* is in the British Museum, and one of *P. ingens* has been set up in Vienna by Hochstetter (Voyage of the Novara). In Madagascar pieces of the tarsal bones of a gigantic bird have been found in the alluvium (*Æpyornis maximus* —the Reek of Marco Polo), and well-preserved. colossal eggs have been discovered in the mud, the contents of which would have been equal to about 150 hen's eggs.

CHAPTER IX.

Class V.—MAMMALIA.*

Warm-blooded, hairy animals with double occipital condyle. They are viviparous and suckle their young with the secretion of milk (mammary) glands.

As opposed to Birds, Mammals are adapted, by the similar structure of the two pairs of extremities, to live principally on land. There are, however, in this class also forms which are fitted in various degrees for an aquatic life, and even live entirely in water, and again forms which move and find their food in the air.

The surface of the skin is rarely quite smooth as in the *Cetacea*, but is traversed by numerous curved, spiral, and partly crossing furrows, and in many places (sole of foot, ischial callosities) is thickened and indurated, so as to form firm, horny plates.

The hairy covering is to Mammalia (named "*Haarthiere*" by Oken), what the plumage is to Birds. Hairs are never entirely absent: even the huge aquatic forms and the largest of the tropical terrestrial species which seem to be naked, possess hairs on certain parts of the body; *e.g.*, the *Cetacea* have short bristles, at least on the lips. Hairs, like feathers, are epidermal structures (fig. 666.) The bulbous root is placed on a vascular papilla (*pulpa*), at the bottom of a pit, which projects into the cutis and is lined by epidermal cells (hair-follicle) while the upper part, or shaft, projects freely on the surface of the skin. Two kinds of hairs may be distinguished, according to the strength and rigidity of the shaft, viz., contour hairs and woolly hairs. Woolly hairs are delicate and curled, and surround in larger or smaller numbers the base of each contour hair. The finer and warmer the fur, the more numerous are the woolly hairs (winter-fur). When the contour hairs have a greater strength they become bristles,

* Joh. Ch. D. v. Schreber, "Die Säugethiere in Abbildungen nach der Natur mit Beschreibungen, fortgesetzt von Joh. Andr. Wagner." Bd. I.—VII.. und Suppl. I.—V. Erlangen und Leipzig. 1775—1855.
E. G. St. Hilaire et Fréd. Cuvier. "Histoire naturelle des Mammifères," Paris. 1819—1835.
C. J. Temmink, "Monographie de mammalogie." Leiden, 1825—1841.
R. Owen. "Odontography," 2 vol. London. 1840—1845.
Blasius. "Die Säugethiere Deutschlands" 1875.
G. Giebel. "Die Säugethiere in zoologisch-anatomischer und paläontologisches Hinsicht." Leipzig, 1850.
A. E. Brehm. "Illustrirtes Thierleben" I., II., und III.
And. Murray. "The Geographical Distribution of Mammalia." London, 1866.

and when still stronger and thicker they constitute spines (Hedgehog, Porcupine.) To the stronger hairs are attached smooth muscles of the dermis, by means of which each one of them can be moved singly, while the striped muscular system of the dermis causes the bristling of the hairy covering and the erection of the spines over larger extents of surface.

The epidermis may also give rise to smaller horny scales as well as to large overlapping scales: the former on the tails of Rodents and Marsupials, the latter upon the whole dorsal and lateral surfaces of the Pangolins (*Manis*), which thus possess a horny epidermal exoskeleton. Another form of exoskeleton is found in the Armadillos; it arises by ossification of the dermis, and consists of suturally united plates, and in the middle of the body of broad, movable, bony girdles. Amongst the dermal ossifications must also be reckoned the antlers of the Deer which are periodically renewed. The horny sheaths of the *Cavicornia*, the horns of the *Rhinocerida* and the various horny coverings of the extremities of the digits are epidermal structures. The latter may be distinguished into nails (*unguis lamnaris, unguis tegularis*), claws (*falcula*), and hoofs (*ungula.*)

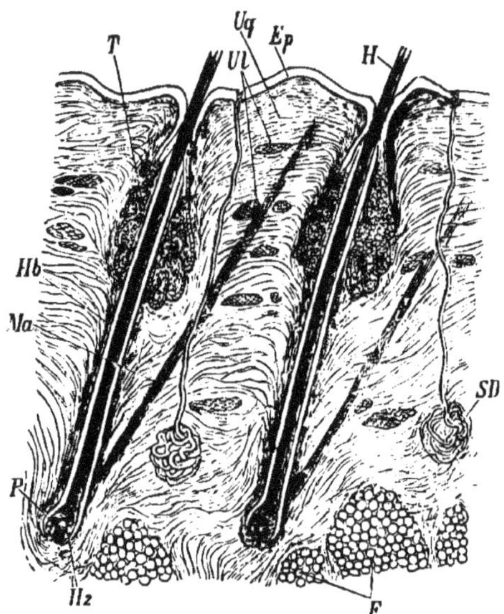

FIG. 666.—Section through the human scalp. *Ep*, Epidermis; *Ul*, transverse bands of the connective-tissue of the cutis; *Uq*, longitudinal bands of the same; *H*, hair; *Hz*, root of hair; *P*, hair papilla; *Hb*, Hair follicle; *Ma*, musculus erector pili; *T*, sebaceous gland; *SD*, sweat gland; *F*, fat body.

Cutaneous glands.—Sweat glands and sebaceous glands (fig. 666) are widely distributed. Sebaceous glands are invariable accompaniments of the hair follicles, but they are also found on naked parts of the skin; they secrete a fatty grease, which keeps the surface of the skin soft. The sweat glands have the form of coiled glandular tubes with sinuous ducts, and are only seldom absent (*Cetacea, Mus, Talpa*).

The larger glands, with strongly smelling secretions, which open on various parts of the integument of many mammals, are to be regarded as modified sebaceous or more rarely sweat glands. As examples of such glands may be mentioned the occipital glands of the Camel, the glands which are placed in a depression of the lachrymal bone of *Cervus*, *Antilope*, *Ovis*, the temporal glands of the Elephant, the facial glands of the Bat, the pedal glands of Ruminants, the lateral glands of the Shrewmouse, the sacral gland of *Dicotyles*, the caudal glands of the Desman, the crural glands of the male Monotremes, etc. These excretory organs are most frequently found near the anus, or in the inguinal region, and are then often placed in special

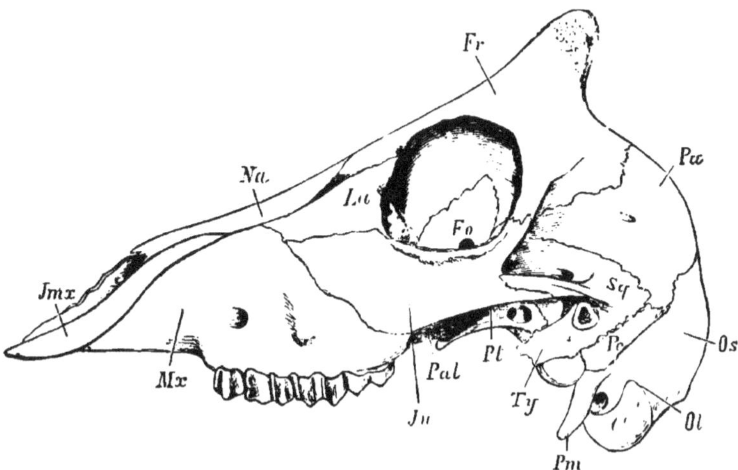

FIG. 667.- Skull of a goat, from the side. *Ol*, exoccipital; *C*, condyle; *Pm*, paramastoid process; *Os*, supra-occipital; *Sq*, squamosal; *Ty*, tympanic; *Pe*, petrous (mastoid portion); *Pa*, parietal; *Fr*, frontal; *La*, lachrymal; *Na*, nasal; *Fo*, optic foramen; *Mx*, maxillary; *Jmx*, præmaxillary; *Ju*, jugal; *Pal*, palatine; *Pt*, pterygoid.

cutaneous pits—*e.g.*, the anal glands of many *Carnivora*, *Rodentia*, and *Edentata*, the civet gland of the *Viverridæ*, the musk pouch of *Moschus moschiferus*, and the preputial glands of the male Beaver.

The **skeleton** is formed of heavy bones containing marrow. The skull (fig. 667) is a spacious capsule, the bony pieces of which are only exceptionally (*Ornithorhynchus*) fused in early life, but as a rule they remain for the most part separated by suture throughout life. There are, however, many cases in which in the adult animals the sutures have partly or wholly vanished (Ape, Weasel). The great extension of the cranial capsule is due not only to the large size of the roof of the skull, but also to the fact that the lateral bones of the skull in place of the interorbital septum extend forward into the ethmoid region. Thus it happens that the ethmoid (*lamina cribrosa*) constitutes the boundary of the anterior and lower part of the skull (fig. 668). The temporal bones also take an essential part in bounding the cranial cavity, since not only the petrous and a part of the mastoid,* but also the large squamosal occupy the space remaining between the alisphenoids and exoccipitals. The occipital always articulates with the first cervical vertebra (*atlas*) by two condyles, and its lateral portions (exoccipitals) frequently present a pyramidal process on each side (*jugular* or *paramastoid process*). The præsphenoid and basisphenoid (fig. 668) often remain separate

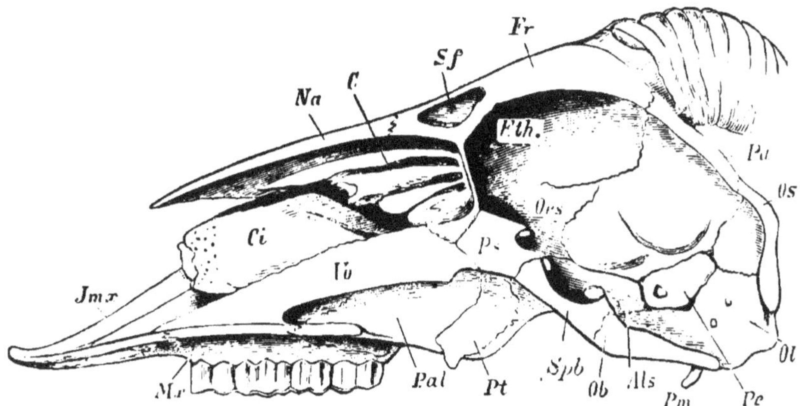

FIG. 668.—Median longitudinal section of sheep's skull, from inside. *Ob*, Basi-occipital *Ol*, exoccipital; *Os*, supra-occipital; *Pe*, petrous; *Spb*, basi-sphenoid; *Ps*, præ-sphenoid; *Als*, alisphenoid; *Ors*, orbito-sphenoid; *Pa*, parietal; *Fr*, frontal; *Sf*, frontal sinus; *Eth*, ethmoid; *Na*, nasal; *C*, ethmoturbinal; *Ci*, inferior turbinal; *Pt*, pterygoid; *Pal*, palatine; *Vo*, vomer; *Mx*, maxillary; *Jmx*, præmaxillary.

for a long time. To the latter are applied the alisphenoids with the parietals, which belong to this region. An interparietal is often developed behind the parietal; it is, however, usually ankylosed with the supra-occipital, more rarely with the parietal. The frontal bones constitute the roof of the skull in the region of the orbitosphenoids; they are less frequently fused than are the parietals. The temporal bone has several constituents (1) The petrous portion, which is composed of the three pieces of the periotic capsule—the *pro-*, *opistho-*, and *epiotic*; (2) the mastoid portion, which is a part of the epiotic; (3) the squamous portion or squamosal, which is a larger bony scale; (4) the tympanic bone, which is attached to the squamosal, bounds

* [The petrous and mastoid together constitute the periotic.]

the external auditory meatus, and is frequently dilated to a projecting capsule (*tympanic bulla*). Postfrontals are absent. The perforated *cribriform plate* (*lamina cribrosa*) of the ethmoid forms the anterior boundary of the cranial cavity. In the Apes and Man only, do the lateral parts of the ethmoid (the part known as *lamina papyracea*) take part in the formation of the inner wall of the orbit. In all other cases the ethmoid is placed in front of the orbit, and its sides are covered by the maxillaries; in such cases it has a considerable longitudinal extension. Two parts may be distinguished in the ethmoid—(1) A median plate—the *lamina perpendicularis*—which is continued in front into the cartilaginous internasal septum, and is underlaid by the vomer; (2) the lateral masses, with the *lamina cribrosa* and the labyrinth (ethmoidal cells and the two upper turbinals); the first corresponds to the unpaired ethmoid, the second to the præfrontals of the lower Vertebrates. Finally, in the anterior part of the nasal cavities there are, as independent ossifications, the inferior turbinals (maxillo-turbinals), which are attached to the inner surfaces of the maxillary bones. On the outer surface of the ethmoid region are placed, as membrane bones, the nasals above and the lachrymals to the sides. The lachrymal (absent as an independent bone in the *Pinnipedia* and most *Cetacea*) is placed in the anterior wall of the orbit; but usually also appears as a facial bone on the outer surface.

The firm fusion of the maxillo-palatine apparatus with the skull and the relation of the mandibular suspensorium to the tympanic cavity, are characteristic of the Mammalia. The lower jaw articulates directly with the temporal bone without the interposition of a quadrate, the morphological equivalent of which is shifted, in the course of development, into the tympanic cavity and transformed into the *incus*, while the upper part of Meckel's cartilage (*articulare* of the lower jaw) becomes the *malleus* (Reichert). The *stapes*, on the contrary, is said to be developed from the upper piece of the hyoid arch (*hyomandibular*). The maxillary, pterygoid, and palatine bones have similar relations to those of the *Chelonia* and *Crocodilia*, but a quadratojugal is always wanting, since the jugal is applied to the squamosal. A palatal roof (hard palate) separating the buccal and nasal cavities is always present; the posterior nares open at its hind end.

The cranial capsule is so completely filled by the brain in the Mammalia that its internal surface presents a relatively accurate impression of the surface of the brain. Owing to the considerable size of the brain the cranial capsule is far more spacious than in any other class of Vertebrates; but it presents great variations in this respect in the individual groups, more especially with regard to the development of the face, the prominence of which in general varies inversely with the development of the intellectual faculties (Camper's facial angle). The hyoid bone is reduced to a transverse, bridge-like piece (body of the hyoid), with two pairs of cornua. In *Mycetes* it is largely developed and excavated.

The vertebral column, except in the *Cetacea*, is divided into five regions, viz., cervical, thoracic, lumbar, sacral and caudal (fig. 669). In the aquatic *Cetacea*, which are without hind limbs, the lumbar

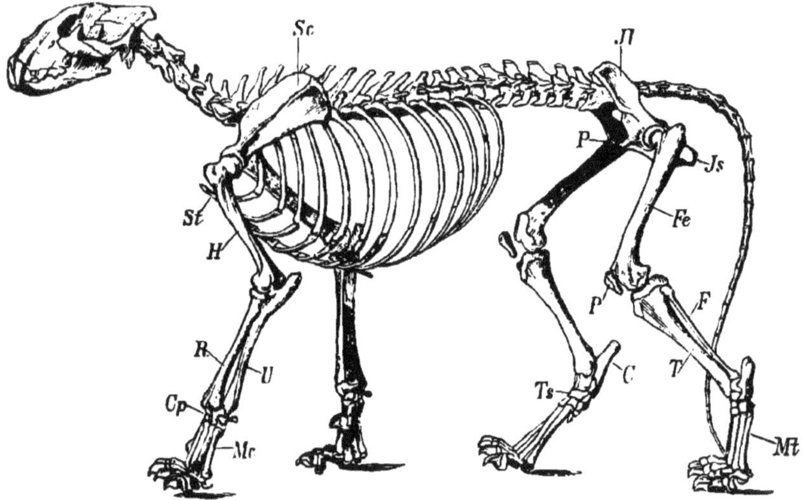

Fig. 669.—Skeleton of the Lion (after Giebel; Bronn's Classen und Ordnungen). *St*, sternum; *Sc*, scapula; *H*, humerus; *R*, radius; *U*, ulna; *Cp*, carpus; *Mc*, metacarpus; *Il*, ilium; *P*, pubis; *Js*, ischium; *Fe*, femur; *T*, tibia; *F*, fibula; *P*, patella; *Ts*, tarsus; *Mt*, metatarsus; *C*, calcaneum.

region passes gradually into the caudal; on the other hand the cervical region is strikingly shortened, and the fusion of its anterior vertebræ renders it rigid and immovable. The vertebral bodies are only exceptionally (neck of Ungulates) connected by articular surfaces, but are usually joined by elastic discs (intervertebral ligaments). The first cervical vertebra (*atlas*) is a bony ring with broad, wing-like, transverse processes, on the articular surfaces of which the two occipital condyles rest and permit of the head being raised and depressed. The turning of the head to the right and left is effected by the movement of the atlas about a median process—the *odontoid process*—of the next vertebra, which is called the *axis* (*epistropheus*).

This process corresponds morphologically to the centrum of the atlas, which is separated from the latter and joined to the centrum of the axis.

The dorsal vertebræ are characterised by high, crest-like, spinous processes, and by the possession of ribs. The anterior ribs are attached by cartilage to the sternum, which is usually elongated and composed of a number of bony pieces arranged one behind another: the posterior ribs (the so-called "false ribs") do not reach the sternum. The ribs articulate with the vertebræ by means of a capitulum and a tuberculum. While the number of cervical vertebræ is almost constantly seven, that of the dorsal vertebræ is subject to a greater variation. As a rule there are thirteen, sometimes twelve dorsal vertebræ; but there is a less number in some Bats and Armadillos, while there are fifteen or more in some animals. The Horse has eighteen, the Rhinoceros and Elephant nineteen to twenty, and the three-toed Sloths have twenty-three to twenty-four. The lumbar vertebræ, which have long lateral processes in place of ribs, are usually seven in number. The number rarely sinks to two as in *Ornithorhynchus* and the two-toed Anteaters, and still more rarely rises to eight or nine (*Stenops*). The sacral vertebræ, which vary in number from two (Marsupials) to four, more rarely nine (Armadillo), are firmly united with one another, and by their transverse processes (with the rudiments of the ribs) with the iliac bones. The caudal vertebræ, which vary considerably in number and mobility, become narrower towards the end of the axis of the body, and often (Kangaroo and Anteaters) possess inferior spinous processes; but all the processes become less and less conspicuous towards the posterior extremity.

The anterior pair of extremities is never absent. The clavicle is absent when the anterior limbs serve only for the support of the anterior part of the body in locomotion, or perform simple, pendulum-like movements, as in swimming, walking, running, jumping, etc., (Whales, Ungulates, Carnivora). Otherwise the scapula is connected with the sternum by a more or less strong, rod-shaped clavicle. The coracoid is almost always reduced to the coracoid process of the scapula; in the *Monotremata* only is it a large bone which reaches the sternum. The posterior extremities are more firmly connected with the body than are the anterior. In the Whales alone is the pelvic girdle rudimentary, and is represented by two rib-like bones which are quite loosely connected with the vertebral column. In all other Mammals the pelvic girdle is fused with the lateral parts of the sacrum, and is closed ventrally by the symphysis of the pubis and sometimes also of the ischium. The appendages articulated to the pectoral and pelvic girdles are considerably shortened in the swimming Mammalia, and either constitute, as in the *Cetacea*, flat fins, the bones of which are immovable upon one another (in the *Sirenia* there is a joint at the elbow), and in which there are a great number of phalanges, or, as in the *Pinnipedia*, have the form of fin-like legs, which can also be used in locomotion on land. In the *Cheiroptera* (Bats), the anterior legs present a large surface in consequence of an expansion of the integument (*patagium*) uniting the fore-limbs with the sides of the body, and extended between the elongated fingers. The fins of the *Cetacea* and the wings of the *Cheiroptera* are, with the exception of the thumb of the latter which projects from the patagium and bears a claw, without nail-like structures.

In the land *Mammalia* the extremities present considerable variations both in their length and special structure. The length of the tubular *humerus* in general varies inversely with that of the metacarpus of the anterior extremity. The *radius* and *ulna* in the fore-limb and the *tibia* and *fibula* in the hind-limb are almost always longer than the humerus and femur respectively. The *ulna* forms the hinge-joint of the elbow, and is prolonged at this point into a process called the *olecranon*; the radius, on the other hand, is connected with the carpus, and can often be rotated round the ulna (*pronation*, *supination*); in other cases it is fused with the ulna, which then constitutes a rudimentary, styliform rod as far as the articular process. In the hind-limb the knee-joint projects forwards, and is usually covered by a knee-cap, the *patella*; the fibula is sometimes (Marsupials) movable on the tibia, but as a rule these two bones are fused, and the fibula which is placed posteriorly and externally is usually reduced. The variations in the terminal parts of the limbs are far more striking (fig. 670). The number of digits is never greater than five, and is often less. The digits disappear in the following order: first, the inner digit or thumb (digit No. 1), which is composed of two phalanges, becomes rudimentary and vanishes; then the small outer digit (digit No. 5) and the second inner digit (digit No. 2) are reduced, sometimes remaining on the posterior surface of the foot (Ruminants) as small accessory claws which do not reach the ground, or totally vanish. Finally the second external digit (No. 4) is reduced or absent, so that the middle digit alone remains for the support of the limb (horse).

This gradual reduction of the digits is accompanied by a simplification and alteration of the carpal and tarsal, metacarpal and metatarsal bones; the metacarpals (metatarsals) of the rudimentary or absent digits are reduced to styliform bones or are entirely absent, while the two middle metacarpals (metatarsals) (3 and 4) are often united to form a strong and long tubular bone. The small carpal and tarsal bones which are employed in the formation of the foot-joint, and serve essentially to diminish the shock produced by the movements of the limbs when used in locomotion, are arranged usually in two, sometimes in three rows; in the tarsus, two bones—the *astragalus* and *calcaneum*—are usually much larger than the rest. The

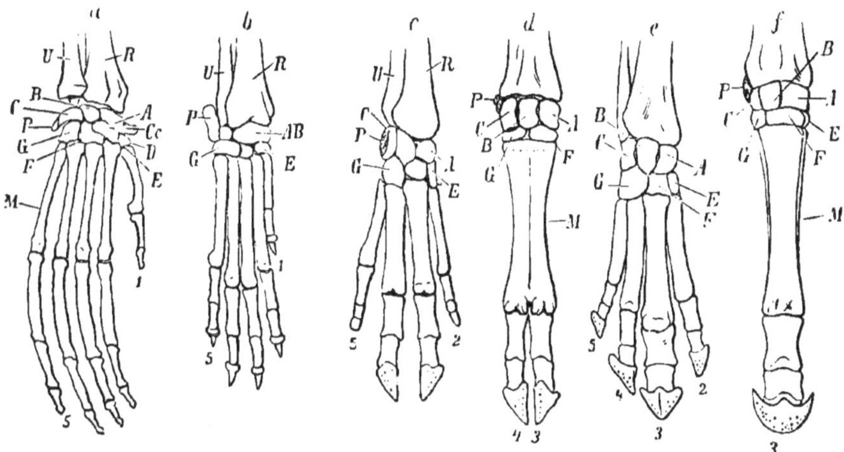

FIG. 670.—Skeleton of hand of—*a*, orang; *b*, dog; *c*, pig; *d*, ox; *e*, tapir; *f*, horse (*b*, *c*, *d*, after Gegenbaur). *R*, radius; *U*, ulna; *A*, scaphoid; *B*, semi-lunar; *C*, triquetrum (cuneiform); *D*, trapezium; *E*, trapezoid; *F*, capitatum (magnum); *G*, hamatum (unciform); *P*, pisiform; *C*, centrale carpi; *M*, metacarpus.

digits of the anterior foot may be called fingers after the analogy of the human hand. The anterior foot becomes a hand when the inner finger or thumb is opposable. The great toe of the posterior foot is also sometimes opposable, but the foot does not on this account become a hand, but only a prehensile foot (Apes); for the hand is characterised by the special arrangement of the carpal bones and muscles. According to the manner in which the foot rests on the ground in movement, animals are distinguished as plantigrades, digitigrades and unguligrades. In the last case the number of digits and metacarpals (and metatarsals) is reduced, and the limb is much elongated by the transformation of the metacarpal (metatarsal) bone into a long tubular bone.

The **nervous system** (fig. 671) is characterised by the size and high development of the brain, the hemispheres of which are so large that they not only fill the anterior part of the cranial cavity, but even

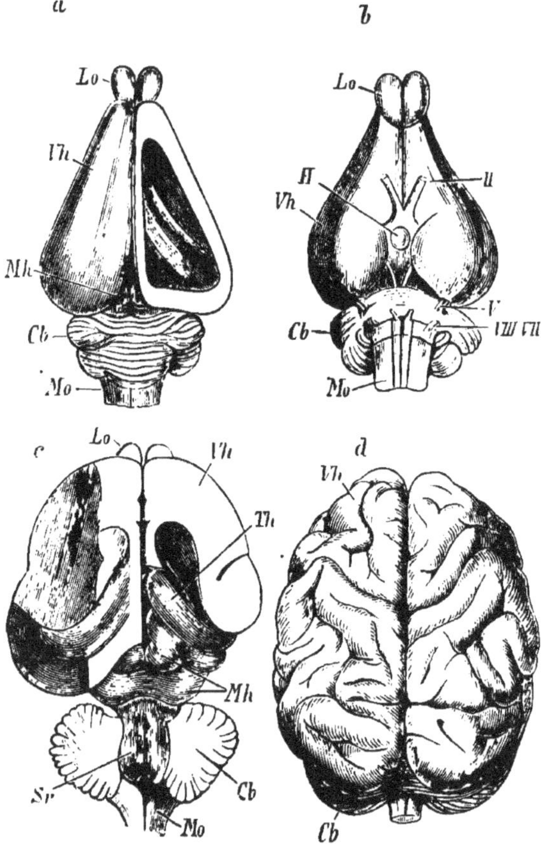

Fig. 671.—Brains of Mammalia. *a*, Brain of rabbit, from above; the roof of the right hemisphere is removed so as to expose the lateral ventricle. *b*, The same from below. *c*, Brain of cat; on the right side the lateral and posterior part of the hemisphere is removed, and almost as much on the left side, and the greater part of the hemispheres of the cerebellum have been removed. *d*, Brain of orang (*a*, *b*, *c*, after Gegenbaur; *d*, from the règne animal). *Vh*, cerebral hemispheres; *Mh*, corpus quadrigeminum; *Cb*, cerebellum; *Mo*, medulla oblongata; *Lo*, olfactory lobe; *II*, optic nerve; *V N*, trigeminal; *VII VIII*, facial and auditory nerves; *H*, hypophysis cerebri; *Th*, optic thalamus; *Sr*, sinus rhomboidalis.

partly cover the cerebellum. In the Marsupials and Monotremes the surface of the hemispheres is still smooth; but in the Edentates, Rodents, and Insectivores it is marked by depressions and ridges, which in the higher forms become regular furrows (*sulci*) and convolutions (*gyri*). A commissure (*corpus callosum*) connecting the two hemispheres is well-developed, and rudimentary only in the *Aplacentalia*. On the other hand the optic lobes, which are known as the *corpora quadrigemina*, and are the equivalents of the *corpora bigemina* of the lower forms, are reduced in size, and are in great part or entirely covered by the posterior lobes of the hemispheres. The pituitary body (*hypophysis*) and the pineal gland are never absent. The cerebellum in the *Aplacentalia* resembles that of the birds in the disproportionate development of its median lobe. There are, however, numerous intermediate stages between such a cerebellum and a cerebellum in which the lateral lobes are largely developed. The *pons Varolii* also is little developed in the lower forms, but in the higher Mammals is increased to a large swelling at the point where the brain is prolonged into the spinal cord. The twelve cranial nerves are completely separated. The spinal cord usually extends only as far as the sacral region, where it ends with a *cauda equina*: there is no posterior rhomboidal sinus.

Sense organs. The olfactory organ presents, on account of the complication of the ethmoidal labyrinth, a greater development of the olfactory mucous membrane than in any other class. The two nasal cavities, which are separated by the median septum, often communicate with spaces in the neighbouring cranial and facial bones (*sinus frontales, sphenoidales, maxillares*), and open externally by paired apertures; in the *Cetacea*, which have no sense of smell, the latter may be fused to form a median opening (*Delphinidæ*). In this case the nasal passages serve only as air-passages. The nasal openings are, as a rule, supported by movable cartilaginous pieces, which in some cases are largely developed and lead to the formation of a proboscis, which is used as a burrowing and tactile organ, and when greatly developed (Elephant) as a prehensile organ. In the diving Mammals the nasal apertures can be closed by muscles (*Phocidæ*) or by valvular apparatuses. A nasal gland is often present on the external wall of the nares, or in the cavity of the upper jaw (*maxillary sinus*). The olfactory nerve is distributed as in the Birds on the superior turbinal bones, and on the upper parts of the nasal septum. The internal nares are always paired and open into the pharynx, far back at the end of the soft palate.

The **eyes** (vol. i., fig. 88) present various degrees of development; they are always small in the Mammals which live beneath the earth, and in some cases (*Spalax, Chrysochloris*) are quite hidden beneath the skin, and are incapable of receiving luminous impressions. They are usually placed at the sides of the head in an incompletely closed orbit (continuous with the temporal fossa). As a rule, each eye has a separate field of vision; a convergence of the optic axes is only possible when the eyes are placed on the front of the head (*Primates*). Besides the upper and lower eyelids there is an internal nictitating membrane (with the Harderian gland), which is, however, not fully developed, and is without the muscular apparatus of the Birds' nictitating membrane; it is sometimes reduced to a small rudiment (*plica semilunaris*) at the inner corner of the eye. The eyeball is more or less spherical (in the *Cetacea*, etc., with shortened axis), and

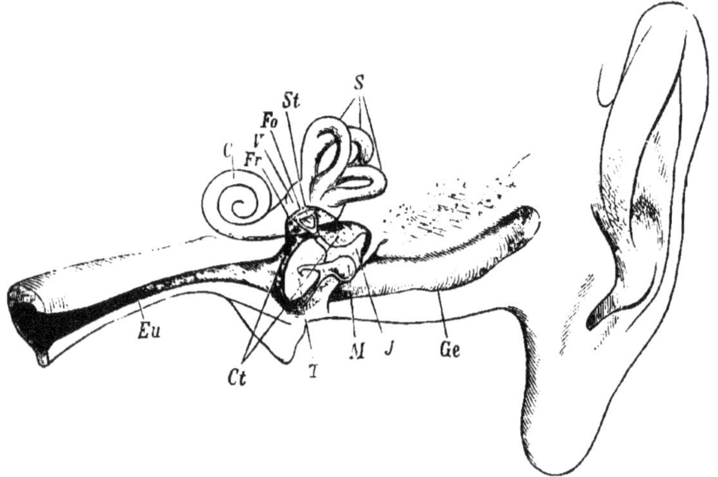

FIG. 672.—The human ear (combined representation) with view of the tympanic membrane from the tympanic cavity. *Ge*, External auditory meatus; *T*, membrana tympani; *Ct* tympanic cavity; *Eu*, Eustachian tube; *M*, malleus; *J*, incus; *St*, stapes closing the fenestra ovalis (*Fo*); *Fr*, fenestra rotunda; *V*, vestibule; *C*, cochlea; *S*, semicircular canals.

can often be retracted into the orbit by a *retractor bulbi*. The lachrymal gland with its duct, which opens into the nasal cavity, lies on the upper and outer side of the orbit. The choroid has a *tapetum* in the Carnivores, Pinnipedes, Dolphins, Ungulates, and some Marsupials.

The **auditory organ** (fig. 672, and fig. 578, iii.) differs from that of the bird principally in the more complicated development of the external ear, in the greater number of sound-conducting bones (*stapes, incus, malleus*), and in the form of the cochlea, which is usually coiled into two or three spiral passages. The tympanic cavity is also more spacious, and is by no means always confined to the space enclosed by the tympanic bone which often projects like a vesicle (tympanic bulla), but is in communication with cavities in the neighbouring cranial bones. The tympanic cavity is largest in the *Cetacea*, in which the sound is not transmitted, as in the terrestrial animals, by the tympanic membrane and the auditory ossicles to the fenestra ovalis of the vestibule, but is conducted mainly by means of the bones of the head through the air of the tympanic cavity to the fenestra (*f. rotunda*) of the unusually large cochlea, and thence by the perilymph of the scala tympani. The three semicircular canals, with the vestibule and cochlea, are very firmly embedded in the petrous bone, which in the *Cetacea* is only connected by ligaments with the neighbouring bones. The Eustachian tubes open in the *Cetacea* alone into the nasal passages, in all other Mammals into the pharynx. An external ear (*pinna*) is wanting in the Monotremes, many Pinnipedes, and in the *Cetacea*, in which the external meatus outside the convex tympanic membrane is represented by a solid cord: it is rudimentary in the aquatic animals which are able to close the external opening of the ear by a valvular apparatus, and in the burrowing Mammalia. In all other cases it consists of a very variously-shaped external appendage, supported by cartilaginous pieces and usually moved by special muscles.

The sense of **touch** is mainly located in the skin of the ends of the extremities (tactile corpuscles on the tips of the fingers and on the surface of the hand of Man and the Apes): also on the tongue, proboscis, and lips, in which long bristle-like tactile hairs (*vibrissæ*), embedded in follicles, with peculiar nervous ramifications, are very generally present.

The sense of **taste** has its seat principally at the root of the tongue (*papillæ circumvallatæ*, compare fig. 89, vol. i), but also on the soft palate, and is far more highly developed than in any other class of animals.

Dentition.—At the entrance to the digestive organs the jaws are almost always armed with teeth. Only individual genera—as *Echidna*, *Manis*, and *Myrmecophaga*—are entirely without teeth, while the whalebone Whales, which bear on the inner surface of the palate vertical horny plates (whale-bone), arranged in transverse rows (fig. 673), possess teeth, at least in the fœtal condition. Horny teeth, produced by hardening of papillæ of the buccal mucous membrane, are present in *Ornithorhynchus* and *Rhytina*.

The dentition of the Mammalia is never so much developed as that of Fishes and Reptiles; and the teeth, which are wedged into alveoli, are confined to the maxillæ, præmaxillæ, and mandible. The external part of the tooth (*i.e.*, the part which projects from the gum, and is called the *crown*, as opposed to the root) is covered with the harder enamel, which consists of prisms arranged at right angles to the cavity of the tooth (pulp cavity). Two kinds of teeth may be distinguished—(1) Simple teeth (*d. simplices*), in which the layer of enamel forms a simple cap; (2) complicated teeth (*d. complicati*), in which the enamel is folded and penetrates into the dentine. When simple or complicated teeth are connected together by osseous tissue (*cement*), they are called composite teeth (*d. compositi*—Hare, Elephant). Rarely, and only in those cases in which the dentition is used, as in the Crocodile, as a prehensile or cutting apparatus, are the teeth in all parts of the jaws alike, having the form and function

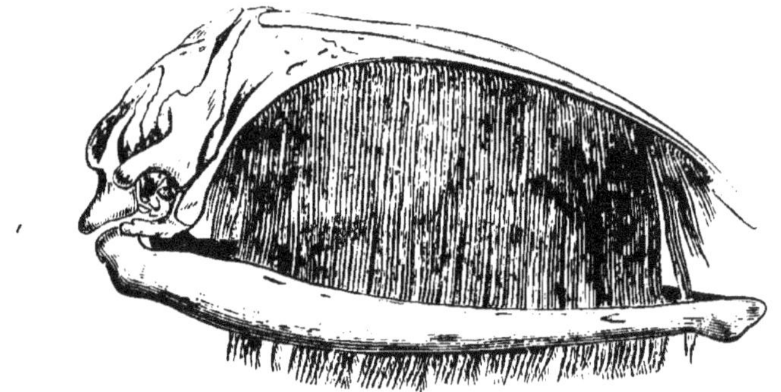

Fig. 673.—Skull of *Balaena mysticetus* with the whalebone (règne animal).

of prehensile conical teeth (Dolphin). As a rule, they are distinguished according to their position in the anterior, lateral, and posterior parts of the jaw as *incisors*, *canines*, and *grinders* (back teeth). The incisors are chisel-shaped, and serve to cut the food; in the upper jaw they belong exclusively to the præmaxillary bones. The canines, which are placed to the sides of the incisors, one in each half of the jaw, are conical or hooked, and serve principally as weapons for attack and defence; not unfrequently, however they are absent (Ruminants, Rodents), and there is a wide gap (*diastema*) between the incisors and the grinders. The latter are specially adapted for the finer mastication of the food, and their crowns are usually provided with a tuberculated or grinding surface. The teeth either last throughout life, and the dentition is not renewed (*Monophyodonta*: Edentates, Cetaceans), or there is a single change of teeth (*Diphyodonta*) (fig. 674). In the latter case the teeth which are changed constitute the *milk dentition*. The anterior grinders, which with the incisors and canines are replaced, are known as the *præmolars*, as opposed to the posterior, true *molars*, which belong to the permanent dentition, and are not replaced. The true molars only appear after the milk teeth have been replaced, and are distinguished by the size and number of their roots, as well as by the extent of their crowns. Formulæ, in which the numbers of incisors, canines, præmolars, and molars in the upper and lower jaws are given, are used to indicate in a simple manner the nature of the dentition, *e.g.*, the dental formula of man is

$$\frac{2}{2}\ \frac{1}{1}\ \frac{2}{2}\ \Big|\ \frac{3}{3}\cdot\left[i.\ \frac{2\cdot2}{2\cdot2}\ c.\ \frac{1\cdot1}{1\cdot1}\ p.m.\ \frac{2\cdot2}{2\cdot2}\ m.\ \frac{3\cdot3}{3\cdot3}\right].$$

Alimentary canal. In addition to the hard structures at the entrance to the digestive cavity, soft, movable lips which bound the mouth opening, and a fleshy tongue which is of very various form and lies on the floor of the buccal cavity, are of special importance for the introduction and preparation of the food (fig. 675). In the Monotremata the lips are replaced by the edges of the beak. The tongue, however, is never absent, but it may be immovable, and completely fused with the floor of the mouth, as in the Whales. Its front part is mainly tactile in function, but in some cases it is used to seize (Giraffe) or capture food (Ant-eaters). Variously shaped papillæ, which are often cornified and bear recurved hooks, project from its upper surface. The *papillæ circumvallatæ* alone have a relation to the sense of taste. The tongue is supported by the hyoid bone and by a cartilaginous rod, which represents the *os interglossum* (*Lytta*). The anterior cornua of the hyoid are attached to the styloid processes of the temporal bone, the posterior bear the larynx. Beneath the tongue there is sometimes (most developed in the *Insectivora*) a single or double projection, which is termed the lower tongue. The sides of the buccal cavity are soft and fleshy, and are not unfrequently in the Rodents, Apes, etc., dilated into wide sacs, — the so-called cheek-pouches. The soft palate (*palatum molle*) must be mentioned as a structure peculiar to the *Mammalia*; it constitutes

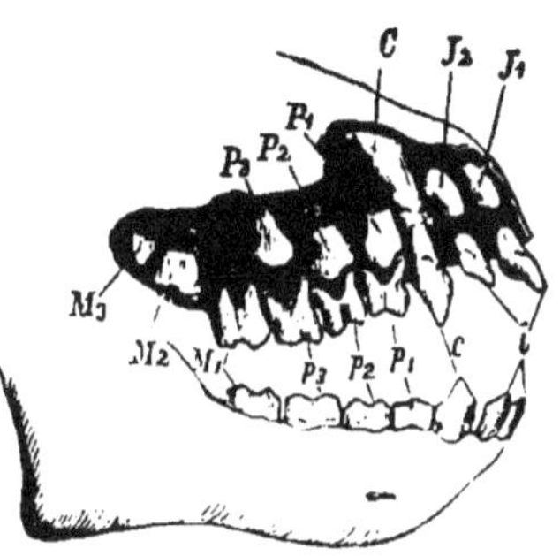

FIG. 674.—Dentition of *Cebus* (while changing the teeth) after Owen. *i*, Incisors; *c*, Canines; *p1 p2 p3*, Præmolars of the milk dentition; *J1 J2* Incisors; *C*, Canine; *P1 P2 P3*, Præmolars of the permanent dentition *M1 M2 M3*, Molars.

the boundary between the buccal cavity and pharynx. All Mammals, with the exception of the carnivorous *Cetacea*, have salivary glands,—a parotid, a submaxillary, and a sublingual,—the fluid secretion of

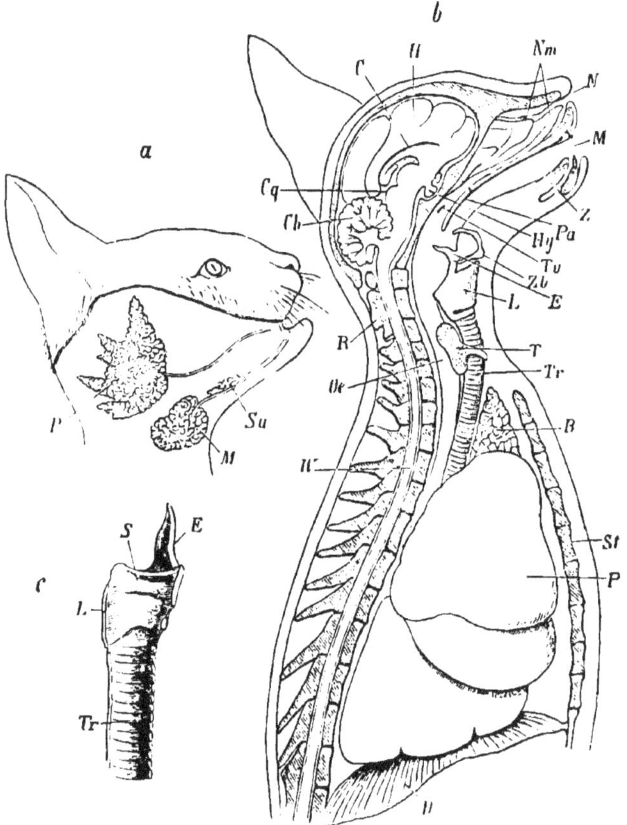

Fig. 675.- Entrance to the digestive apparatus and the respiratory organs of the Cat (after C. Heider). *a*, head with exposed salivary glands. *P*, Parotid; *M*, Sub-maxillary; *Su*, Sub-lingual. *b*, Longitudinal section through the Head and Thorax; the Respiratory organs are seen from the side. *N*, Nasal aperture; *Nm*, Turbinal bones; *M*, Mouth; *Z*, Tongue; *Pa*, Velum palati; *Oe*, Oesophagus; *L*, Larynx; *E*, Epiglottis; *Zb*, Hyoid; *Tr*, Trachea; *P*, Lung; *D*, Diaphragm; *T*, Thyroid; *B*, Thymus; *Tu*, Opening of Eustachian tube into the Pharynx; *H*, Cerebral hemispheres; *C*, Corpus callosum; *Cp*, Corpora quadrigemina; *Cb*, Cerebellum; *R*, Spinal cord; *Hy*, Hypophysis; *W*, Vertebral column; *St*, Sternum. *c*, Longitudinal section through the Larynx (*L*) and the first part of the Trachea (*Tr*). *S*, Vocal cord; *E*, Epiglottis.

which is poured out in large quantities, especially in the *Herbivora*. The œsophagus, which follows the wide gullet, only exceptionally presents crop-like dilations; it is usually of considerable length, and opens into the stomach behind the diaphragm (vol. i., fig. 50). The stomach is, as a rule, a simple transversely placed sac, but is frequently divided by the gradual differentiation and constriction of its anterior, lateral, and posterior regions into a number of parts, which are most completely separated in the Ruminants and distinguished as four separate stomachs. The pyloric region is principally distinguished by the presence of gastric glands, and is more or less sharply separated from the beginning of the small intestine by a sphincter muscle and by an inwardly projecting fold (*pyloric valve*). The intestine is divided into a small and a large intestine, the boundary between which is indicated by the presence of a valve and a caecum, which is especially developed in herbivorous animals. The anterior part of the small intestine, or *duodenum*, contains the so-called Brunner's glands in its mucous membrane, and receives the secretion of the large liver and of the pancreas. The liver is multi-lobed, and is sometimes without a gall bladder. When a gall bladder is present the bile duct (*d. cysticus*), and the hepatic duct (*d. hepaticus*) unite to form a common duct (*d. choledochus*). The small intestine is longest in animals which eat grasses and leaves, and is characterised by the numerous folds (*valvulæ conniventes*) and villi of its mucous membrane, and by the possession of a great number of groups of glands (Lieberkühn's, Peyer's glands). The terminal region of the large intestine or rectum opens, except in the *Monotremata* which are characterised by the possession of a cloaca, behind the urogenital opening, though the two openings are sometimes surrounded by a common sphincter (*Marsupialia*).

The **heart** (fig. 676) of Mammalia, like that of Birds, is divided into a right venous and a left arterial portion, each with a ventricle and auricle (sometimes as in *Halicore* the division is marked externally). It is enclosed in a pericardium, and sends off an arterial trunk, which forms a left aortic arch, from which two vessels frequently arise, viz., (1) a right anonyma, with the two carotids and right subclavian; and (2) the left subclavian; or, as in man, three vascular trunks, viz., (1) a right anonyma, with the right carotid and right subclavian; (2) the left carotid; and (3) the left subclavian, all close to one another. As a rule, a superior and an inferior vena cava open into the right auricle; more rarely, as in the Rodents, Monotremes, and Elephants, there are two superior venæ cavæ. *Retia mirabilia* have been recognised principally for the arterial vessels, and are found on the extremities of burrowing and climbing animals (*Stenops*, *Myrmecophaga*, *Bradypus*, etc.); on the carotids round the hypophysis, and on the ophthalmic arteries in the orbit in Ruminants:

finally on the intercostal arteries and the iliac veins of the Dolphin. A renal-portal system is always absent.

The **lymphatic system** is provided with numerous lymphatic glands, and its main trunk (*ductus thoracicus*), which is placed on the left, opens into the superior vena cava.

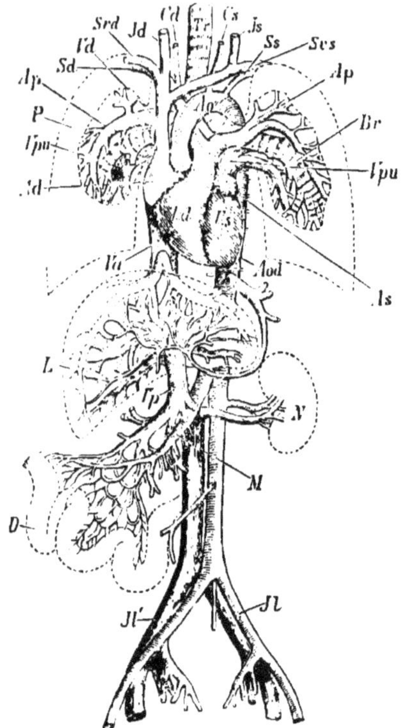

Fig. 676. - Circulatory apparatus of Man (from Owen after Allen Thomson). *Vd*, Right Ventricle; *Vs*, left Ventricle; *Ad*, right Auricle; *As*, left Auricle; *Ao*, Aortic arch; *Aod*, descending Aorta; *Cd*, right Carotid; *Cs*, left Carotid; *Sd*, right subclavian Artery; *Ss*, left subclavian Artery; *M*, Mesenteric Artery; *Jl*, common Iliac Artery; *Va*, inferior Vena Cava; *Vd*, superior Vena Cava; *Jl'*, common Iliac vein; *Vp*, Vena portæ; *Jd*, right Jugular; *Js*, left Jugular; *Svd*, right subclavian Vein; *Svs*, left subclavian; *Ap*, pulmonary Artery; *Vpu*, pulmonary Vein; *Tr*, Trachea; *Br*, Bronchi; *P*, Lungs; *L*, Liver; *N*, Kidney; *D*, Intestine.

Of the so-called vascular glands the **spleen**, the **thymus**, and the **thyroid**, which is especially developed in the young, are very generally present (fig. 675).

The paired lungs (fig. 675) are freely suspended in the thoracic cavity, and are distinguished by the numerous ramifications of the bronchial tubes, the finest branches of which end with conical, funnel-shaped dilatations (*infundibula*), which are provided on their lateral surfaces with swellings. Respiration is mainly effected by the movements of the diaphragm, which forms a complete, usually transversely placed, septum between the thoracic and abdominal cavities: by the contraction of its muscular parts it acts as an inspiratory muscle; that is, it dilates the thoracic cavity. The elevation and depression of the ribs also have an effect in dilating the thorax. The trachea is, as a rule, straight, without coils, and divides at its lower end into two bronchi leading to the lungs. There may be, in addition, a small accessory bronchus on the right side. The trachea is supported by cartilaginous half-rings which are open behind, and only exceptionally by complete rings of cartilage. The first part of the trachea, or larynx, is placed at the lower end of the pharynx, behind the root of the tongue; it is supported by the posterior horns of the hyoid bone, possesses lower vocal cords, complicated pieces of cartilage (cricoid, thyroid, and arytenoid cartilages) and muscles, and constitutes a vocal organ.

In the *Cetacea* alone is the larynx, which projects in the base of the pharynx as far as the posterior nares, used exclusively for respiration. A movable epiglottis (almost tubular in the *Cetacea*), attached to the upper edge of the thyroid cartilage, projects over the glottis. When food is being swallowed it sinks, and closes the glottis. Accessory cavities, with membranous or cartilaginous walls, are sometimes attached to the larynx. These sometimes function as air reservoirs, *e.g.*, the air-sacs of *Balæna*, sometimes as a resonating apparatus for the strengthening of the voice, as in many Monkeys (*Mycetes*).

The **kidneys** (fig. 677) still sometimes consist (Seals, Dolphins) of numerous lobes united together at the pelvis of the kidney. As a rule, however, they are compact bean-shaped glands, lying in the lumbar region, outside the peritoneum. The ureters arise from the so-called pelvis of the kidney, and always open into a urinary bladder, placed in front of (ventral to) the intestine. The duct of the bladder, or *urethra*, enters into a more or less close relation with the ducts of the generative organs, and leads into a *sinus* or *urogenital canal* opening in front of the anus. Above the kidney there is a glandular organ termed the *suprarenal body*.

The **male sexual organs** (fig. 677) of most *Mammalia* are characterised by the change in the position of the testes. In the *Monotremata* and *Cetacea* alone do the testes remain in their original position near the kidneys, in all other cases they descend in front of the pelvis, and, pushing the peritoneum before them, enter the inguinal canal (many Rodents), or, still more frequently, pass through the inguinal canal into a double cutaneous fold, which is transformed into the scrotum. Not unfrequently (Rodents, Bats. Insectivores) they pass back through the open inguinal canal into the abdominal cavity after the breeding season: this is effected by the *cremaster*, a slip of muscle separated from the oblique abdominal muscle. The *scrotum*, as a rule, lies behind the penis; but in the Marsupials it is formed by an invagination of the integument directly at the entrance of the inguinal canal in front of the male copulatory organ. The coiled excretory ducts of the testes, which are derived from the Wolffian body, constitute the *epididymis*, and lead into the two *vasa deferentia*, which, after forming glandular dilatations (seminal vesicles), open close together into the urethra. At this point open the ducts of the *prostates*, which differ much in form, and are often divided into several groups of glands. Further down a second pair of glands, known as *Cowper's glands*, opens into the urethra. Remains of the Müllerian ducts, which in the female are used as the oviducts, frequently persist between the openings of the vasa deferentia. They are called the organ of Weber (*uterus masculinus*), and in the so-called Hermaphrodites their parts are much enlarged, and may be differentiated in the manner peculiar to the female sex. In all cases the end of the urethra, which functions as a urogenital canal, is in connection with external copulatory organs: these always have the form of an erectile *penis*, which, in the *Monotremata*, is concealed in a pouch in the cloaca. The penis is supported by cavernous erectile bodies, which in the Monotremata are confined to paired *corpora cavernosa urethræ*; but in all other Mammalia there are, in addition to the *corpus cavernosum urethræ* (*c. spongiosum*) which is unpaired and surrounds the urethra, two upper *corpora cavernosa penis*, which are attached to the ischium, and only rarely fuse with one another. A cartilaginous, or bony support, the so-called *os penis* (*Carnivora*, *Rodents*), may also be developed, especially frequently in the *glans*, which is formed by the *corpus cavernosum urethræ* (fig. 677). The *glans*, which is bifid only in exceptional cases (*Monotremata*, *Marsupials*).

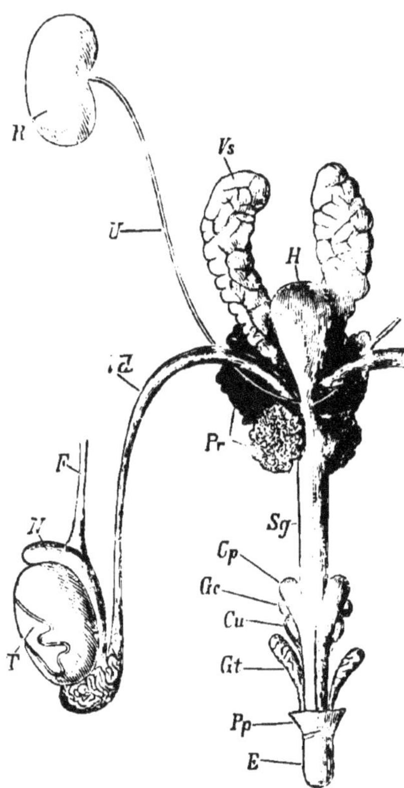

FIG. 677.—Urinary and sexual organs of *Cricetus vulgaris* (after Gegenbaur). *R*, Kidney; *U*, Ureter; *H*, Urinary bladder; *T*, Testis; *F*, Funiculus spermaticus (Spermatic cord); *N*, Epididymis; *Vd*, Vas deferens; *Vs*, Vesiculæ seminales; *Pr*, Prostate; *Sg*, Urogenital sinus (Urethra); *Gc*, Cowper's glands; *Gt*, Tyson's glands; *Cp*, Corpora cavernosa penis; *Cu*, Corpus cavernosum urethræ; *E*, Glans penis; *Pp*, Prepuce.

varies greatly in its form, and lies retracted in a reduplication of the skin (*foreskin* or *prepuce*) which is richly glandular (*gl. Tysonianæ*).

Female sexual organs. The ovaries (fig. 678) are unsymmetrical only in the *Monotremata*, in consequence of the reduction of the right ovary. In all other cases they are equally developed on either side; they are placed in folds of the peritoneum, close to the funnel-shaped dilated mouths of the oviducts, by which they are sometimes com-

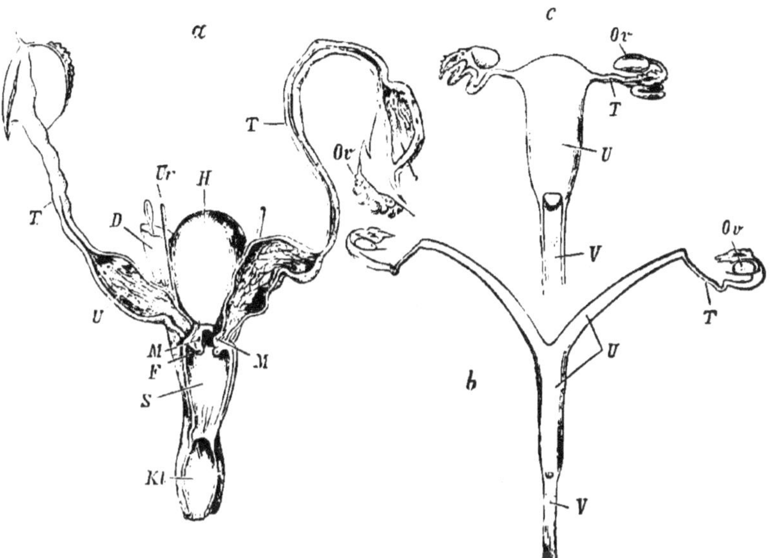

FIG. 678.—Female generative organs. *a*, of *Ornithorhynchus* (after Owen); *b*, of *Viverra genetta*; *c*, of *Cercopithecus nemestrinus*; *Ov*, Ovary; *T*, Oviduct; *U*, Uterus; *V*, Vagina; *H*, Urinary bladder; *Ur*, Ureter; *M*, Mouth of Uterus; *F*, opening of Ureter; *S*, urogenital Sinus; *Kl*, Cloaca; *D*, Intestine. A style is passed through the opening of the latter into the Cloaca.

pletely surrounded. The oviduct is divided into the *Fallopian tube*, which is always paired and begins with a free ostium; the dilated, sometimes paired, more frequently unpaired, middle portion—the *uterus*; and the terminal part, or *vagina*, which is unpaired, except in Marsupials, and opens behind the opening of the urethra into the short urogenital sinus, or vestibule. In the *Monotremata* the two tubular uteri open, without forming a vagina, on papilliform prominences into the urogenital sinus, which is still connected with the cloaca (fig. 678 *a*).

According to the different degrees of duplicity of the uterus (when a vagina is present), we may distinguish: the *uterus duplex*, with more or less complete external separation and double os uteri (Rodents, Marsupials); the *uterus bipartitus*, with single os uteri, but almost complete internal partition (Rodents): the *uterus bicornis* (fig. 678 *b*). in which the upper parts, or horns of the uterus are separate (*Ungulata, Carnivora, Cetacea, Insectivora*); and finally the *uterus simplex* (fig. 678 *c*) with single cavity and very muscular walls (*Primates*).

The vestibule, with its glands of Duvernoy (Bartholin), which correspond to the Cowperian glands of the male, is separated from the vagina by a constriction, and sometimes also by a fold of the mucous membrane, called the *hymen*. The external generative organs consist of the *labia majora* and *labia minora*, at the sides of the sexual opening. and of the *clitoris*. The *labia majora* are two external folds of skin, and are equivalent to the two halves of the scrotum: the *labia minora* are two smaller internal folds, and are not always present. The *clitoris* possesses erectile tissue and a *glans*, and is the equivalent of the penis. The clitoris may sometimes (as in *Ateles*) reach a considerable size, and be perforated by the urethra (Rodents. Moles, Lemurs). In such cases of perforated clitoris. there is, of course, no common urogenital sinus. Morphologically, the female genitalia represent an earlier stage of development of the male organs, which, in the cases of the so-called hermaphrodite formation, may in consequence of arrest of development preserve a more or less female structure. As a rule the two sexes are easily distinguished by the different form of the external generative organs. Frequently there is a marked dimorphism in the whole external appearance; the male being larger, having a different hairy covering, being possessed of a louder voice, and provided with stronger teeth or special weapons (horns). On the other hand, the milk glands, which are situate in the inguinal region, on the abdomen, and on the thorax, and which almost always project into teats or nipples, are rudimentary in the male sex.

The breeding time (rut) is usually in the spring, rarely towards the end of summer (Ruminants), or even in the winter (*Sus, Carnivora*). An important phenomenon, which accompanies the rut in the female, and is independent of copulation, is the passage of one or more ova from the Graafian follicles of the ovary into the oviduct. The ova of the Mammalia were first discovered by C. E. von Baer. They are extraordinarily small ($\frac{1}{20}$ to $\frac{1}{10}$ line in diameter) and are surrounded by a strongly refractile membrane (*zona*

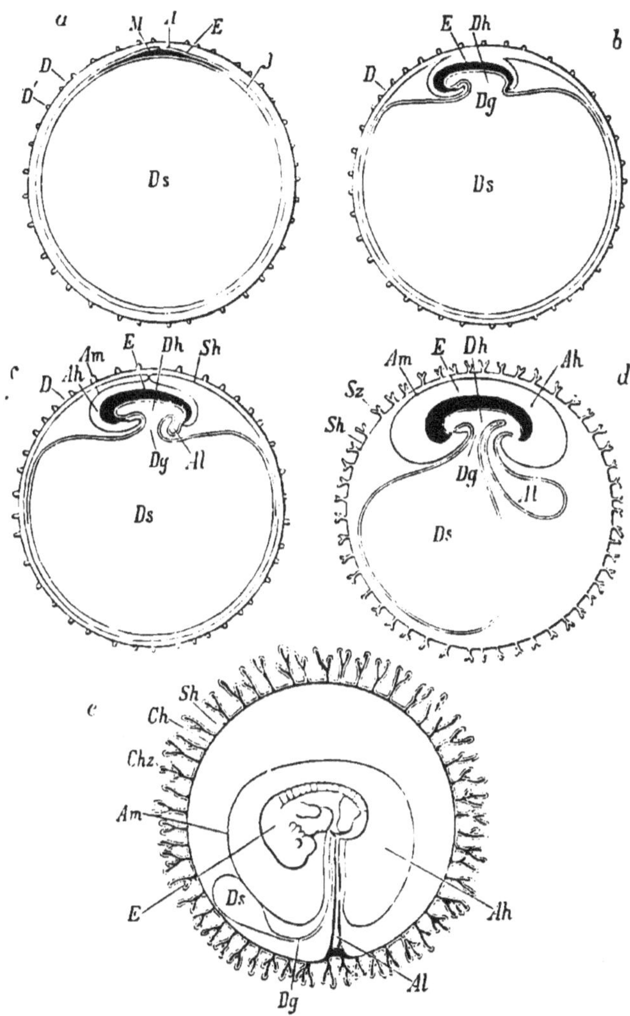

FIG. 679.—Diagramatic figures illustrating the formation of the fœtal membranes of a Mammal (after Kölliker). *a*, Ovum with first rudiments of embryo; *b*, Ovum with yolk-sac and developing amnion; *c*, Ovum with amnion closing and developing allantois; *d*, Ovum with villous serous envelope, embryo with mouth and anus; *e*, Ovum in which the vascular layer of the allantois is applied to the serous envelope and has grown into the villi of the latter, yolk-sac reduced, the amniotic cavity is increasing; ***D***, Zona radiata; ***D'***, Villi of zona; ***Sh***, subzonal membrane (serous envelope); ***Sz***, Villi of subzonal membrane; ***Ch***, Chorion (vascular layer of Allantois); ***Chz***, Chorionic villi (consisting of chorion and subzonal membrane); ***Am***, Amnion; ***Ah***, Amniotic cavity; ***E***, Embryo; ***A***, Embryonic thickening of the external layer; ***M***, of the middle layer; ***J***, of the inner layer; ***Ds***, cavity of the embryonic (blastodermic) vesicle, later of the yolk-sac (umbilical vesicle); ***Dh***, Intestinal cavity; ***Dg***, Umbilical stalk; ***Al***, Alla.

pellucida), round which a layer of albumen is often deposited in the oviduct.

The fertilization and total segmentation* of the ovum always take place in the oviduct (Fallopian tube). *Amnion* and *allantois* are present in *Mammalia*. In the uterus the ovum acquires a villous coat (*chorion*), derived from the original zona and from the subzonal membrane (so-called serous envelope), which is developed within the zona. It becomes attached to the uterine wall by means of the chorion (fig. 679). Later on, the peripheral part of the allantois also becomes applied to the chorion, and, as a rule, penetrates with its vessels into the villi (*secondary chorion*), so that there is developed a relatively large surface, permeated with branches from the fœtal vessels, the blood of which is in intimate endosmotic connection with the blood of the uterine wall. This connection of the allantois and chorion of the fœtus with the uterine walls gives rise to the **Placenta**, by means of which the nourishment and respiration of the fœtus are provided for in the body of the mother. The placenta is wanting only in the *Monotremata* and *Marsupialia*, which, therefore, are known as *Aplacentalia*, as opposed to the rest of the Mammalia, which have a placenta, and are called *Placentalia*. The placenta presents great variations in the individual orders, in its special development and in the mode of its connection with the uterine walls. Either the villi of the placenta are loosely connected with the uterine walls, and separate from the latter at birth (*Adeciduata*), or they become so intimately united with the glands of the uterine mucous membrane that the latter comes away with the embryo at birth, as the *decidua* or *after-birth* (*Deciduata*).† In the first case the allantois may grow completely round the ovum, and the villi be numerous and uniformly distributed over the whole chorion (*diffuse placenta* of *Ungulata*, *Cetacea*), or be aggregated in special places, forming small tufts, the so-called *cotyledons* (Ruminants). In the other case, the placenta with its villi is confined either to an annular zone on the chorion (*Pl. annularis*, or *zonary placenta* of *Carnivora*, *Pinnipedia*), or to a discoidal area (*discoidal placenta* of Man, Apes, Rodents, Insectivores, Bats).

* [According to Caldwell's recent discovery, which was communicated to the British Association at Montreal in September of the present year (1884), but of which no details have as yet come to hand, the Monotremata are oviparous and their ova meroblastic.]

† [For a fuller account of the structure and development of the various kinds of placenta, the reader is referred to Balfour's *Comparative Embryology*, vol. ii., p. 193.]

In the fœtus, respiration is effected through the placenta, and the lungs are functionless. In correspondence with this the circulation of the fœtus differs from that of the animal after birth (fig. 680). From the heart the blood is driven into the descending aorta, which sends off behind two large vessels to the placenta (*umbilical* or *allantoic* arteries). The blood, returning from the placenta in the allantoic vein, passes in great part through a connecting vessel (*ductus venosus Arantii*) into the inferior vena cava, and thence in part passes into the right auricle, but the greater part passes, in consequence of a special arrangement of valves, directly into the left auricle through an opening in the interauricular septum, called the *foramen ovale.* The blood which reaches the right ventricle passes through a vessel (*ductus arteriosus Botalli*), connecting the pulmonary artery with the aorta, directly into the systemic circulation, except a small portion which goes to the lungs. From this condition of the circulation, it results that all the arterial vessels, except the allantoic vein. contain mixed blood.

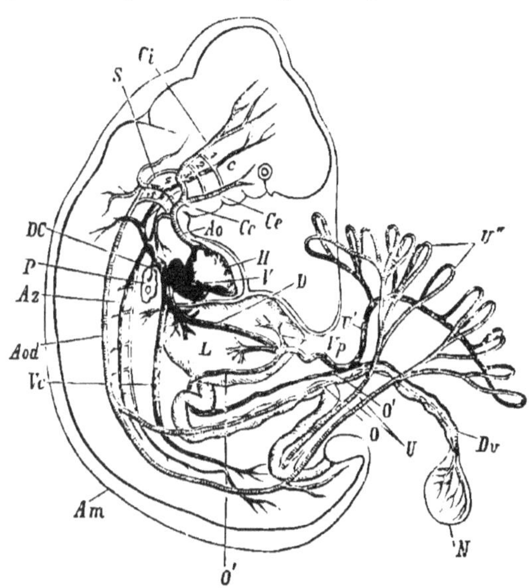

FIG. 680.—Diagram of the arrangement of the principal vessels in a human fœtus (after Huxley). *H*, Ventricle; *V*, Auricle; *Ao*, Aortic trunk; *Cc*, common Carotid; *Ce*, external Carotid; *Ci*, internal Carotid; *S*, subclavian artery; 1, 2, 3, 4, 5, the arterial arches—the persistent left aortic arch is not visible; *Aod*, descending aorta; *O*, Omphalomeseräic (vitelline) Artery; *O'*, Omphalomeseräic (vitelline) Vein; *U*, Umbilical (allantoic) arteries with their placental ramifications (*U''*); *U'*, Umbilical (allantoic) vein; *Vp*, Portal vein; *Vc*, Vena cava inferior; *C*, anterior cardinal vein; *D*, Ductus venosus Arantii; *DC*, Ductus Cuvieri; *Az*, Azygos Vein; *P*, Lungs; *L*, Liver; *N*, Umbilical vesicle (yolk-sac); *Dv*, Vitelline duct (*Ductus omphalomeseraicus*); *Am*, Amnion.

As remains of the first stage of the circulation before the development of the placenta, the *omphalomeseräic vessels*—an artery and a vein—which belong to the umbilical vesicle, still persist.

The duration of gestation depends on the size of the body and the stage of development at which the young are born. It is longest in the large terrestrial, and the colossal aquatic animals (*Ungulata, Cetacea*), which live under favourable conditions of nourishment. The young of these animals are so far advanced in their bodily development at birth, that they are able to follow the mother (to a certain extent like *præcoces*). The period of gestation is relatively shorter in the *Carnivora*, the young of which are born naked and with closed eyes and, like *altrices*, are for a long time completely helpless, and need the care and protection of the mother. It is, however, shortest in the aplacental Monotremes* and Marsupials. In these animals the young, which are born at a very early stage (in the Kangaroo they are no larger than a nut), pass into a pouch formed by cutaneous folds in the inguinal region, and here adhere firmly to the nipples of the mammary glands. In this pouch, as in a second uterus, they are nourished by the secretion of the mammary glands, which assume at this early stage the nutrient function of the absent placenta. The number of the young, which are born, also varies very greatly in the different genera. The large *Mammalia*, of which the period of gestation is longer than six months, as a rule bear only one, more rarely two young; but in the smaller Mammals and some domestic animals (Pig) the number is considerably larger, so that twelve to sixteen, or even twenty young may be born at one time. The number of teats on the mother usually indicates the greater or smaller number of the progeny.

Many Mammals live a solitary life, and pair only at the breeding time; they are principally such carnivorous animals as find their subsistence by hunting in definite hunting grounds, like the Mole in its subterranean passages. Others live united in companies, in which the oldest and strongest males frequently undertake the care of protection and leadership. Most Mammals seek their food by day. Some, *e.g.*, the Bat, leave their hiding places in the twilight and at night. Most *Carnivora* and numerous *Ungulata* also sleep in the daytime. Some *Rodentia*, *Insectivora*, and *Carnivora* fall, during the cold season of the year when food is scarce, into an interrupted (Bear, Badger, Bat), or continuous (Dormouse, Hedgehog, Marmots) winter sleep in their hiding places which are often carefully protected, or in nests formed in the earth. During this time the temperature is lowered, the respiration is less active, the heart-beat is slowed, and they take up no food, but consume the fat masses which were stored up in the autumn. The following animals are known to migrate:

* *Vide* note on p. 296.

the Reindeer, the South American Antelopes, and the North-American Buffalo, the Seals, Whales, and Bats, but more especially the Lemmings, which migrate in enormous herds from the northern mountains southwards to the plains, are stopped by no obstacles on their journey, and even cross rivers and arms of the sea.

The intellectual faculties are more highly developed than in any other class of animals. The *Mammalia* possess the faculty of discrimination and memory: they form ideas, judgments, and conclusions; they exhibit affection and love to their benefactors, dislike, hate, and anger to their enemies; each individual has a definite character. Further, the intellectual faculties of Mammals are capable of being developed and improved, but to a relatively small extent on account of the absence of articulate speech. The more docile and intelligent of the Mammalia have been chosen by man as domestic animals, and in this capacity have played an important and indispensable part in the history of civilisation (Dog, Horse). Instinct, however, always occupies an important place in the life of Mammals. It leads many of them to construct spacious passages and ingenious nests above or below the earth, in which they rest and bring up their offspring. Almost all Mammals make special places for their brood, which they often line with soft materials; some even construct true nests, like those of birds, of grass and stalks on the earth. Many of those which inhabit subterranean holes and passages store up winter-provisions, which they consume in the sterile season, sometimes only in autumn and spring (winter-sleepers.)

Geographical distribution. Some orders, as the Rodents and Bats, are represented in all parts of the world. Of the *Cetacea* and *Pinnipedia* most species belong to the Polar regions. In general, the Old and New Worlds have each their own fauna. The mammalian fauna of Australia consists almost exclusively of Marsupials. The oldest fossil remains (lower jaw) of Mammals are found in the Trias (Keuper Sandstone and Oolite, Stonesfield slate) and are probably Marsupial. But it is not until the tertiary period that the mammalian fauna presents a rich development.

I.—APLACENTALIA.

Order 1.—MONOTREMATA.*

The jaws are elongated to the form of a beak; the feet are short, five-toed, and furnished with strong claws. Marsupial bones and a cloaca are present [*Oviparous* :† *with meroblastic ovum.*]

The most important character of the Monotremes is the presence of a cloaca. The dilated end of the rectum receives the openings of the generative and urinary ducts (fig. 678 *a*). In addition to this character we must mention the simple condition of the female generative organs, the absence of teeth in the jaws, the possession of a large coracoid, and the slight development of the corpus callosum.

FIG. 681.—*Echidna hystrix.*

FIG. 682.—*Ornithorhynchus paradoxus.*

The form of the body and the mode of life of the Monotremes partly recall the Anteaters and Hedgehog (*Echidna hystrix*, fig. 681) and partly the Otters and Moles (*Ornithorhynchus*); in fact, *Ornithorhynchus* received the appropriate name of "Watermole" from the Australian settlers (fig. 682). *Echidna* is covered with strong spines, and possesses an elongated edentulous snout, with a vermiform, protrusible tongue. The short five-toed legs end with powerful scratching claws, which are excellently adapted for rapid burrowing. *Ornithorhynchus*, on the contrary, has a close, soft fur, a flattened

* R. Owen, Article "Monotremata," in Todd's "Cyclopaedia of Anatomy," vol. iii., 1843.

† *Vide* note on p. 296.

body, and, as in the Beavers, a flat tail. The jaws, like the beak of a Duck, are adapted for burrowing in mud, but are armed on both sides with two horny teeth, and are surrounded by a horny integument, which, at the base of the beak, projects in a peculiar manner, so as to form a kind of shield. The legs of *Ornithorhynchus* are short; the five-toed feet end with strong claws, but are also furnished with very extensible webs, and are, therefore, equally well adapted for swimming and burrowing. Both sexes have, like the Marsupials, in front of the pubis the so-called marsupial bones, which in the female *Echidna* support a pouch. The testes remain inside the body cavity (*i.e.*, do not descend into a scrotum). The males in both the genera possess a hollow spur on the hind foot, which receives the duct of a gland, to which for a long time poisonous properties were erroneously attributed. It appears more probable that this spur serves only as a stimulant during copulation, since it fits into a pit in the thigh of the female. The embryos * are born at an early stage, and in *Echidna* pass into the marsupial pouch of the mother. On the abdomen of the latter there are two mammary glands, which are without a projecting nipple. Fossil remains are as yet unknown.

Ornithorhynchus paradoxus Blumb., The Duck-bill Platypus, Australia and Van Diemen's Land: *Echidna hystrix* Cuv., in the mountainous regions of south-east Australia: *E. setosa* Cuv., Van Diemen's Land.

Order 2.—MARSUPIALIA.†

Mammalia with various dentition, with two marsupial bones supporting a marsupial pouch, which encloses the teats of the mammary glands.

The principal characteristic of the Marsupials is the possession of a sac, or pouch (*marsupium*), which is supported by two marsupial bones (fig. 683), encloses the teats of the mammary glands, and receives the helpless young after birth. In the absence of a placenta, birth, as in the Monotremes, takes place at a very early stage. Even in *Macropus giganteus*, the males of which attain almost the height of a man, the period of gestation does not last more than thirty-nine days, and the embryo at birth is blind and naked, its extremities are

* Vide note, p. 296.

† R. Owen, "Marsupialia," in Todd's Cyclopædia of Anatomy, vol. iii., 1842.
G. R. Waterhouse, "A Natural History of the Mammalia," vol. v., "Marsupialia," London, 1846.

scarcely visible, and it is not much more than an inch in length. It is placed in the pouch by the mother, sucks firmly on to one of the two or three teats, and remains in the pouch for eight or nine months.

In their external appearance, in their mode of nourishment, and in their habits, the Marsupials differ extraordinarily from each other. Many of them are herbivorous, and in their dentition approach the Rodents or the Ungulates; others are omnivorous; others, like true *Carnivora*, prey on Insects, Birds and Mammals. In their general appearance and mode of locomotion they repeat a series of types of different mammalian orders. The Wombats represent the Rodents,

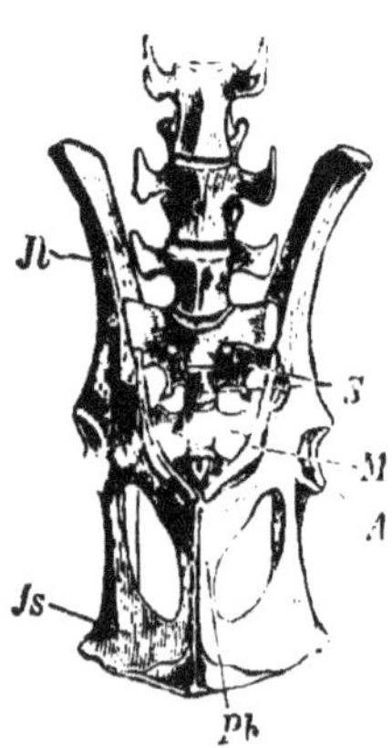

FIG. 683.—The pelvis and adjoining parts of the vertebral column of *Macropus*. *Jl*, Ilium; *Pb*, Pubis; *Js*, Ischium; *M*, Marsupial bones; *A*, Acetabulum; *S*, the two sacral vertebræ.

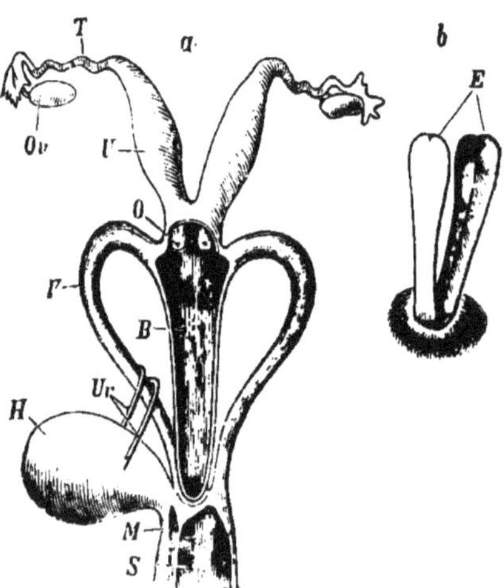

FIG. 684.—*a*, Female generative organs of *Halmaturus* (after Gegenbaur); *Ov*, Ovary; *T*, Oviduct; *U*, Uterus; *O*, Mouth of uterus; *V*, Vagina; *B*, Cæcum of vagina; *Ur*, Ureter; *H*, Urinary bladder; *M*, opening of bladder into the urogenital sinus (*S*). *b*, Bifid penis of *Didelphys philander* (after Otto, from Gegenbaur); *E* the two halves of the glans.

the fleet Kangaroos, which move by huge bounds, correspond to the Ruminants, and represent, in a certain degree, game, which is absent in Australia. The flying Marsupials (*Petaurus*) resemble the flying Squirrels (*Pteromys*); the climbing Phalangers (*Phalangista*), in their shape and mode of life, recall the Lemurs (*Lemur*); while others, as the *Peramelidæ*, show a likeness to the Shrews (*Soricidæ*) and Insectivores. Finally, the carnivorous Marsupials approach in their dentition to the true *Carnivora* as well as the *Insectivora*, to which they scarcely yield in the large number of their small incisor teeth and tuberculated molars.

The female generative organs frequently present racemose ovaries: the two oviducts are prolonged into two completely separate uteri, which are followed by the peculiarly formed and likewise double vagina (fig. 684*a*); the two vaginæ are united at the point where they receive the openings of the uteri into a common portion, which gives off a long cæcal diverticulum, usually divided by a septum. From this part arise the two vaginal canals, which curve round and open into the urogenital canal. Since the external opening of the latter coincides more or less closely with the anus (the two openings are surrounded by a common sphincter), the Marsupials may be said to have a kind of cloaca. The penis ends as a rule with a bifid glans (fig. 684*b*), corresponding to the double vagina of the female.

Most Marsupials live in Australia, many also in islands of the Pacific Ocean and the Moluccas; *Didelphys* in South America. In Europe they are wholly absent at the present time, though they were distributed there in the tertiary period.

Tribe 1. **Glirina** (Rodent-like Marsupials). Unwieldy, heavy animals of the size of a Badger, with close soft fur, with Rodent dentition, short limbs, and rudimentary tail. The rudimentary inner toe of the hind foot alone is without the curved claw.

Fam. **Phascolomyidæ.** Dentition $i. \frac{1}{1}\, c. \frac{0}{0}\, p.m. \frac{1}{1}\, m. \frac{4}{4}$. *Phascolomys Wombat* Per. Les. (*fossor*). Van Diemen's Land and New South Wales.

Tribe 2. **Macropoda** (Jumping Marsupials). With small head and neck, weak, small, five-toed front legs, and hind part of the body unusually developed. The very long hind legs serve for jumping, and are aided by the long tail, the root of which is thickened. The powerful hind feet end with four toes with hoof-like claws; the two internal toes are united, and the median one is very long and powerful. The dentition recalls that of the Horse, though the number of incisors in the lower jaw (2) is smaller. The stomach is colon-like in shape, the cæcum is long. They feed on grass and plants.

Fam. **Halmaturidæ** (Kangaroos). Dental formula $i. \frac{3}{1}\, c. \frac{0\,(1)}{0}\, p.m. \frac{1}{1}\, m. \frac{4}{4}$ *Macropus giganteus* Shaw. Great Kangaroo. *Hypsiprymnus rufescens* Gould. Kangaroo Rat.

Tribe 3. **Scandentia (Carpophaga).** Climbing Marsupials. The second and third toes of the hind foot are fused, but the inner toe is without nail and opposable. The long tail is prehensile in accordance with the arboreal life. With respect to the dentition these animals are intermediate between the *Glirina* and *Halmaturidæ*.

Fam. **Phascolarctidæ.** Body stout, unwieldy, head thick, ears large, and tail quite rudimentary. *Phascolarctus cinereus* Goldf., Koala. Dentition $c. \frac{3}{1}\, i. \frac{1}{0}\, p.m. \frac{1}{1}\, m. \frac{4}{4}$. New South Wales.

Fam. **Phalangistidæ.** Of slender form, with prehensile tail. *Petaurus flaviventer* Desm.: *P. pygmaeus* Desm., scarcely 4 inches long: *Phalangista ursina* Temm., Celebes: *P. (Trichosaurus) vulpina* Desm. (fig. 685): *P. viverrina*, New South Wales; *Tarsipes rostratus* Gerv.

Fig. 685.– *Trichosurus vulpinus.*

Tribe 4. **Rapacia** (carnivorous Marsupials). The dentition presents the characters of that of the *Insectivora* and *Carnivora*. Stomach without glandular apparatus. The cæcum is but slightly developed. Some are climbers, some jumpers and runners.

Fam. **Peramelidæ (Entomophaga).** Bandicoots. With elongated hind legs and pointed snout, as in the *Insectivora*. They dig holes in the earth. *Perameles nasuta* Geoffr., New South Wales.

Fam. **Dasyuridæ.** Native cats, devils, etc. With distinct carnivorous characteristics and hairy tail, which is not prehensile. *Myrmecobius fasciatus* Waterh., Marsupial Ant-eater; *Phascogale penicillata* Temm. Bloodthirsty and bold carnivorous animal of the size of a squirrel, in a certain sense the Weasel of South and West Australia. *Ph. flavipes* Waterh., yellow-footed Marsupial mouse: *Dasyurus viverrinus* Geoffr., Dentition: $i. \frac{4}{3}\, c. \frac{1}{1}\, p.m. \frac{2}{2}\, m. \frac{4}{4}$: New South Wales; *Thylacinus cynocephalus* A. Wagn., Tasmanian Wolf.

Fam. **Didelphyidæ (Pedimana).** Opossums. Dentition: $i. \frac{5}{4}\, c. \frac{1}{1}\, p.m. \frac{3}{3}\, m. \frac{4}{3}$: with tolerably pointed snout, large eyes and ears, and usually long prehensile tail. The feet have five toes. On the hind foot the inner toe is opposable. *Didelphys virginiana* Shaw: *D. cancrivora* Gm., Brazil, with completely prehensile tail: *D. opossum* L.; *D. philander* L.; *D. dorsigera* L., Surinam.

II. PLACENTALIA.

I.—Adeciduata.

Order 3.—Edentata* (Bruta).

Mammals with incomplete dentition, usually with numerous grinders without roots, and scratching or curved claws on the extremities.

This group which includes but few genera is characterised by the relatively low grade of development of all the systems of organs and especially by the incomplete dentition, teeth being in exceptional cases altogether wanting. Except in the case of a single Dasypod the incisors are always absent (fig. 686). When canine teeth are present they are small, blunt, and conical. The grinders also are weak and of simple structure, being without roots and enamel. Many (*Vermilinguia* and *Dasypoda*) are insectivorous, others (*Bradypoda*) phytophagous. They are all sluggish, stupid animals, with small brain without convolutions; they climb or dig holes, and at the present time only inhabit the southern zones. Except the African *Orycteropus* and the genus *Manis*, which lives in Africa and Asia, they are all confined to South America.

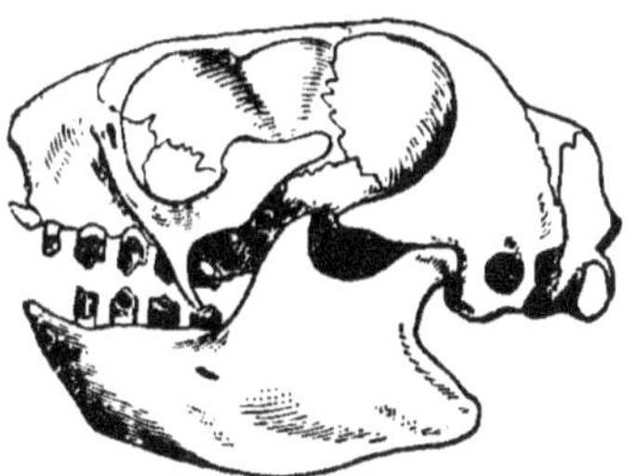

Fig. 686.—Skull of *Bradypus torquatus*.

Fam. **Vermilinguia** (Ant-eaters). With very elongated, pointed snout, narrow mouth, from which the thin, vermiform tongue can be protruded to a great distance. The jaws are weak. Teeth are altogether absent, except in *Orycteropus*, which possesses numerous grinders formed of longitudinal hollow fibres and scarcely of the hardness of bone. The legs are short, strong, and adapted for digging; they are used for scraping up the nests of Ants and Termites. They extend their long viscous tongue into the ant-heaps thus broken into: the insects bite firmly hold of it, and by the rapid retraction of the tongue become the prey of the Ant-eater. *Myrmecophaga jubata* L.; *M. tetradactyla* L., (*tamandua* Desm.). *didactyla* L., South America. *Manis*, Pangolin; *M. macrura* Erxl., West coast of Africa; *M. brachyura* Erxl. and *javanica* Desm., both found in the East Indies. *Orycteropus capensis* Geoffr., Aardvark of South Africa.

Fam. **Dasypoda** (Armadillos). The body is covered with bony plates which

* Th. Bell, Article "Edentata," in Todd's "Cyclopædia of Anatomy," vol. ii. 1836.

W. v. Rapp. "Anatomische Untersuchungen über die Edentaten." Tübingen. 1852.

are arranged in transverse rows on the back and tail, so as to form a movable dermal armour (fig. 687). The limbs are short, and with their powerful scraping claws are well adapted for burrowing. Incisor teeth are absent, except in *Dasypus sexcinctus* and in the fossil *Chlamydotherium*. Both jaws have small cylindrical grinding teeth, the number of which varies in the different forms. They inhabit South America. *Dasypus novemcinctus* L., the long-tailed Armadillo, with eight to ten bands; *D. gigas*, with upwards of a hundred teeth; *Chlamydophorus truncatus*, Hart, the Pichyciego, in the neighbourhood of Mendoza.

Fam. **Bradypoda** (Sloths). With rounded head (fig. 686) and anteriorly directed eyes, with very long anterior limbs and pectoral mammæ. The incisor teeth, and sometimes also the canines, are absent; there are three to four grinders in each half of the jaw. The large process on the jugal, descending over the lower jaw, is worthy of remark. The Sloths are exclusively arboreal: they use the curved claws at the end of the two or three closely connected digits for hanging on to branches during their strong but slow movements. On the ground they can only drag themselves along extremely awkwardly and helplessly. The

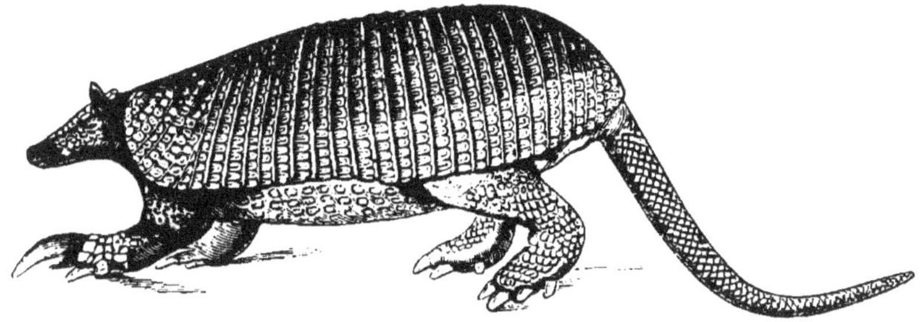

FIG. 687.—*Dasypus gigas.*

body is covered with long and coarse hair, like dry hay. They live in the forests of South America. *Bradypus tridactylus* Cuv., Aï, or three-toed Sloth; *Br. torquatus* Ill., *Cholæpus didactylus* Ill., Unau, or two-toed Sloth.

Order 4.—CETACEA.*

Aquatic Mammalia with spindle-shaped body which is not covered with hair; with fin-like front limbs and horizontal caudal fin. The posterior limbs are absent.

The Whales repeat the piscine type in the form of their body and in the articulation of their skeleton (Fig. 688). By their whole organisation they are true Mammals with warm blood and pulmonary respiration, and they are most nearly allied to the Ungulates, which they approach through the *Sirenia*. Some species attain a colossal

* D. F. Eschricht, "Zoologisch-anatomisch-physiologische Untersuchungen über die nordischen Walthiere." Leipzig, 1849.
D. F. Eschricht og J. Reinhardt, "Om Nordhvalen," Kjöbenhavn, 1861.

size, that only water can carry them, and the sea supply them with food. The cervical region is not visible externally, and the head passes directly into the cylindrical trunk, while the caudal end develops a horizontal fin, in addition to which there is often a fatty fin on the dorsal surface. Hairs are almost entirely absent in the larger forms; being only represented by bristle-like hairs on the upper lip, which are present during the whole of life, or only during the fœtal period. In the smaller species, and in the *Sirenia* there is a sparse covering of bristles. On the other hand, there is developed beneath the thick leathery skin in the subdermal cellular tissue a considerable layer of fat, which to a certain extent takes the place of fur, and serves both to prevent the loss of heat and to lower the specific gravity. The head is often elongated into the form of a snout, and is without an external ear. The eyes are strikingly small and are often placed near the angle of the mouth; the nasal apertures are shifted on to the forehead. The anterior limbs are represented by short, externally unjointed swimming fins, which can

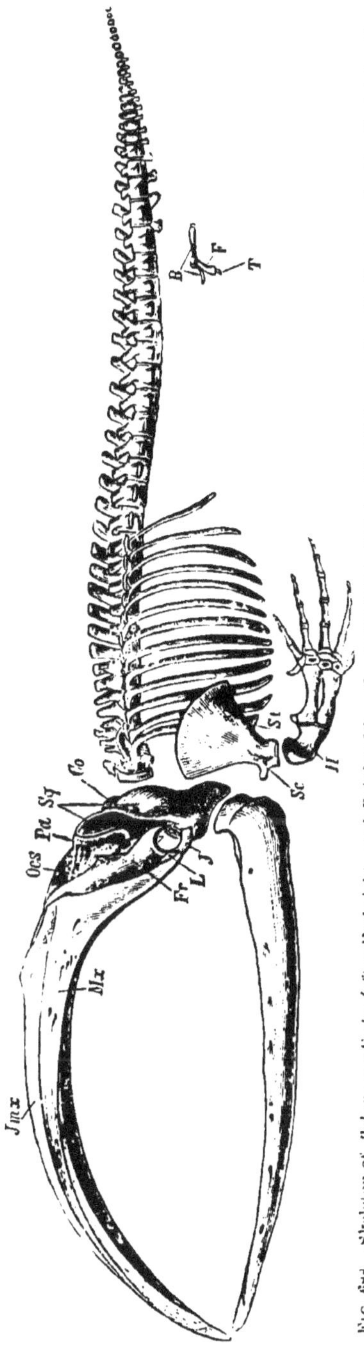

FIG. 684.—Skeleton of *Balæna mysticetus* (after Eschricht and Reinhardt). *Ocs*, Occipital bone; *Co*, Occipital condyle; *Sq*, Squamosal; *Pa*, Parietal; *Fr*, Frontal; *Jmx*, Præmaxillary; *Mx*, Maxillary; *J*, Jugal; *L*, Lachrymal; *St*, Sternum, which is only connected with the first rib; *Sc*, Scapula; *H*, Humerus; *B*, rudiment of pelvis, *F*, of femur, *T*, of tibia.

only be moved as a whole. The hind limbs are wholly wanting as external appendages.

The skull, as compared with the large facial part of the head, which is often elongated like a beak, is small and often asymmetrical, the right side being the largest. Its bones are separated by sutures and loosely connected. The parietals early fuse with the interparietals to a single bone. The hard petrous bone remains isolated from the other parts of the temporal bone. The nasal cavity, in connection with the great development of the premaxillaries, is shifted entirely on to the skull. Except in the *Sirenia*, the nasal bones are rudimentary. The jaws are frequently altogether without teeth. A change of dentition takes place only in the *Sirenia*. In the true *Cetacea* the dental germs are developed in fœtal life: but the teeth either fall out before birth (Whalebone Whales), or develope into the permanent teeth (Dolphins). Of the hind limbs traces are only sometimes found, as small bones which are interpreted as the rudiment of a pelvis: in *Balæna mysticetus* rudiments of a femur and tibia are also present (fig. 688). The single or double nasal aperture is placed more or less high up on the skull, and leads straight down into the nasal cavities, which descend as a paired, but posteriorly single, nasal canal, which at the soft palate can be shut off from the pharynx by a sphincter muscle. The view that the Whale spouts water through its nasal apertures has been proved to be erroneous. The expired aqueous vapour condenses into a cloud, and gave rise to the illusion that a column of water was ejected from the nostrils. The lungs are very spacious; they extend, like the swimming bladders of Fishes, far backward, and play an essential part in the maintainance of the horizontal position of the body in water; the diaphragm also has a corresponding horizontal position. Saccular dilatations on the aorta and pulmonary arteries, as well as the so-called arterial networks may serve as aids to respiration during diving.

The females bear a single (the smaller species rarely two) relatively far advanced young, which, however, need the care of the mother for a long time after birth. The two teats of the mammæ lie in the inguinal region, in the *Sirenia* on the thorax.

The Whales usually live together in herds. The smaller species frequent the coasts, and even enter the mouths of rivers. The larger species prefer the open sea and colder climates. They swim with great strength and speed, usually keeping near the surface. The gigantic whalebone Whales, which are entirely without teeth, but possess whalebone on the palate, feed on small marine animals, nudibranchiate Molluscs and Medusæ. The Dolphins, with their uniform carnivorous dentition, feed on larger fishes; the *Sirenia*, which are intermediate, so far as their form is concerned, between the Whales and Seals, are herbivorous. Fossil remains are found in the older tertiaries.

Sub-order 1. **Cetacea carnivora** (True Whales). Either with conical teeth in the jaws or with whalebone on the palate. The nasal apertures are placed on the forehead. The larynx projects like a pyramid into the posterior nares. The mammæ are placed on the inguinal region. The skin is devoid of hairs, and beneath it is a thick layer of fat. The limbs are movable at the shoulder joint only; their constituent bones are rigidly and immovably connected.

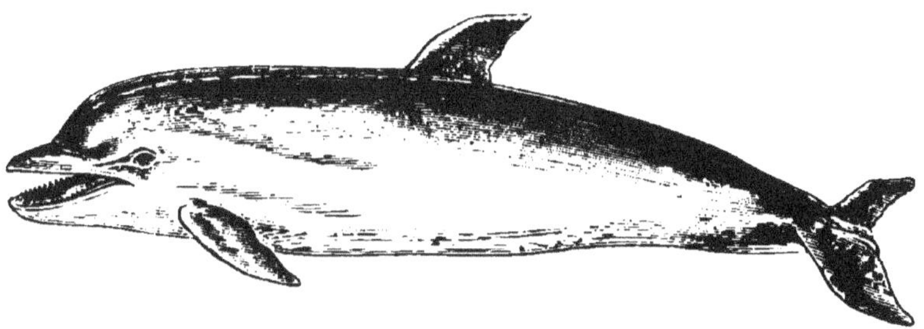

FIG. 689.—*Delphinus delphis* (règne animal).

Tribe 1. **Denticete** (Toothed Whales). Carnivorous Whales which feed principally on fish, with conical teeth in both or only in one jaw. Dentition *monophyodont*. Head of proportionate size. Nasal apertures often united to a single semilunar opening.

Fam. **Delphinidæ.** Both jaws with similar conical teeth, but not always armed along their whole length. Nasal apertures united to a semilunar spiracle. *Phocæna communis* Less., Common Porpoise, four to five feet long, ascends the mouths of rivers, lives on Fishes. European seas. *Beluga* (*Delphinapterus*) *leucas* Gray, White Fish: *Globiocephalus globiceps* Cuv., Black Fish, North Atlantic Ocean; *Delphinus delphis* L., Common Dolphin (fig. 689).

Fam. **Monodontidæ** (Narwhals). Upper jaw with only two anteriorly directed teeth which in the female are small: but in the male one of them (usually that of the left side) becomes a colossal, spirally grooved tusk. The other small teeth of both jaws fall out early. *Monodon monoceros* L., Narwhal, North Polar Sea, twenty feet long.

Fam. **Hyperoodontidæ.** With elongated beak-like snout, only one or two fully developed teeth on each side in the lower jaw. Facial bones, especially præmaxillaries, often asymmetrical. Spiracle semilunar. *Hyperoodon bidens* Flem., more than twenty feet long. North Atlantic Ocean.

Fam. **Catodontidæ**—**Physeteridæ** (Sperm-whales). Head of enormous size, being one third of the length of the body. Swollen to the extremity by the accumulation of fluid fat (spermaceti). Upper jaw without teeth. Rami of the lower jaw applied to one another and armed with a row of conical teeth. Spiracles separate. They live on Cephalopoda. *Catodon macrocephalus* Lac., Cachelot, forty to sixty feet long, North Sea; *Physeter tursio* Gray, North Atlantic Ocean.

Tribe 2. **Mysticete** (Whalebone Whales). Head very large, jaws without teeth, with whalebone (fig. 673). Œsophagus narrow; spiracles separate.

Fam. **Balænidæ.** *Cetacea* of considerable size, with enormous head, wide slit-like mouth without teeth, and double nasal openings; with very small eyes near the angle of the mouth. Two rows of horny transverse plates, frayed out at their lower edges, arise from the palate and upper jaw. These are the whalebone plates. They project vertically into the mouth, are closely packed together one behind the other, and decrease in size anteriorly and posteriorly They form a kind of sieve, which when the huge mouth is closed retains the small *Medusæ*, Nudibranchs, etc., which are taken in with the sea water, while the water flows out. *Balænoptera*, Rorquals; *B. rostrata* Fabr., North Sea; *Balæna mysticetus*, Greenland Whale, reaches a length of sixty feet.

Sub-order 2. **Cetacea herbivora, Sirenia.** With thick, sparsely bristled skin, swollen lips, and anterior nasal apertures, with pectoral mammæ. The large fins are movable at the elbow joint, and end like hands with traces of nails. Neck distinct. Dentition and internal organisation approximate to those of the Ungulates. The incisors are replaced. The grinders have a flat crown, and are always well developed in both jaws. There are no canine teeth. In the Dugong there are two tusk-like incisors in the upper jaw, while the lower incisors fall out early. They feed especially on fuci and seaweed on the sea coast.

Fam. **Sirenia.** Nasal openings placed far forward. *Manatus Australis* Tils., American Manatee. Found at the mouths of the Orinoco and Amazon: *M. senegalensis* Desm., African Manatee; *Halicore indica* Desm., Dugong, Indian Ocean and Red Sea: *Rhytina Stelleri* Cuv., Steller's Sea-cow, extinct.

Order 5.—Perissodactyla (Odd-toed Ungulates).*

Large Ungulates usually of unwieldy build; the middle digit is more developed than the others. The stomach is simple and the cæcum is very large. The dentition is usually complete.

In the earlier tertiary times the Ungulates were already a well-defined group, the smaller species of which presented approximations to the *Insectivora* (*Microchœrus*) and Rodents. The *Ungulata* are either herbivorous or omnivorous. The dentition is highly differentiated; the grinders are traversed by folds of enamel, with transverse ridges and short tubercles, which are usually worn down to an even, masticating surface. Large chisel-shaped incisors, which, however, may fall out or in the lower jaw be completely absent, are often present. There is always a gap between the incisors and the præmolars. The canines are often absent, or only present in the upper jaw, principally in the males, and then are transformed into tusk-like weapons. Even when both upper and lower canines are present, they have this significance, and are much larger in the male sex.

Among the many differences which the Ungulates present in their whole organisation and mode of life, the difference in the number of hoofs (which corresponds with that of the toes) was held to have a special value, and accordingly Multiungulates, Biungulates, and Uniungulates were distinguished as separate orders. This division was, however, by no means a natural one, since it led not only to the union of widely divergent forms as Multiungulates, but also to the separation of the Uniungulates and Biungulates from their near allies. The progress of palæontological knowledge has shown that this division is untenable. Remains of extinct forms, which partly fill up the gaps between members of the supposed orders, have been discovered. Accordingly the order of Multiungulates has been recently broken up, and two members of it—the *Proboscidea* and the *Hyracoidea*—have been placed among the *Deciduata*: and further, two orders founded upon the odd or even number of the toes—a character which had already been used by

* G. Cuvier, "Recherches sur les ossements fossiles." Third edition, Paris, 1846.

T. Rymer Jones. Article "Pachydermata" in Todd's "Cyclopædia," with Supplement by F. Spencer Cobbold, 1859.

W. Kowalevski, "Monographie des Genus Anthracotherion Cuv., und Versuch einer natürlichen Classification der fossilen Hufthiere. Palæontographica." 1873.

Cuvier—were established; these are the *Perissodactyla* (Pachydermes à doigts impaires Cuv., and Solidungula Aut.) with an odd number of toes, and the *Artiodactyla* with an even number of toes. The names do not correspond strictly with the number of the toes since there are Perissodactyles—as the *Tapir* and *Eohippus*—which have four toes on the front feet; and on the other hand, there are Artiodactyles—as *Anoplotherium tridactyle*,—which have three toes on both front and hind feet. But, when applied in a limited sense with reference to the number of the pillar-like supporting bones of the middle digit or digits, the names are in all cases suitable. In the *Perissodactyla* the unpaired central digit serves as the principal support, which in the *Artiodactyla* is afforded by the third and fourth digits which are symmetrical and similar. In most *Perissodactyla* there are three digits of which the middle one is specially strongly developed. The forms which exist at the present time are confined to the families of the *Tapiridæ*, the *Rhinoceridæ*, and the *Equidæ*, of which the last were represented as far back as the eocene epoch (*Anchitherium*) by forms which constitute connecting links between the *Palæotheridæ* and *Tapiridæ* on the one hand, and the ancestral forms of living horses on the other.

Fam. **Tapiridæ.** Short-haired Ungulates of medium size, with movable proboscis. Dentition: $i. \frac{3 \cdot 3}{3 \cdot 3}$ $c. \frac{1 \cdot 1}{1 \cdot 1}$ $p.m. \frac{4 \cdot 4}{3 \cdot 3}$ $m. \frac{3 \cdot 3}{3 \cdot 3}$. The moderately-long front legs end with four (fig. 670 *c.*), the hind legs with three digits. *Tapirus indicus* Desm., East India; *T. americanus* L., South America.

Fam. **Rhinoceridæ.** Large unwieldy Pachyderms with one or two epidermal horns on the strongly-arched nasal bones. Dentition: $i. \frac{1 \cdot 1}{1 \cdot 1}$ $c. \frac{0 \cdot 0}{0 \cdot 0}$ $p.\ m. \frac{4 \cdot 4}{4 \cdot 4}$ $m. \frac{3 \cdot 3}{3.3}$. The four incisor teeth are rudimentary, and sometimes fall out in old age. Rhinoceroses appeared in the miocene, and are also found in the pliocene and diluvium of Europe. *Rhinoceros javanus* Cuv., Java; *Rh. sumatrensis* Cuv.; *Rh. africanus* Camp.; *Rh. tichorhinus* Cuv., with bony nasal septum, and hairy skin; diluvial, found well-preserved in ice. *Rh. leptorhinus* Cuv., upper tertiaries, in Italy and south of France.

Fam. **Equidæ (Solidungula** Aut.). Long-limbed, slender Ungulates of considerable size. The three-jointed middle digit alone treads upon the ground, and its strong terminal joint is surrounded by a broad hoof (fig. 690). The second and fourth digits are either present as small accessory digits (in fossil horses), or are reduced to the metatarsal (metacarpal) bones (splint bones).

The dentition (fig. 691) consists of six upper and six lower, large chisel-shaped incisors, which are arranged in a curved line, and are distinguished by the transversely oval pit on their biting surfaces. Canine teeth are as a rule present only in the male sex in both jaws, and are small and conical.

In the fossil forms there are seven grinders on each side in each jaw; in the recent species of the *Equidæ* the number is reduced to six, but there is a small tooth in front of the first præmolar which soon falls out (Wolfszahn, Bojanus). Fossil forms first appear in the eocene (*Orohippus* still with a rudimentary fifth digit, as well as the three other digits which rested on the ground, and *Anchitherium*). They persisted in the miocene and pliocene (*Hipparion*) and then pass into the diluvial genus *Equus*, to which the domestic horses of the present day belong. *Anchitherium Dumasii* Gerv. Feet with three digits, middle

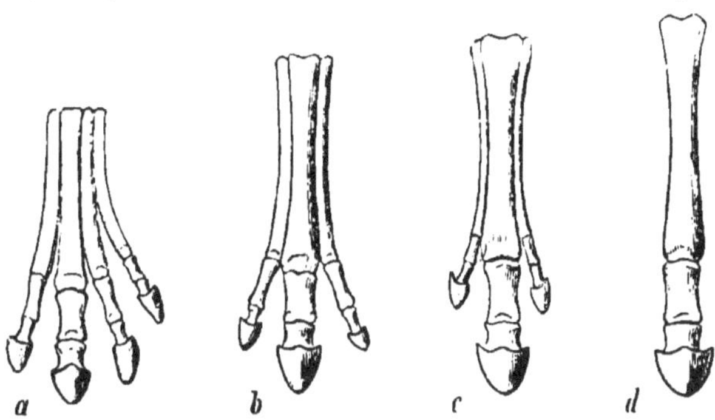

FIG. 690.—Pedal skeleton of different genera of *Equidæ* (after Marsh). *a*, Foot of *Orohippus* (Eocene); *b*, Foot of *Anchitherium* (Lower Miocene; *c*, Foot of *Hipparion* (Pliocene); *d*, Foot of the recent genus *Equus*.

digit large, remains of fifth metacarpal on the anterior limbs. The grinders are $\frac{7}{7}\cdot\left[p.\,m.\ \frac{4\cdot4}{4\cdot4},\ m.\ \frac{3\cdot3}{3\cdot3}\right]$ *Hipparion gracile* Kp., miocene. Of the seven grinders, the anterior is a simple prism with a semilunar transverse section: it is lost with the milk dentition. *Equus caballus*. Foot composed of one digit, with remains of metatarsals (metacarpals) of second and fourth digits (splint

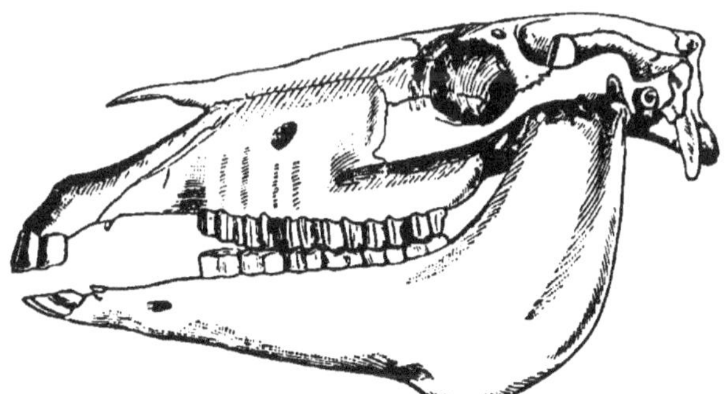

FIG. 691. Skull of *Equus caballus*.

bones): grinders are $p.\,m.\ \frac{3\cdot3}{3\cdot3},\ m.\ \frac{3\cdot3}{3\cdot3}$, with remains of an anterior seventh grinder in the milk dentition. This genus is only known in the domestic state, but is probably descended from one or several of the species of horses which lived in the diluvial period. *Asinus tæniopus* Hengl., the wild ass of South-East Asia, the ancestral form of the domestic ass (*E. asinus*); *A. hemionus* Pall., Dziguetai; *A. onager* Pall., Kulan, Mongolia. The African species (placed with the sub-genus *Hippotigris* Sm.) are *E. quagga* Gm.; *E. zebra* L., Zebra; *E. Burchelli* Fisch., Burchell's Zebra.

Order 6.—ARTIODACTYLA (PARIDIGITATA).

Ungulates with paired digits, of which the two outer are usually rudimentary, the two middle of equal size and rest on the ground; often without canine and incisor teeth in the upper jaw; grinders always with folds of enamel.

Some of the *Artiodactyla* are unwieldly and heavily built, some are slender and graceful animals, sometimes with short, sometimes with long limbs: the former with thick naked skin and a covering of stiff bristles, the latter with a thick fur. They walk mainly on the third and fourth digits, which are always larger than the two external digits, and touch the ground with their hoofs (fig. 670 *c d*). The second and fifth digits, when present, may also take part in supporting the body, but are usually rudimentary, placed behind, and do not reach the ground: they may be reduced to the remains of their metatarsals (metacarpals), and not be visible as external digits; this is the case with both of them in *Anoplotherium* and with the external one in the posterior three-toed limb of *Dicotyles*.

The animals belonging to this order may be arranged in two series:—the *Pachydermata* and the *Ruminantia*. The *Artiodactyla* were represented in the older tertiaries by forms, which with the *Palæotheridæ*, and perhaps descended from the same source as the latter, were the forerunners of the Suidæ and the Ruminants.

Sub-order 1. **Artiodactyla pachydermata.** With complete dentition, always with canine teeth and simple stomach. The metatarsal bones of the middle digits are never ankylosed.

Fam. **Anoplotheridæ.** Dentition with all three kinds of teeth which are arranged in a continuous row (*i.e.*, without diastema). *Anoplotherium commune* Cuv. Fossil.

Fam. **Suidæ** * (Setigera). With close covering of bristles, and a short proboscis-like snout. The dentition (fig. 692) includes all the kinds of teeth, but the rows of teeth are not perfectly continuous. The 4—6 incisors are placed in an obliquely horizontal position, and fall out in old age. Canines usually much elongated and triangular, and in the male as powerful weapons (tusks). There are 6—7 grinders with folded enamel in each jaw (on each side). Only the two middle digits rest on the ground, while the smaller external

* Herm. v. Nathusius, "Vorstudien für Geschichte und Zucht der Hausthiere, zunächst am Schweineschädel." Berlin, 1864; and "Die Racen des Schweines," 1860.

digits are placed behind (fig. 670 *e*). *Phacochœrus æthiopicus* Cuv., Wart-hog, South Africa; *Ph. Ælianus* Rüpp. (*Sus africanus* L.), found from Abyssinia to Guinea; *Porcus babyrussa* L., The Babyrussa, Moluccas; *Dicotyles*, Peccaries. *D. torquatus* Cuv.; *D. labiatus* Cuv., America; *Potamochœrus africanus* Schreb. (*larvatus* Fr. Cuv.), South-West Africa; *Sus europœus* Pall. (*S. scrofa* L.), Wild Boar. Dentition $i.\frac{3 \cdot 3}{3 \cdot 3},\ c.\frac{1 \cdot 1}{1 \cdot 1},\ p.m.\frac{4 \cdot 4}{4 \cdot 4},\ m.\frac{3 \cdot 3}{3 \cdot 3}$, is widely distributed from India to Western Europe and North Africa. Is the ancestral stock of a great number of races of our domestic pig; though on the other hand the pigs of China, Cochin-china, and Siam, and the Neapolitan, Hungarian, and Andalusian pigs, the small Bündtner pig and the Peat pig from the more recent stone period (neolithic) of the Swiss lake dwellings are derived (Nathusius) from a special ancestral species (*S. indicus*), which is not known with certainty in the wild state, but is allied to the *S. vittatus* Müll. Schl. from Java and Sumatra.

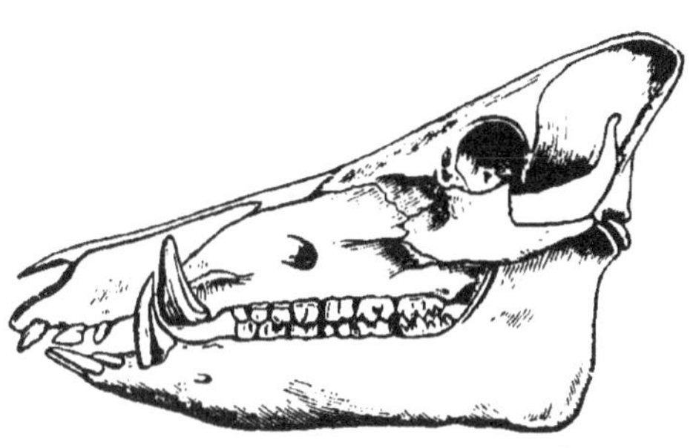

FIG. 692.—Skull of *Sus scrofa fera*.

Fam. **Obesa.** Of unwieldy shape, with large massive head, and broad truncated, swollen snout. *Hippopotamus amphibius* L. Dentition :—

$$i.\frac{2 \cdot 2}{2 \cdot 2},\ c.\frac{1 \cdot 1}{1 \cdot 1},\ p.\ m.\frac{4 \cdot 4}{4 \cdot 4},\ m.\frac{3 \cdot 3}{3 \cdot 3}.$$

H. major Cuv., diluvium of Central and Southern Europe.

Sub-order 2. **Artiodactyla ruminantia.*** With incomplete dentition (fig. 693), in which the upper incisors and canines are usually not developed. On the other hand there are eight, rarely only six shovel-shaped incisors in the lower jaw. The general form of the grinders affords tolerably constant characteristics. The quad-

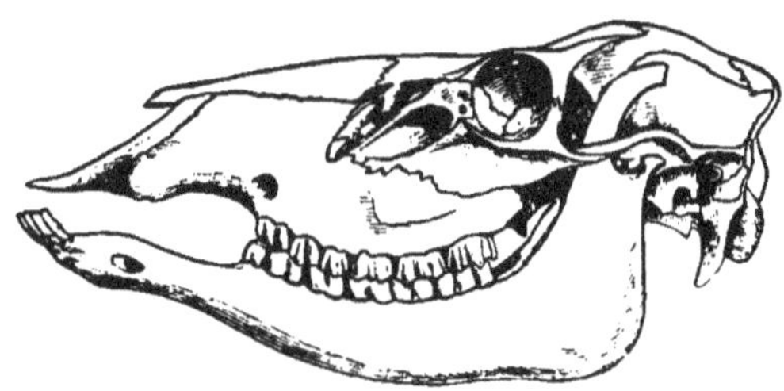

FIG. 693.—Skull of *Cervus canadensis*.

* Cf. especially G. J. Sundevall, "Methodische Uebersicht über die wiederkauenden Thiere." 2 Theile. 1847.

Rütimeyer. "Fauna der Pfahlbauten."

Rütimeyer. "Versuch einer natürlichen Geschichte des Rindes." In den *Denkschr. der Schweizer naturforsch. Gesellsch.*, Bd. 22 and 23.

rangular crown has four chief prominences, which are separated by deep valleys, which are not filled with cement, but are sometimes furnished with small accessory protuberances. The præmolars are small, and have usually only one or two protuberances. The metatarsal and metacarpal bones are always ankylosed, to form a cannon bone (fig. 670*d*).

The Ruminantia are characterised physiologically and anatomically by rumination and by the structure of the stomach and dentition which is correlated with this peculiarity. The food always consists mainly of vegetable substances, which contain only a small portion of albuminous matter, and must, therefore, be eaten in great quantities. In this relation, the division of labour between the acquisition and reception of food on the one hand, and its mastication on the other, is an advantageous arrangement, which is foreshadowed by the structure of the stomach of other Mammalia. The animal plucks and swallows its food while moving freely from place to place, and chews and masticates it when at rest. The act of rumination depends upon the complicated structure of the stomach, which is divided into four, more rarely into three, peculiarly connected divisions (fig. 694). The superficially masticated, coarse food passes through the lateral opening of the œsophageal groove, the lips of which are separate from one another, into the first and largest division of the stomach—the *paunch*, or *rumen* (fig. 694 *Ru*). Thence it passes into the small *reticulum* (*R*), a small rounded appendage of the rumen, which receives its name from the net-like folds of its inner surface. After the food is softened by the secretion which is poured into this division of the stomach, it ascends by a process resembling vomiting through the œsophagus into the mouth, and there undergoes a second, more thorough mastication: it is then returned in a semi-liquid form through the œsophageal groove, which is now closed by the coming

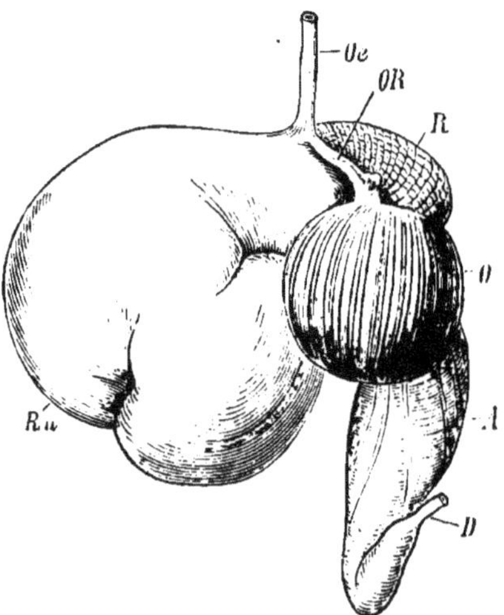

FIG. 694.—Stomach of a Calf. *Ru*, Paunch or rumen; *R*, Reticulum; *O*, Manyplies or psalterium; *A*, Abomasum or rennet stomach; *Oe*, End of œsophagus; *OR*, Œsophogeal groove; *D*, beginning of intestine.

together of its lips, into the small third division of the stomach, which is called the *psalterium* on account of the numerous leaf-like folds of its inner surface. From the psalterium the food enters the fourth stomach—the longitudinally folded *rennet stomach*, or *abomasum*, in which the digestion takes its further course under the influence of the secretion of the numerous peptic glands. In only a few cases,—in the Java Musk-deer and the *Tylopoda* (Camels and Llamas) is the psalterium absent as a separate division.

Fam. **Tylopoda**. Ruminants without accessory digits, with a callous sole covering all three phalanges behind the small hoofs. The præmaxillaries bear two, in the young animal four or six incisor teeth, while the number of the lower incisors is reduced by two. There are also strong canines in both jaws. There is no separate psalterium. *Auchenia lama* L., Llama; *A. huanaco* H. Sm.; *A. Alpaca* Gm.; *A. vicugna* Gm. All on the west coast of South America. *Camelus dromedarius* L., Dromedary, Grinding teeth. $\frac{6}{5}$: *C. bactrianus* L. two-humped Camel of Tartary, Mongolia.

Fam. **Devexa = Camelopardalidæ.** Giraffes. With very long neck, long front legs; the hind legs are much shorter, and, therefore, the back slopes backwards *Camelopardalis giraffa* Gm., wooded plains of Central Africa.

Fam. **Moschidæ**. Small, slender Ruminants, without horns, with tusk-like, strongly-developed upper canine teeth in the male. The male has between the navel and the penis a glandular sac, in which strong-smelling musk accumulates. *Moschus moschiferus* L., high mountains of Central Asia, from Thibet to Siberia; *Tragulus javanicus* Pall., without musk-bag, Island of Sunda.

Fam. **Cervidæ** (Deer). Of slender build, with horns in the males, and two rudimentary digits. In almost all cases there is a brush of hairs on the inside of the hind foot, which affords a good means of distinguishing deer from the antelopes. Upper canines often present in the male. Grinding teeth :—$\frac{6}{6}$. The horns, which, except in the Reindeer, are confined to the male, are of systematic importance; they are solid dermal bones, which are attached to a bony process of the forehead, and are detached at regular periods from the thickened circular base, cast off, and renewed. They feed on leaves, buds, and shoots. The females have four mammæ, but usually bear only one young. Australia and South Africa only are without *Cervidæ*. Fossil species first appear in the middle tertiaries. *Cervus capreolus* L., Roe-deer; *C. elaphus*, L. Red-deer; *C. canadensis* Priss., North America; *C. campestris* Cuv.; *Dama vulgaris* Brook, Fallow-deer; *Megaceros hibernicus* Ow. (*euryceros*), extinct Irish elk of the diluvium; *Alces palmatus* Klein = *C. alces* L., Moose or Elk, in North Europe, Russia, and North America; *Rangifer tarandus* H., Sm., Reindeer, antlers in both sexes, with numerous broadly-projecting prongs; they are used as beasts of burden, and for draught and riding, by the Laps.

Fam. **Cavicornia**. Without canine teeth, with $\frac{6}{6}$ grinders and hollow horns in both sexes. All are gregarious, and most polygamous.

Sub-fam. **Antilopinæ**. *Antilope dorcas* Licht., Gazelle, Africa; *Saiga saiga* Wagn., steppes of Asia; *Hippotragus equinus* Geoffr., Blaubock of South Africa; *H. oryx* Blainv.; *H. addax* Wagn., Africa; *Strepsiceros Kudu* Gray,

Africa; *Bubalis pygarga* Sundv., Buntbock, South Africa: *Catoblepas gnu*, the Gnu, plains of South Africa: *Rupicapra rupicapra* Pall., Chamois, Pyrenees and Alps.

Sub-fam. **Ovinæ.** *Ovis aries* L., domestic sheep, of which numerous races are distributed over the whole earth (German sheep, Haideschnucke, Merino, Zackelschaf, Fat-tailed sheep). There was a domesticated race of sheep in the stone age. The Moufllon, *O. musimon* Schreb., and the Argali, *O. argali* Pall., living in Northern and Central Asia have been often regarded as the wild ancestral species. *Capra*, Goats, and Ibexes. *C. ibex* L., Steinbock of the Alps; *C. ægagrus* L., Bezoar-goat, Caucasus; *C. hircus* L., Domestic Goat, numerous races, distributed everywhere.

Sub-fam. **Bovinæ.** *Ovibos moschatus* Blainv., Musk-ox of North America; *Bison europæus* Ow. (improperly called Aurochs): *B. americanus* Gm.: *Bubalus buffelus* L., Indian buffalo: *B. caffer* L.: *Pœphagus grunniens* L., Yak, Thibet and Mongolia, domesticated: *Bos gaurus* H., Sm., Gaur, East Indies: *B. indicus* L., Zebu; *B. primigenius* Boj., diluvial, lived in Cæsar's time in Germany (called "Ur" in the Nibelungen-Liede), still preserved in a semi-wild condition in Chillingham Park. Cuvier regarded them as the ancestral form of the domestic ox. *B. taurus* L.—and there can be no doubt that the Friesland or Holstein ox is to be referred to *B. primigenius*. Rütimeyer has recently shown that a second species, which existed in the diluvial period, *B. brachyceros* Ow., is to be regarded as the ancestral species of the domestic ox.

2. DECIDUATA.

Order 7. Proboscidea.

Multiungulates of very large size, with long proboscis, which functions as a prehensile organ; with compound grinding teeth, and tusks in the præmaxillæ.

The thick hide is folded, and is only sparsely covered with hairs. There is a tuft of hairs on the tail. The head is short and deep, is swollen by chambers in the frontal and parietal bones, and possesses a long movable proboscis. The occipital region descends abruptly, and almost perpendicularly. The perpendicularly-placed præmaxillæ with their large rootless tusks, are enormously developed. In the *Mastodonta* there are also two incisor teeth in the lower jaw, which soon fall out in the female, but in the male are retained as tusks. There are no canines. According to the age one, two, or sometimes even three grinding teeth are present in each jaw; they are composed of a number of parallel dental plates placed behind one another. In the genus *Elephas* these plates are connected with cement, and present on the masticatory surface transverse rhombic spaces, bounded by enamel substance. In the *Mastodonta* the cement is absent, and there are mammillary prominences on the masticatory surface. According to Owen, there are three præmolars, and the same number of molars. There are, however, never more than three, usually only two grinding teeth above the gum at the same time; for the hinder teeth, which increase in size and number of lamellæ, only appear after the anterior have fallen out. At first each half of the jaw has one grinder, behind which a second is soon developed. Later on the front one is worn out, and falls out, and then a new tooth makes its appearance behind the second. The cylindrical limbs end with five digits, which are connected as far as the small hoofs. The females have a two-horned uterus, and two thoracic mammæ. The placenta is zonary. Elephants live together in herds, and inhabit damp, shady places in the hot parts of Africa and India. They possess great intelligence, and when tamed are extremely useful animals. They were used even in antiquity as beasts of burden in war, and in the chase.

Fam. **Elephantidæ.** *Elephas indicus* Cuv. The transverse spaces of the molars in the form of narrow bands, with almost parallel, finely folded edges. Head very deep, with concave forehead and relatively small ears. Attains a height of ten to twelve feet. Ceylon and India. The Elephant of Sumatra, according to Temmink, belongs to a special species (*E. sumatranus*). *E. primigenius* Blumb., Mammoth, diluvial; *E.* (*Loxodon*) *Africanus* Blumb. The transverse spaces of the molars are lozenge-shaped and less numerous. Skull less deep. Ears very large. Central and South Africa. *Mastodon giganteum* Cuv. diluvial in North America.

The miocene genus **Dinotherium** Kp. is, according to its skull, closely allied (and therefore included with) the *Proboscidea.* Its extremities, however, have not yet been found, and the view that it is allied to the *Sirenia* cannot be directly contradicted. In the dentition there are no incisors in the præmaxillæ, while there are two large downwardly curved tusks in the lower jaw. Grinding teeth $\frac{5}{5}$, with two to three rows of transverse tubercles. *D. giganteum* Kp. Eppelsheim.

The **Lamnungia** are usually separated as a distinct order, and are placed near the Elephants. They are small and resemble the *Agouti*; in their dentition they are intermediate between the Rodents and Pachyderms, and in the formation of their feet show resemblances to the Tapirs and have, therefore, often been placed with the Pachyderms. The body is closely haired, the front feet have four digits, the posterior three, all of which are provided with small hoofs.

Hyrax. Dentition: $\frac{1\ 0\ 6\ (8)}{2\ 0\ 6\ (7)}\left[i.\ \frac{1\cdot1}{2\cdot2}\ c.\ \frac{0}{0}\ p.m.\ \frac{4\cdot4}{4\cdot4}\ m.\ \frac{3\cdot3}{3\cdot3}\right]$. *H. capensis* Schreb., Daman, Dassy, Rock-rabbit: *H. syriacus* Schreb. (Fig. 695): probably the Coney (Saphan) of the Old Testament.

Order 8. Rodentia = Glires.

With freely movable, clawed digits. Dentition with $\frac{1\ (2)}{1}$ *chisel-shaped incisors, grinding teeth with transverse enamel folds, and without canines.*

The *Rodentia* are a large group of small, active Mammalia. They are easily recognizable by the dentition and structure of the teeth. The order, nevertheless, includes many forms transitional to the *Insectivora*. Rodents are plantigrade animals, with freely movable digits, which are usually armed with claws, only rarely with arched nails, or even hoof-like nails. They all feed on vegetable, usually hard substances, especially on stalks, roots, seeds, and fruits. Only a few are omnivorous.

Fig. 695.—*Hyrax syriacus.*

There are two large chisel-shaped, somewhat curved incisors (fig. 696), which possess enamel only on their anterior surfaces. The posterior surface is, therefore, quickly worn away by use, and the more so since the arrangement of the narrow, laterally compressed glenoid cavity necessitates an antero-posterior movement of the lower jaw during mastication. The wearing away is compensated by a proportionate, continous growth of the tooth. The grinders, which are separated from the incisors by a wide gap, possess usually transversely arranged folds of enamel, and are only tuberculated when the animal is omnivorous. When these teeth are being used the lower jaw is drawn so far back that the incisors are not rubbed against one another, and the lower jaw is moved backwards and

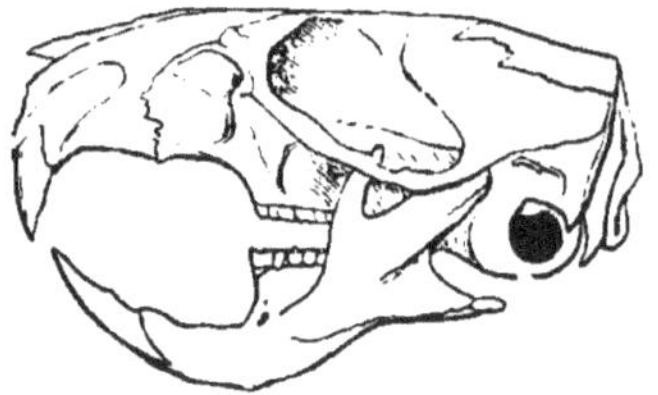

Fig. 696.—Skull of *Cricetus vulgaris* (after Giebel; Bronn's Classen und Ordnungen).

forwards in the longitudinal direction, in correspondence with the position of the transverse folds of enamel. Many of them build nests, dig out complicated burrows, and lay up stores for the winter. The latter usually possess cheek-pouches. Some fall into a deep winter sleep at the cold time of the year, others migrate in large flocks. They produce numerous young, some of them four or six litters in the year, and possess, accordingly, a great number of abdominal and thoracic mammæ. Uterus usually completely divided; placenta discoidal.

Fam. **Leporidæ.** With long ears, powerful hind legs, and short tail. Dentition: $\frac{1}{1}\ \frac{0}{0}\ \frac{5\ (6)}{5}\left[i.\ \frac{2\cdot2}{1\cdot1}\ c.\ \frac{0\cdot0}{0\cdot0}\ p.\ m.\ \frac{3\cdot3}{\ }\ m.\ \frac{3\cdot3}{3\cdot3}\right]$. In the præmaxillæ there are two posterior accessory incisors (*Duplicidentata*). *Lepus timidus* L., Hare: *L. variabilis* Pall. Alpine hare: *L. cuniculus* K., Rabbit; *Lagomys*, Pikas. *L. alpinus* F. Cuv., barely a foot in length, Siberia: *L. princeps* Richards, Rocky Mountains.

Fam. **Subungulata.** Grinders, $\frac{4}{4}$. The feet have naked soles, and end in front with four, and behind usually with three toes. *Cavia aperea* L., Aperea, in Brazil and Paraguay; *C. cobaya* Schreb., the tame Guinea-pig; *Cœlogenys paca* L., the Paca, Brazil: *Dasyprocta aguti* L., the Agouti: *Hydrochœrus capybara* Erxl., the Capybara, four feet in length, the largest of living rodents.

Fam. **Aculeata.** With short, obtuse snout, and spines on the dorsal side of the body. *Cercolabes prehensilis* L., the Kuandu, Brazil: *Erethizon dorsatus* L., North America; *Hystrix cristatus* L., Porcupine, Italy and Spain.

Fam. **Octodontidæ.** *Octodon Cumingii* Benn., Chili; *Myopotamus coypus* Geoff., the Coypu, distributed from Brazil to Patagonia.

Fam. **Lagostomidæ,** Chinchillas. *Eriomys laniyera* Benn., the Chinchilla Chili; *Lagidium Cuvieri* Wagn.; *Lagostomus trichodactylus* Brookes, Viskatscha,

Fam. **Dipodæ.** Jerboas. With very long hind legs, which serve for jumping, and large, usually tufted, jumping tail. *Jaculus labradorius* Wagn., Hüpfmaus; *Dipus Ægyptius* Hempr. Ehrnb., Arabia: *D. sagitta* Schreb., Sea of Aral: *Pedetes caffer* Ill., Cape jumping hare (Springhase), South Africa.

Fam. **Muridæ.** Mice. Grinders: $\frac{3}{3}$. With large eyes and ears, and long sometimes hairy, sometimes ringed, scaly tail. *Cricetus frumentarius* Pall., the Hamster; with internal cheek pouches; constructs subterranean passages and chambers, in which it accumulates winter provisions. It passes through a short winter sleep, and is very hurtful to corn-fields. *Mus rattus*, L., House-rat, Black Rat: *M. decumanus* Pall., Grey Rat: *M. musculus* L., House Mouse; *M. minutus* Pall. (*pendulinus*); *Hydromys chrysogaster* Geoffr., Australia.

Fam. **Arvicolidæ.** Voles. With thick, broad head, rootless grinders, short, hairy ears and tail. *Arvicola amphibius* L., Water-rat: *A. arvalis* Pall., Field-mouse; *A. agrestis* L.: *Hypudæus glareolus* Schr.: *Myodes lemmus* L., the Lemming, on high mountains of Norway and Sweden, known by its migrations in immense flocks before the approach of the cold weather. *Fiber zibethicus* L., North America.

Fam. **Georhychidæ.** *Spalax typhlus* Pall., Blindmouse. South East Europe; *Georhychus capensis* Pall.

Fam. **Castoridæ.** Beavers. Grinders: $\frac{4}{4}$. With flat, scaled swimming-tail. Two glandular sacs which secrete the castoreum open into the prepuce. *Castor fiber* L., the common Beaver.

Fam. **Myoxidæ.** Dormice. Connecting links between the Mice and Squirrels. *Myoxus Glis* Schreb., Dormouse; *M.* (*Muscardinus*) *avellanarius* L.; *M.* (*Eliomys*) *nitela* Schreb.

Fam. **Sciuridæ.** Squirrels. Grinders: $\frac{5\,(4)}{4}$. *Sciurus vulgaris* L., Europe and North Asia; *Tamias striatus* L.; *Pteromys volans* L., Flying Squirrel, Siberia; *Spermophilus Citillus* L., East Europe; *Arctomys marmota* Schreb., the Marmot, Alps; *A. bobac* Schreb., Poland.

Order 9. Insectivora.

Plantigrade Mammals with clawed digits, with complete dentition, small canines, and sharp-pointed grinders.

Small Mammals, which resemble in their appearance different types of Rodents, but in structure and mode of life lead to the Carnivora. The head ends with a pointed snout, which is often elongated like a proboscis. The external ears are sometimes large, and sometimes reduced; the eyes are always small and reduced, and sometimes hidden beneath the fur. The dentition (fig. 697) is especially important, and resembles that of the insectivorous Bats. All the three kinds of teeth are present. The incisors are usually of considerable size, but of variable number. The canines are not always clearly distinguished from the incisors and the front grinders. The grinders are numerous, and have sharply-tuberculated crowns, and are divided into anterior præmolars, of which the posterior corresponds to the carnassial tooth of the true Carnivora, and into posterior molars, which are characterized by being composed of prismatic divisions. All are plantigrades, with naked soles, and usually five-toed feet, armed with strong claws. The mammæ are abdominal; the placenta is discoidal. They feed on small animals, principally on Insects and Worms, which they destroy in great numbers, thereby benefitting man.

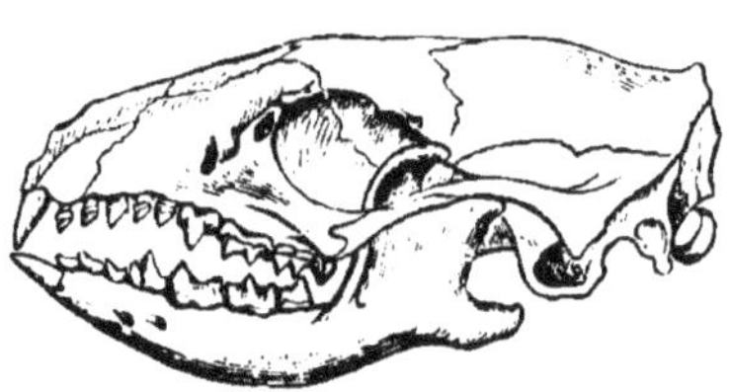

Fig. 697.—Skull of *Erinaceus europæus*.

Fam. **Erinaceidæ** (*Hedgehogs*). Back covered with stiff bristles and spines, which afford a complete protection to the animal when the body is rolled into a ball by the action of the strongly-developed cutaneous muscles. *Erinaceus europæus* L., Hedgehog. Urchin, with 36 teeth: $\frac{3\ 7}{3\ 5}\left[i.\frac{3\cdot3}{3\cdot3}\ c.\frac{0\cdot0}{0\cdot0}\ p.m.\frac{4\cdot4}{2\cdot2}\ m.\frac{3\cdot3}{3\cdot3}\right]$. Digs holes with two exits about a foot deep in the earth and hibernates. *E. fossilis* Schreb., Cave Hedgehog; *Centetes ecaudatus* Wagn., Tanrec, Madagascar; snout elongated like a proboscis.

Fam. **Soricidæ** (Shrews). With proboscis-like snout, soft fur, and tail covered with short hairs. Peculiar glands on the sides of the body or at the root of the tail give the true Shrews an unpleasant musty smell. *Cladobates tana* Wagn.: *Cl. murinus* Müll. Schl., Borneo; *Macroscelides typicus* Smith. South Africa. *Sorex*; with 28 to 33 teeth: *S. vulgaris* L., Common Shrew-mouse; *S. fodiens* Pall., Water Shrew-mouse: *S. pygmæus* Pall. *Myogale moschata* Pall., the Desman, as large as the Hamster. South East Russia.

Fam. **Talpidæ** (Moles). With short, laterally-directed digging feet, soft velvety fur, and proboscis. *Talpa*. Dentition: $\frac{3\ 1\ 3\ |\ 4}{4\ 1\ 2\ |\ 4}\left(i.\frac{3\cdot3}{4\cdot4}\ c.\frac{1\cdot1}{1\cdot1}\ p.m.\frac{3\cdot3}{2\cdot2}\ m.\frac{4\cdot4}{4\cdot4} = 44\right)$. *T. europæa* L., Mole, constructs an ingenious subterranean dwelling, which communicates by a long gallery with the daily multiplying burrows which the animal makes in hunting for food. The nest consists of a softly-lined central chamber and two circular passages, of which the upper one is the smaller, and communicates by three passages with the central chamber, while the lower and larger lies in the same plane as the chamber. Five or six communicating passages pass from the upper circular passage into the lower, from which a number of horizontal passages radiate, and usually curve round and open into the common gallery. *T. cæca* L., the Blind Mole of South Europe; *Chrysochlorys inaurata* Schreb., Cape Golden Mole; *Condylura cristata* L., the North American Star-nosed Mole; *Scalops aquaticus* L., Water Mole, North America.

Order 10.—PINNIPEDIA.

Hairy aquatic Mammalia with five-toed fin-like feet, of which the posterior are directed backwards; with complete dentition; without caudal fin.

The body is elongated, spindle-shaped, possesses four fin-like feet, and ends with a short conical tail. The head is very small in proportion to the body, of globular shape, with swollen lips, and usually without external ears. The surface of the body is covered with a short, but close, smooth fur. The short limbs end with broad swimming fins, which possess five digits, armed with blunt or sharp claws. The movements on land are effected in the following way: the animal raises the anterior part of its body, and throws it forward; it employs the two front feet as supports to fix the body, and then by bending its back drags the hinder part forward. In swimming the anterior extremities are applied to the body, and are used as rudders, while the hind feet serve as swimming fins.

The dentition, which is usually complete, indicates a predatory mode of life, resembling that of the true *Carnivora*, to which order the *Pinnipedia* are also allied in other anatomical characters, as in the possession of a two-horned uterus and a zonary placenta. With regard to the dentition, however, there are essential differences between the families of the Walruses and Seals. The Seals have $\frac{3}{2}$, more rarely $\frac{2}{1}$ chisel-shaped incisors, small canines in each jaw, and $\frac{6-5}{5}$ jagged grinders, of which one or two are true molars. The Walruses have a complete dentition only in the young stage; the incisors, which at first are $\frac{3}{3}$, are soon reduced to $\frac{1}{1}$ in the præmaxilla. The canines in the upper jaw are transformed into huge tusks, which are used when the animal crawls on land to fix the anterior part of the body. There are five grinders in the upper jaw and four in the lower, with masticatory surfaces which wear away, in course of time, obliquely from within outwards. The change of teeth usually takes place during embryonic life. The Seals live principally on fish; the Walruses on sea-weeds, Crustacea, and Molluscs, the shells of which they crush with their grinders.

Fam. **Phocidæ** (Seals). Pinnipedes with complete dentition, short canines, and jagged molars. *Halichærus grypus* Nilss., Utsel. *Phoca vitulina* L., Common Seal, $\frac{3}{2}\ \frac{1}{1}\ \frac{5}{5}$; *Ph. grænlandica* Nilss., Northern Seas; *Cystophora cristata* Fabr., Greenland; *Otaria jubata* Forst., Sea-lion of South America; *O.* (*Callorhinus*) *ursina* Per., Sea-bear, Greenland.

Fam. **Trichechidæ** (Walruses). The upper canines are large, rootless, downwardly-directed tusks. The grinders are at first bluntly pointed, but are gradually worn down, and eventually reduced to three in each ramus, while in the upper jaw there is an internally placed incisor. *Trichechus rosmarus* L., Walrus of the Polar Seas. Dentition: $\frac{2\,(1)}{2\,(0)}\ \frac{1}{0}\ \frac{3\,(4)}{3\,(4)}$.

Order 11.—Carnivora = Feræ.

Carnivorous Mammalia with predatory dentition, without or with a rudimentary clavicle, and with strongly-clawed digits.

The Carnivora are distinguished from the Insectivora by their larger size, and by their genuine carnivorus dentition (fig. 698). The dentition contains all three kinds of teeth: above and below six small incisors with single roots, and at their sides a long, conical pointed canine tooth; then a number of grinders, which are distinguished into præmolars (*d. spurii*), a carnassial tooth (*d. sectorius*), and molars (*d. molares*). We never find prismatic grinders with needle-shaped points on the crown, as in the *Insectivora*. The compressed and sharp-edged præmolars are the least developed; the characteristic carnassial teeth are distinguished by the size of their cutting, usually two-or three-toothed crown, and often by the possession of a posterior bluntly-tuberculated lobe (upper carnassial tooth). The lower carnassial tooth is always the first molar, while the upper is the last præmolar. The true molars have several roots; they possess bluntly-tuberculated crowns, and vary in size and number. The external form of the skull and dentition, the high temporal crest of the skull for the attachment of the large temporal muscle, and the marked curvature of the zygomatic arch for the passage of the same, the transverse articular cavity (glenoid cavity) of the temporal bone, and the cylindrical articular head of the lower jaw, which restricts the motion of the jaw to the vertical plane and excludes lateral movements,—are characters which are common to all the Carnivora, and coincide with the form of the dentition.

FIG. 698.—Skull of *Felis Leo*.

The limbs end with four or five freely-movable digits, which are armed with strong cutting claws (accessory to the dental apparatus), and in the front limbs are also used for seizing the prey. Only a few Carnivora, as the Bears, are true plantigrades resting the whole sole of the foot on the ground; others, as the *Viverridæ*, only place the anterior part of the sole (the digits and metacarpals) on the ground; the most agile of the Carnivora, on the other hand, are digitigrade, *e.g.*, the *Felidæ* (fig. 699.) The uterus is two-horned, the placenta zonary. Most *Carnivora* have peculiar anal glands, which emit an intense odour. The Carnivora are found in all parts of the world, except Australia, where they are replaced by the carnivorous Marsupials. Fossil remains first appear in the Eocene.

Fam. **Ursidæ** (Bear-like Carnivora). Plantigrades of unwieldy form, with elongated snout, and broad, usually quite naked soles, and five digits. *Ursus* L., Bear. Of unwieldy build, with very short tail. Grinders: $\frac{3}{4}\ \frac{1}{1}\ \frac{2}{2}$ [*i.e. p.m.* $\frac{3\cdot3}{4\cdot4}$ *carnassial*, $\frac{1\cdot1}{1\cdot1}$ *m.* $\frac{2\cdot2}{2\cdot2}$]. The front grinders fall out early. *U. maritimus* Desm., Polar Bear. Northern Polar Sea; *U. arctos* L., Brown Bear; *Procyon lotor* L., Washing Racoon, is wont to dip its food in water, North America: *Nasua rufa* Desm., the Coatimondi (Rüsselbär), Brazil; *Cercoleptes caudivolvulus* Ill., the Kinkajou (Wickelbär), Guiana and Peru.

Fam. **Mustelidæ** (Marten-like Carnivora). Some are plantigrade (Badger), some semiplantigrade; body elongated, with short legs, and five-toed feet with non-retractile claws; only one molar behind the large carnassial. *Meles taxus* Pall., Badger. *Mephitis mesomelas* Licht., Skunk (Stinkthier), North America. *Gulo borealis* Briss., Glutton; *Mustela martes* L., Pine-marten, grinders: $\frac{3}{4}\ \frac{1}{1}\ \frac{1}{1}$; *M. foina* Briss., House-marten: *M. zibelina* L., Sable-marten, Siberia; *Putorius putorius* L.; *P. vulgaris* L., Weasel; *P. erminea* L., Ermine; *P. lutreola* L., (Nörz); *Lutra vulgaris* Erxl., Common Otter; *L. canadensis* Schreb., North America; *Enhydris marina* Erxl., Sea otter, West islands of North America.

Fam. **Viverridæ** (Civets). Body elongated, sometimes cat-like, sometimes marten-like in form; with pointed snout and long tail, which is sometimes rolled up into the form of a ring; they are either plantigrade, semiplantigrade, or digitigrade. The feet have five digits, and the claws are usually entirely, or half retractile. Half the foot, or only the toes, are placed on the ground. *Viverra zibetha* L. Grinders: $\frac{3}{4}\ \frac{1}{1}\ \frac{2}{2}$. With large glandular sac between the anus and external generatives, in which the oily secretion known as "Civet," and used as perfume and for external application in medicine, accumulates. *V. civetta* Schreb., the African Civet-cat, domesticated in Egypt and Abyssinia; *V. genetta* L., the Genet, South Europe; *Herpestes ichneumon* L., the Mongoose or Ichneumon (Pharaonsratte), Egypt and South Europe.

Fam. **Canidæ** (Dogs). Digitigrades, with non-retractile claws, five-toed front feet, and four-toed hind feet. *Canis lupus* L., Wolf. Grinders: $\frac{3}{4}\ \frac{1}{1}\ \frac{2}{2(1)}$. In Europe, especially Norway and Sweden, also in Asia; *C. latrans* Sm., the Prairie Wolf; *C. aureus* L., the Jackal: *C. familiaris* L., Dog (*cauda sinistrorsum recurvata* L.). The numerous races, which are known only in the domesticated and run-wild state, have certainly been derived from more than one ancestral species. *C. vulpes* L., Fox; *C. lagopus* L., Polar Fox, gray in summer, white in winter.

Fam. **Hyænidæ** (Hyæna-like Carnivora). Digitigrades with sloping back which bears a mane of elongated hairs. The dentition resembles that of the Cats in the small development of the molars, of which there is only one in the upper jaw. *Hyæna striata* Zimm., the striped Hyæna of Africa and parts of India. Grinders $\frac{3}{3}\ \frac{1}{1}\ \frac{1}{0}$. *H. crocuta* Zimm., the spotted Hyæna of South Africa.

Fam. **Felidæ** (Cats). Digitigrades of slender build, adapted for jumping; with short jaws, and only few grinders—four in the upper, three in the lower jaw. Molars absent, except one small tooth above projecting transversely inwards. The canines and carnassials are, however, so much the more powerfully developed. The anterior of the two præmolars of the upper jaw is reduced. In walking, the last phalanx of each digit is raised vertically, so that it does not touch the ground, and the claws are protected from wear. *Felis leo* L., Lion; grinders, $\frac{2}{2}\frac{1}{1}\frac{1}{0}$. *F. concolor* L., Puma; *F. tigris* L., Tiger, Asia: *F. onca* L., Jaguar, Paraguay and Uraguay: *F. pardalis* L., Panther-cat, South America; *F. pardus* L., Panther or Leopard, Africa and West Asia; *F. catus* L., Wild Cat, grey, with stripes, and transverse bars and vertical pupils, Central and Northern Europe; *F. maniculata* Rüpp., Nubian Cat: *F. domestica* L., the Domestic Cat, only known in domesticated state, probably descended from several species; *Cynailurus guttata* Herrm., and *jubata* Schreb. (Guepardc); *F. serval* L., Serval, Senegal: *Lynx lynx* L., with a tuft of hairs on the ear: *L. caracal* Schreb., Asia and Persia.

Order 12.—Chiroptera.

Mammals with complete dentition: with a flying membrane (patagium) extending between the limbs and the sides of the body, and between the elongated fingers of the fore-limb: with two thoracic mammæ.

Amongst the Marsupials (*Petaurus*), the Rodents (*Pteromys*), and the *Prosimiæ* (*Galeopithecus*), there are a number of forms which are assisted in jumping by a kind of parachute, which consists of a cutaneous expansion—the *patagium*—stretched between the limbs on each side. The patagium is much more completely developed in the Bats; in these animals it is continued over the extraordinarily elongated fingers of the hand, and in virtue of its enormous size and its great elasticity constitutes a true organ of flight, which, however, differs considerably from that of birds. The tail is included in the patagium, but the thumb and the foot are separate from it (fig. 699). The thumb has two phalanges, and is armed with a claw, as also are the five digits of the foot. Peculiar outgrowths of the skin of the head, lobe-like appendages of the nose and ear, often give the face a very strange expression. Except upon these appendages, and on the thin elastic patagia, both of which have a large supply of nerves and a delicate sense of touch, the surface of the body is closely covered with hair. The skeleton (fig. 699) is light, and displays in its structure the Mammalian type: it is, however, distinguished from that of other Mammalia by the rigidity of the thoracic framework, and by the length of the strongly developed sacrum, with which the ischia are united. The possession of a *crista sterni*, and the ossification of the sternocostal cartilages, and some other peculiarities recall the skeleton of the birds. The femur and crus (middle division of leg) are, unlike the corresponding parts of the arm, very short. A spurlike process, called the *calcar*, projects from the inner side of the ankle-joint, and serves for the support of the femoral and caudal part of the patagium. Of the sense organs the eyes are relatively slightly developed, but on the contrary, the senses of smell, of hearing, and of touch, are, in correspondence with the nocturnal habits, of great importance. Spallanzani has shown that Bats which have been made blind are able to avoid all obstacles in their flight with great skill. The sense of hearing is not less developed; it is essentially assisted by a large

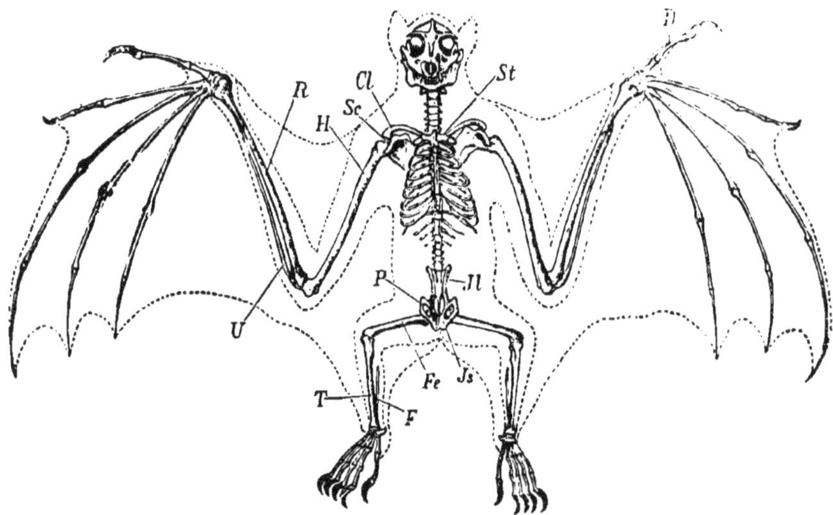

FIG. 699.—Skeleton of *Pteropus* (after Owen, slightly altered). *St*, Sternum; *Cl*, Clavicle; *Sc*, Scapula; *H*, Humerus; *R*, Radius; *U*, Ulna; *D*, Thumb; *Jl*, Ilium; *P*, Pubis; *Js*, Ischium; *Fe*, Femur; *T*, Tibia; *F*, Fibula.

pinna, which is provided with special lobes, and can be closed by a valve.

Bats are nocturnal animals, and feed on Insects. Amongst the exotic species there are some which attack Birds and Mammals, and suck their blood (Vampire); other, and especially the larger species live on fruit. Many fall into a winter sleep. They bear only one or two young at a birth, suckle them with their pectoral mammary glands, and carry them about during their flight.

Sub-order 1. **Frugivora** (Fruit-eating Bats). With elongated dog-like head, small ears, and short rudimentary tail. The index finger, which has three phalanges, often bears a claw as well as the thumb (fig. 699). The other fingers have two phalanges, and are without claws. The dentition has four or two incisors, which often fall out, one canine, and four or six grinders with flat, bluntly-tuberculated crowns. The præmaxillæ are loosely united with one another, and with the maxillæ. The tongue is beset with a number of backwardly-directed, horny spines. They inhabit the forests of the hot regions of Africa, East India, and Australia. Many of them are eaten on account of their well-flavoured flesh.

Fam. **Pteropidæ** (Flying Foxes). The small ears and the nose are without the cutaneous appendages and valves. *Pteropus edulis* Geoff., Kalong, East Indies. Dentition: $\frac{2}{1}\frac{1}{1}\frac{2}{3}\Big|\frac{3}{3}$. *Harpyia cephalotes* Pall., Amboina.

Sub-order 2. **Insectivora** (Insect-eating Bats). The snout is short; the ears are large, and frequently covered with valves. Grinding teeth sharply tuberculated or cutting, and composed of three-sided pyramids. The thumb alone bears a claw. Some of them live on insects: some on the blood of warm-blooded animals.

FIG. 700.—Head of *Phyllostoma* (*Vampyrus*) *spectrum* (règne animal).

Tribe 1. **Gymnorhina.** The nose is smooth, and without foliaceous appendages. The præmaxillæ are firmly ankylosed with the maxillæ. The ears sometimes meet one another on the top of the head, and are sometimes widely separate; the valves of the ears vary considerably.

Fam. **Vespertilionidæ.** The long and slender tail is entirely included in the interfemoral membrane. *Plecotus auritus* L., Long-eared Bat: *Synotus barbastellus* Schreb., the Barbastelle: *Vespertilio murinus* Schreb. Dentition: $\frac{2}{3}\frac{1}{1}\frac{3}{3}\Big|\frac{3}{3}$. *Vesperugo noctula* Schreb.; *V. pipistrellus* Schreb.

Fam. **Taphozoidæ.** Tail shorter than the interfemoral membrane. The base of the thumb is within the patagium. *Taphozous leucopterus* Temm., South Africa; *Mystacina tuberculata* Gray, New Zealand.

Tribe 2. **Phyllorhina.** Cutaneous appendages are spread on and over the nose. They consist of a horseshoe-shaped anterior leaf (*ferrum equinum*), a medium saddle (*sella*), and a posterior, usually vertical lancet-shaped leaf (*lancet*) (fig. 700). The præmaxillæ are not ankylosed with the maxillæ. Ears separated. Feed partly on the blood of warm-blooded Vertebrates, which they attack while sleeping.

Fam. **Rhinolophidæ.** Ears separated without tragus. *Rhinolophus hipposideros* Bechst., small Horseshoe-nose. *Rh. ferrum equinum* Shreb., large Horseshoe-nose; *Phyllorhina gigas* Wagn., Guinea.

Fam. **Megadermidæ.** The large ears approximated, with long tragus. *Megaderma lyra* Geoffr.; *Rhinopoma microphyllum* Geoffr., Egypt.

Fam. **Phyllostomidæ.** With thick head and long truncated tongue. Nasal apparatus usually with upright lancet. Ears almost always separate, with ear-valve. *Phyllostoma hastatum* Pall., Brazil. Dentition: $\frac{2}{2}\frac{1}{1}\frac{5}{5}$. *Vampyrus spectrum* L., the Vampire of Central America.

Order 13.—Prosimiæ (Lemurs).

Arboreal animals of the Old World, with complete insectivor-like dentition, with hands and prehensile feet, without a closed orbit, and with thoracic and abdominal mammæ.

Fig. 701.—*Chiromys madagascariensis* from Vogt and Specht.

The dentition holds a position intermediate between that of the Carnivora and that of the Insectivora. There are usually four incisors, of which the upper are separated by a wide gap, while the lower project more or less horizontally; there are projecting canines, and numerous sharply-tuberculated grinders. The lower jaw is relatively weak, and its two rami remain permanently separate at the symphysis. The orbits are, indeed, completely surrounded by a bridge of bone, but are not shut off from the temporal fossa as they are in the Apes. In many Lemurs the clitoris is perforated by the urethra. Uterus two-horned or double. There are usually several pairs of teats. The anterior limbs are shorter than the posterior. The great toe, like the thumb, is opposable except in *Galeopithecus*. They thus have the hands and prehensile feet of Apes, and also flat nails on the extremities of the fingers and toes, except in *Galeopithecus* and *Chiromys* (fig. 701), which have claws on all the fingers and toes.

The second toe of the foot alone forms an exception, being armed with a long claw. There may, however, be a claw on the middle toe. The tail presents great variations in size and development, but can never be used as a prehensile tail. The *Prosimiæ* inhabit exclusively the hot regions of the Old World, principally Madagascar, Africa, and South Asia. They are almost all nocturnal in their habits, climb with great skill, but slowly and lazily, and feed on Insects and small Vertebrates.

Fam. **Galeopithecidæ** = **Dermoptera**. With closely-furred patagium, which they use as a parachute in jumping. Lower incisors pectinated and inclined forwards. They are most nearly allied to the Makis; they are nocturnal in their habits, and live partly on fruit and partly on Insects. During the day they sleep in their hiding-places, suspended like Bats. *Galeopithecus volans* L., the flying Maki. Island of Sunda.

Fam. **Chiromyidæ**. With rodent-like dentition, and with claw-like nails on the fingers and toes. The large opposable great toe of the hind foot alone has a flat nail. In the præmaxillæ and in the lower jaw there are two large rootless incisors, which project obliquely forwards, and, unlike those of Rodents, are covered with enamel on both sides. *Chiromys madagascariensis* Desm., the Aye-aye, permanent dentition $\frac{1\ 0\ 1}{1\ 0\ 3}$ (Fig. 701).

FIG. 702.—*Otolicnus galago* from Vogt and Specht.

Fam. **Tarsiidæ** (Long-footed animals). With thick head, large ears and eyes, short snout, much elonged proximal tarsal bones (calcaneum and naviculare), and long tail. The middle toe as well as the second may be armed with a claw (*Tarsius*). In their appearance they resemble the Hazel-mice (*Myoxus avellanarius*), in their movements the Squirrels. *Tarsius spectrum* Geoffr., (Gespenstmaki).

Fam. **Lemuridæ**. The lower incisors directed horizontally forwards. Only the second hind toe has a claw-like nail. *Stenops gracilis* v.d. Hoev., the slender Lori, Ceylon; *Nycticebus tardigradus* L., the unwieldy Lori, East Indies and Island of Sunda; *Lichanotus brevicaudatus* Geoffr., Indri of Madagascar; *Propithecus diadema* Wagn., Vlissmaki of Madagascar. *Lemur catta* L., *macaco* L., *mongoz* L., Makis, Madagascar; Dentition $\frac{2(0)\ 1\ 3\ \ 3}{2\ \ 1\ 3\ \ 3}$. *Otolicnus senegalensis* Geoffr., the common Galago (Fig. 702), Africa.

Order 14.—PRIMATES L., PITHECI* (APES).

With complete dentition and $\frac{2}{2}$ chisel-shaped incisors in closed series on each side; usually with prehensile feet on the hind limbs, and as a rule, with hands on the front limbs; with closed orbit and two pectoral mammæ.

The Apes, as a rule, are of slender build, corresponding with their quick and easy movements as arboreal animals; there are, however, heavy unwieldy forms, which, as the Baboons (*Cynocephalidæ*), avoid forests and inhabit rocky mountain regions. The body is more or less closely covered with hairs, except on the face, which is naked in parts, and the callous parts of the buttocks (ischial callosities). The hair is often longer in places on the head and trunk, forming tufts and manes. The human look of the face depends mainly upon the slight prominence of the jaws, and is greatest in the early part of life. The facial angle of the adults rarely exceeds 30°; but in one case, viz., in *Chrysothrix sciurea*, is almost twice that size. With the increase in size of the brain, the cranial capsule becomes rounder, and the *foramen magnum* gradually moves from the posterior part on to the lower surface. The pinna of the ear also has a human look, as has also the position of the anteriorly directed eyes. The orbits are completely closed towards the temporal fossæ. Further, the mammæ are two in number and pectoral in position, as in Man. Finally, the dentition and the structure of the extremities (fig. 703) are so similar to those of Man, that he has been placed in the same order as the Apes. There are in each jaw four chisel-shaped incisors, which, as in Man, are placed close to one another without any interval. There are projecting conical canines, and in the Apes of the Old World five, in those of the New World six bluntly-tuberculated grinders, the form of which indicates that the diet is mainly vegetable. The size of the canines, which project almost as much as those of the *Carnivora*, occasions the presence of a considerable gap between the canines and the first præmolars of the lower jaw.

The anterior limbs are usually longer than the posterior. A clavicle is always present. The forearm permits of a rotation of the radius round the ulna, and accordingly of the pronation and supination of the hand, the fingers of which, except in the

* Vrolik, "Recherches d'anatomie comp. sur le Chimpanzé," Amsterdam, 1841.
G. L. Duvernoy, "Des caractères anatomiques des grands Singes pseudo-anthropomorphes." *Arch. du Muséum*, Tom. VIII., 1855.
R. Owen, "Osteology of the Anthropomorphidæ," *Transact. Zool. Soc.*, vol. i., 1835; vol. ii., 1841; vol. iii., 1849; vol. iv., 1853.

Arctopitheci, have flat or arched nails. In structure and function the hand is considerably inferior to that of man. The pelvis is long and extended, but in the *Anthropomorpha* it is shorter and more like that of man, though it is always flatter. The tibia and fibula are always separate and movable. The posterior extremity ends in all cases in a well-developed, prehensile foot, which, according to the osseous structure and muscular arrangements, we are not justified in calling a hand. The opposable hallux always has a nail, while the other toes may be armed with claws (*Arctopitheci*). By the arrangement of the hind limbs the Apes are admirably adapted for climbing and jumping. On the other hand they are less fitted for walking or running upon the four limbs, in consequence of the position of the foot; the leg is directed obliquely inwards, so that only the external edge of the foot rests on the ground. The gait is, therefore, clumsy, except in the *Arctopitheci*.

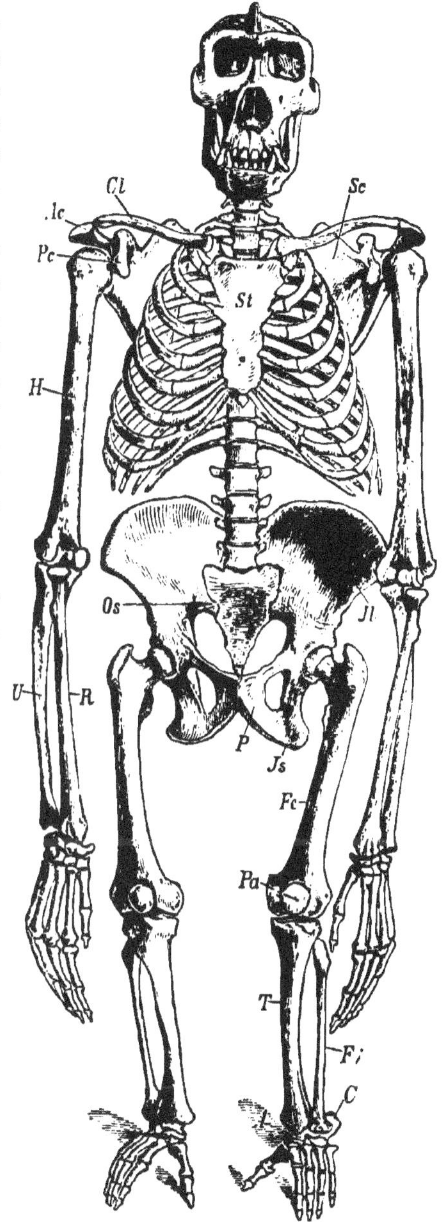

FIG. 703.—Skeleton of *Gorilla engena*. *St*, Sternum; *Sc*, Scapula; *Ac*, Acromion; *Pc*, Coracoid process; *Cl*, Clavicle; *H*, Humerus; *R*, Radius; *U*, Ulna; *Os*, Sacrum; *Jl*, Ilium; *Js*, Ischium; *P*, Pubis; *Fe*, Femur; *Pa*, Patella; *T*, Tibia; *Fi*, Fibula; *C*, Calcaneum; *A*, Astragalus.

Their movements on the boughs and branches of trees, which are effected with great ease and safety, are often assisted by the long tail, which may even act as an accessory prehensile organ.

Most Apes are gregarious, and live in forests in hot countries. In Europe, the precipices of Gibraltar are the single resort of a species of Ape, the Barbary Ape (*Inuus ecaudatus*), which is probably African in origin, and has elsewhere completely vanished from Europe. Only a few Apes lead a solitary life; most of them live together in large companies, which are led by the largest and strongest male. They feed chiefly on fruits and seeds, but also on insects, eggs and birds. The females produce only one young (more rarely two) at a birth; they protect and tend their offspring with great love. Intellectually the Primates take with the dogs and elephants the highest place amongst Mammalia.

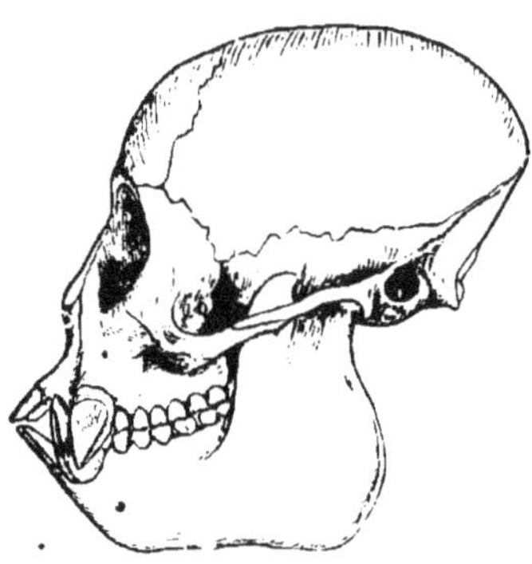

FIG. 704.—Skull of *Pithecia Satanas*.

Suborder 1. **Arctopitheci.** Marmosets. South American Apes of small size, with long hairy tail, and claw-like nails. The great toe is opposable, and has a flat nail. The thumb is not opposable. In the number of teeth (thirty-two) they resemble the apes of the Old World, from which, however, they differ in the fact that the præmolars (three) are more numerous than the molars (two). They produce two or even three young at a birth, and feed on eggs, insects, and fruit.

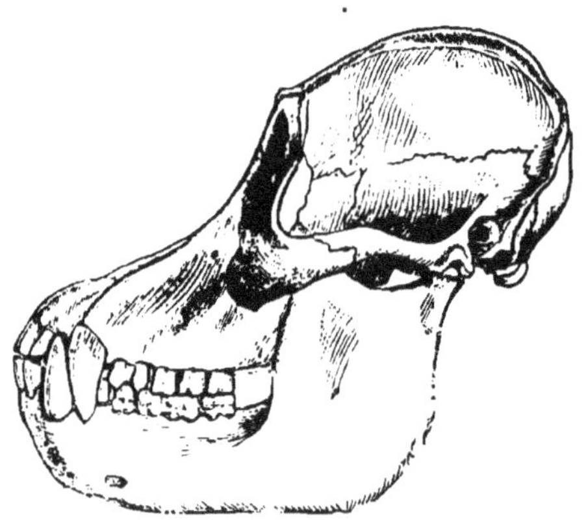

FIG. 705.—Skull of *Satyrus orang*.

Fam. **Hapalidæ.** Dental formula $\frac{2}{2}\,\frac{1}{1}\,\frac{3}{3}\,\Big|\,\frac{2}{2}$. Without prehensile tail. *Hapale Jacchus* Geoffr., the Sahui or Ouistiti; *Midas Rosalia* L.

Suborder 2. **Platyrrhini.** Apes of the New World, with broad nasal septum, and with thirty-six teeth $\left(\frac{2}{2}\,\frac{1}{1}\,\frac{3}{3}\,\Big|\,\frac{3}{3}\right)$. (Fig. 704

The tail is sometimes used as a prehensile organ. The fingers and toes have arched or flat nails. The thumb is sometimes reduced, and is never opposable to the same extent as the great toe. Cheek pouches and ischial callosities always absent.

Fam. Pithecidæ. Apes with long hairy tail, which cannot be used for prehension; *Pithecia Satanas* Hoffms., Brazil; *Nyctipithecus trivirgatus* v. Humb., New Granada; *Chrysothrix sciurea* L., Saimir, Squirrel Monkey, Guiana; *Callithrix personata* Geoffr., east coast of Brazil.

Fig. 706.—*Gorilla engena*, from Vogt and Specht.

Fam. Cebidæ. Apes with prehensile tail covered with hair or naked at the end. *Cebus capucinus* L., Sai, Capuchin; *Ateles paniscus* L., Koaita, Spider-monkey, Brazil; *A. Belzebuth* Geoffr., Guiana; *Lagothrix Humboldtii* Geoffr., Peru. *Mycetes* Ill., Howling Monkeys. *M. niger* Geoffr., Brazil; *M. seniculus* L.

Suborder 3. **Catarrhini.** Apes of the Old World, with narrow nasal septum, and approximated, downwardly directed nostrils, with thirty-two teeth $\left(\frac{2}{2}\ \frac{1}{1}\ \frac{2}{2}\ \frac{3}{3}\right)$. (Fig. 705). The tail is never prehensile, but is in some cases rudimentary, or, as in the *Anthropomorphidæ*, entirely absent as an external appendage.

Fam. **Cynocephalidæ.** Baboons. Of stout, unwieldy build, with dog-like projecting snout. The canine teeth are large like those of the *Carnivora*. There are cheek pouches and large ischial callosities. *Cynocephalus hamadryas* L.; *C. Babuin* Desm., Abyssinia; *C. Gelada* Rüpp., Gelada: *Papio mormon* L., Mandrill, Africa.

Fam. **Cercopithecidæ.** Of slender, light build, with cheek pouches, ischial callosities, and tail of various length, without terminal tuft. *Macacus sinicus* L., and *silenus* L., India: *M. cynomolgus* L., the Java Ape; *Rhesus nemestrinus* Geoffr., Borneo and Sumatra; *Inuus sylvanus* L., *ecaudatus* Geoffr., Barbary Ape, North Africa and Gibraltar: *Cercopithecus sabæus* F. Cuv., West Africa.

Fam. **Semnopithecidæ.** With small ischial callosities, without true cheek pouches. The thumb is short. *Semnopithecus entellus* L., reverenced in the Indies as the holy ape of the Hindoos; *S. nasicus* Cuv., Borneo.

The African genus *Colobus* is allied to the *Semnopithecidæ*, from which it is distinguished principally by the thumb, which is rudimentary, or wanting. *Colobus Guereza* Wagn., with long pendent white mane and caudal tuft, Abyssinia.

Fam. **Anthropomorphæ.** Without tail, with long front limbs, without ischial callosities [except in the Gibbons] and cheek pouches. The body is closely covered with hair on the under side of the trunk and the limbs. *Hylobates Lar* Ill., *H. syndactylus* Cuv., Siamang, Gibbon. The front limbs are very long, reaching to the ground. *Satyrus Orang* L., Orang-Utang, Pongo. Lives in the swampy forests of Borneo. *Gorilla engena = gina* J. Geoffr., Gorilla (fig 706). Lives gregariously in forests on the west coast of Africa (on the Gaboon River) and reaches a height of five and a half to six feet. *Troglodytes niger* L., the Chimpanzee; lives in great companies in the forests of Guinea, and is said to build a nest with a roof upon trees.

Man.*

With reason and articulate speech, with upright gait, with hands and broad-soled, short-toed feet.

Although the view, which formerly was so widely held, that Man belongs to a special natural kingdom, above and outside the animal

* J. F. Blumenbach. "De generis humanis varietate nativa," Gottingæ, 1795. And, "Decas Collectionis suæ craniorum diversarum gentium illustrata," Gottingæ, 1790-1820.

J. C. Prichard, "Researches into the Physical History of Mankind." 2nd ed. London, 1826.

A. Retzius, "Anthropologische Aufsätze," übersetzt in Müller's Archiv.

Huxley, "On the zoological relations of man with the lower animals," *Nat. Hist. Rev.*, 1861.

Huxley, "Evidence as to Man's Place in Nature," London, 1863.

C. Vogt, "Vorlesungen über den Menschen," etc., Giessen 1863.

M. L. Bischoff, "Ueber die Verschiedenheit in der Schädelbildung des Gorilla, Chimpansé und Orang-Utang," etc., München, 1867.

Quetelet, "Anthropométrie," 1870.

Friedrich Müller, "Allgemeine Ethnographie," Wien, 1879.

kingdom, may now be completely put on one side as incompatible with the spirit and method of natural science, yet there are still differences of opinion as to the position of Man in the class of Mammalia, according to the value attributed to the peculiarities of his bodily structure. While Cuvier, and more recently Owen and others, establish a special order (*Bimana*) for Man, other investigators, as Huxley and his followers, attach a much less importance to the characters which separate Man from the anthropoid Apes, and, in agreement with Linnaeus, who included Man with the Apes in his family of *Primates*, regard them only as of family value. The most important anatomical differences between Man and the Apes depend upon the configuration of the skull and the face, the structure of the brain, the dentition and the formation of the extremities, the arrangement of which, in connection with certain peculiarities of the vertebral column, permit of the upright posture of the body in walking.

The rounded arched form of the spacious cranial capsule, the considerable preponderance of the skull over the face, which is not placed in front of the skull as in the anthropoid Apes and in other animals, but almost at right angles beneath it, are essential human characters, as are the relatively large mass of the brain, the great size of the anterior and posterior lobes, and, finally, the great development of the cerebral convolutions, which, however, in the Apes are arranged on the same type.

All these peculiarities, which are of the greatest importance for the intellectual development of Man, cannot be regarded as fundamental distinctions, but must rather be ascribed to gradual deviations, since there are still greater differences between the highest and lowest Apes. Efforts have been vainly made to show that certain parts, which are always present in Apes and other Mammals, are absent in Man (*præmaxilla* Blumenbach—Goethe); and the attempts to prove the converse of this, viz., that there are parts of fundamental value in the human organism (*pes hippocampi minor* Owen—Huxley), which are found in no other Mammal, have as completely failed. Further, the completely continuous row of teeth, interrupted by no gap for the opposed canines, a character by which the human dentition is distinguished from that of the *Catarrhina*, is not an exclusive human character, but is known in a fossil Ungulate (*Anoplotherium*); while on the other hand similar gaps have been observed, certainly only in exceptional cases (Kaffir skull in the Erlangen collection) in the human dentition. The prominent chin of Man has indeed the value of a characteristic feature, although even this is less conspicuous in the negroes; nevertheless it is obvious that this feature cannot be regarded as a character of fundamental importance.

Far more important are the differences between the limbs of Man and those of the anthropoid Apes. The proportions of the individual regions are essentially different, although the differences are not greater than those which exist between the three species of anthropoid Apes in this respect. While in Man the legs constitute the sole support of the body, and greatly surpass the arms in length and weight, in the Apes the arms are longer, in various degrees, than the legs; the brachium in Apes being relatively shorter, the antebrachium and hand, on the other hand, much longer than in Man. In none of the three anthropoid Apes does the hand attain the perfection of the human hand: that of the Gorilla approaches it most nearly, but is clumsier, heavier, and has a shorter thumb. The foot also of the Apes is relatively very long, and is prehensile, the sole being more or less turned inwards. With regard to the arrangement of the bones and muscles, the human foot differs essentially from a true hand, but not from the prehensile foot of the Apes. The human foot presents a number of features, which are peculiar to Man, and are an essential condition of the maintenance of the upright posture; these are the large and strong, but non-opposable, great toe, the arched instep (articulation of the tarsal and metatarsal bones), and the horizontal position of the sole upon the ground. With these peculiarities of the foot are correlated the large development of the calf muscle, the form of the broad shovel-shaped pelvis, the form of the thorax, and the double curvature of the vertebral column. However high a value we may concede to the form of the head, the development of the brain and the upright position of the body, and the upright gait, yet it is undeniable that Man and the Apes are built upon the same type.

The most important consideration which induced the older naturalists to assign to man an entirely special place outside the animal kingdom, is his high intellectual development, which, founded on the possession of articulate speech, has elevated him to a reasonable being, capable of almost unlimited perfection. In fact, it were foolish to deny the great gap which, in this relation, separates Man from the highest beast. Nevertheless, if we examine without prejudice the intellectual development of the individual in early life, and of civilised humanity since the first dawn of culture, and if we subject the intellectual peculiarities of the higher animals to a comparative investigation, we shall reach with Wundt the conclusion that the mind of Man differs from that of the beasts only in the degree of development which it has attained.

The origin of Man and his early history are hidden in complete obscurity, but the view that he has existed for only a few centuries on the earth is completely contradicted by antiquarian and geological investigations. The simultaneous appearance of the remains of human bones (skulls of Engis and of the Neanderthal) and of stone implements, with the skeletal remains of extinct animals (*Mammoth, Rhinoceros tichorhinus*) of the diluvial period, proves the great antiquity of the human race. Man certainly existed in the pleistocene period, but possibly also at the beginning of the tertiary epoch. There are, however, at the present time, no definite facts with regard to his origin; * the view, that the highest form of life has also originated by the process of natural selection from one of the lower forms of Primates, is only a deduction from the Darwinian theory.

The question as to the unity of the species of Man,* which may be answered in different ways according to the different conceptions of species, may remain undiscussed, since from the impossibility of drawing a distinct line between species and race, a definite conclusion is impossible. Blumenbach, at the end of the last century, distinguished five races of men, and characterised them by the form of the head and skulls, by the colour of the skin, and the structure of the hair.

1. The **Caucasian** race, with white skin, fair or dark hair, globular skull, high forehead, vertically placed teeth, narrow nose, and long oval face. Inhabitants of Europe, West Asia, and North Africa. To this race belong the *Indogermanic* peoples (Germans, Celts, Hindoos, Iranians, etc.); the *Semitic* peoples (Jews, Arabs, Berbers, etc.); and the *Slavs*.

2. The **Mongolian** race, with yellowish skin, almost quadrangular short head, narrow, flat forehead, small nose, projecting cheekbones, and broad face, with oblique eyes (from above and outside to below and inside), stiff, black hair. They inhabit parts of Asia, Lapland, and North America (*Eskimos*).

3. The **Ethiopian** race, with black skin and close, crisp, curly hair, with narrow elongated skull, and prominent jaws with oblique

* Cf. Ch. Darwin, "The descent of Man, and selection in relation to sex." London: John Murray, vol. i. and ii., 1871.

Cf. Th. Waitz, "Anthropologie der Naturvölker," fortgesetzt von Gerland, Leipzig, 1859-1872.

alveolar portion; with thick lips, short flat nose, retreating forehead and chin (facial angle only 75°). They inhabit Central and South Africa (Negroes, Kaffirs, etc.).

4. The **American** race, with brown-yellow or copper-red skin, with stiff black hair, deep-set eyes, projecting cheek bones, and broad face. The forehead is narrow, the nose short, but projecting. They inhabit America.

5. The **Malayan** race, with yellow-brown to black skin, with close black loose hair, broad thick nose, turned-up lips and projecting jaws. They inhabit Australia and the East Indian Archipelago.

Cuvier recognised only the white, or Caucasian, the yellow, or Mongolian, and the black, or Ethiopian races as such; and in distinguishing them he laid great stress on differences in language and capability of civilization. The efforts of the modern Anthropologists to found a better and more natural division of races and stocks are based, according to the works of Retzius, principally on the value of the dimensions of the skulls, for the measurement of which a number of methods have been invented. Retzius distinguishes according to the different forms of the face and skull, the long-heads, or *Dolichocephali* (9 : 7), and short heads or *Brachycephali* (8 : 7) : and further, according to the disposition of the jaws and teeth, the *Orthognathi* and the *Prognathi*. The people of Europe are orthognathous, and in great part, the Celts and Germans excepted, brachycephalic.

INDEX.

www.ingramcontent.com/pod-product-compliance
Lightning Source LLC
Chambersburg PA
CBHW021404210326
41599CB00011B/1006
9783337278298